Research on Characterization and Processing of Table Olives

Research on Characterization and Processing of Table Olives

Editors

Beatriz Gandul-Rojas
Lourdes Gallardo-Guerrero

MDPI • Basel • Beijing • Wuhan • Barcelona • Belgrade • Manchester • Tokyo • Cluj • Tianjin

Editors
Beatriz Gandul-Rojas
Instituto de la Grasa (CSIC)
Spain

Lourdes Gallardo-Guerrero
Instituto de la Grasa (CSIC)
Spain

Editorial Office
MDPI
St. Alban-Anlage 66
4052 Basel, Switzerland

This is a reprint of articles from the Special Issue published online in the open access journal *Foods* (ISSN 2304-8158) (available at: https://www.mdpi.com/journal/foods/special_issues/table_olives).

For citation purposes, cite each article independently as indicated on the article page online and as indicated below:

LastName, A.A.; LastName, B.B.; LastName, C.C. Article Title. *Journal Name* **Year**, *Volume Number*, Page Range.

ISBN 978-3-0365-0514-5 (Hbk)
ISBN 978-3-0365-0515-2 (PDF)

Contents

About the Editors . . . . **vii**

Preface to "Research on Characterization and Processing of Table Olives" . . . . **ix**

Beatriz Gandul-Rojas and Lourdes Gallardo-Guerrero
Characterization and Processing of Table Olives: A Special Issue
Reprinted from: *Foods* **2020**, *9*, 1469, doi:10.3390/foods9101469 . . . . **1**

Giorgia Perpetuini, Roberta Prete, Natalia Garcia-Gonzalez, Mohammad Khairul Alam and Aldo Corsetti
Table Olives More than a Fermented Food
Reprinted from: *Foods* **2020**, *9*, 178, doi:10.3390/foods9020178 . . . . **9**

Paola Conte, Costantino Fadda, Alessandra Del Caro, Pietro Paolo Urgeghe and Antonio Piga
Table Olives: An Overview on Effects of Processing on Nutritional and Sensory Quality
Reprinted from: *Foods* **2020**, *9*, 514, doi:10.3390/foods9040514 . . . . **25**

Dimitrios A. Anagnostopoulos, Vlasios Goulas, Eleni Xenofontos, Christos Vouras, Nikolaos Nikoloudakis and Dimitrios Tsaltas
Benefits of the Use of Lactic Acid Bacteria Starter in Green Cracked Cypriot Table Olives Fermentation
Reprinted from: *Foods* **2020**, *9*, 17, doi:10.3390/foods9010017 . . . . **61**

Antonio Paba, Luigi Chessa, Elisabetta Daga, Marco Campus, Monica Bulla, Alberto Angioni, Piergiorgio Sedda and Roberta Comunian
Do Best-Selected Strains Perform Table Olive Fermentation Better than Undefined Biodiverse Starters? A Comparative Study
Reprinted from: *Foods* **2020**, *9*, 135, doi:10.3390/foods9020135 . . . . **83**

Antonio Benítez-Cabello, Francisco Rodríguez-Gómez, M. Lourdes Morales, Antonio Garrido-Fernández, Rufino Jiménez-Díaz and Francisco Noé Arroyo-López
Lactic Acid Bacteria and Yeast Inocula Modulate the Volatile Profile of Spanish-Style Green Table Olive Fermentations
Reprinted from: *Foods* **2019**, *8*, 280, doi:10.3390/foods8080280 . . . . **99**

Lucía Sánchez-Rodríguez, Marina Cano-Lamadrid,Ángel A. Carbonell-Barrachina, Esther Sendra and Francisca Hernández
Volatile Composition, Sensory Profile and Consumer Acceptability of HydroSOStainable Table Olives
Reprinted from: *Foods* **2019**, *8*, 470, doi:10.3390/foods8100470 . . . . **117**

Antonio López-López, Antonio Higinio Sánchez-Gómez, Alfredo Montaño, Amparo Cortés-Delgado and Antonio Garrido-Fernández
Panel and Panelist Performance in the Sensory Evaluation of Black Ripe Olives from Spanish Manzanilla and Hojiblanca Cultivars
Reprinted from: *Foods* **2019**, *8*, 562, doi:10.3390/foods8110562 . . . . **133**

Barbara Lanza, Sara Di Marco, Nicola Simone, Carlo Di Marco and Francesco Gabriele
Table Olives Fermented in Iodized Sea Salt Brines: Nutraceutical/Sensory Properties and Microbial Biodiversity
Reprinted from: *Foods* **2020**, *9*, 301, doi:10.3390/foods9030301 . . . . **155**

Antonio López-López, José María Moreno-Baquero and Antonio Garrido-Fernández
In Vitro Bioaccessibility of Ripe Table Olive Mineral Nutrients
Reprinted from: *Foods* **2020**, *9*, 275, doi:10.3390/foods9030275 . . . . **169**

Marta Berlanga-Del Pozo, Lourdes Gallardo-Guerrero and Beatriz Gandul-Rojas
Influence of Alkaline Treatment on Structural Modifications of Chlorophyll Pigments in NaOH—Treated Table Olives Preserved without Fermentation
Reprinted from: *Foods* **2020**, *9*, 701, doi:10.3390/foods9060701 . . . . **187**

Antonio Valero, Elena Olague, Eduardo Medina-Pradas, Antonio Garrido-Fernández, Verónica Romero-Gil, María Jesús Cantalejo, Rosa María García-Gimeno, Fernando Pérez-Rodríguez, Guiomar Denisse Posada-Izquierdo and Francisco Noé Arroyo-López
Influence of Acid Adaptation on the Probability of Germination of *Clostridium sporogenes* Spores Against pH, NaCl and Time
Reprinted from: *Foods* **2020**, *9*, 127, doi:10.3390/foods9020127 . . . . **205**

Antonio Bevilacqua, Barbara Speranza, Daniela Campaniello, Milena Sinigaglia and Maria Rosaria Corbo
A Preliminary Report for the Design of MoS (Micro-Olive-Spreadsheet), a User-Friendly Spreadsheet for the Evaluation of the Microbiological Quality of Spanish-Style Bella di Cerignola Olives from Apulia (Southern Italy)
Reprinted from: *Foods* **2020**, *9*, 848, doi:10.3390/foods9070848 . . . . **223**

About the Editors

Beatriz Gandul-Rojas (Dr.) graduated in Chemistry from Seville Univ., and did her Ph. D thesis in the Chemistry and Biochemistry Pigments Group at the Instituto de la Grasa (IG) of the Spanish National Research Council (CSIC). She was a Postdoctoral Fellow in the private industrial sector and, since 1996, she is a Senior Scientist at the IG-CSIC. Her research is part of the axis of food quality and safety, and covers aspects from the suitability of the raw materials and food processing and preservation, to health in relation to its consumption, and focuses on studying (1) modifications of chlorophylls and carotenoids in the processing of green table olives, (2) the pigment composition of virgin olive oil, including quality and authenticity parameters, (3) catabolism of the chloroplastic pigments and enzyme systems involved, and (4) the bioavailability of chlorophyll compounds. She has 58 SCI scientific contributions and 12 book chapters, and she has participated in 44 funded research actions, being leader in 8 projects, 4 private research contracts, and 2 patent license agreements. She has been part of Scientific-Technical Thematic Networks and the Scientific Committee of the International Congress on Pigments in Food. She has been Professor in high specialization courses, doctoral programs, postgraduate Master's degrees, and directed 2 Ph. D theses. She is also the inventor of 4 patents, sells her services to food companies and public control bodies, and generates social value by participating in outreach events.

Lourdes Gallardo-Guerrero (Dr.) is a Researcher at Instituto de la Grasa (Food Phytochemistry Department) of the Spanish National Research Council (CSIC) in Seville (Spain). Her main research interests are focused on the study of chlorophyll and carotenoid pigments for determining food quality at the nutritional, sensorial, and authenticity levels. Most of her research has been carried out on olive fruits and their products, table olives, and virgin olive oil, although other foods have also been addressed. Her interests covered by the research work include basic knowledge of the constituents responsible for food color; studies on chlorophyll stability and transformation mechanisms during food processing; enzymes involved in pigment degradation; food color alteration and adulteration; color improvement of green vegetable products in general, and table olives in particular; and bioaccessibility and bioavailability of pigments.

Preface to "Research on Characterization and Processing of Table Olives"

Table olive is recognized as an essential component of the Mediterranean diet, having been explicitly included in the second level of its nutritional pyramid as an aperitif or culinary ingredient. The olive fruit is extremely bitter and must be processed to be edible as table olive. There is a wide range of production styles of table olives aimed at hydrolyzing the bitter glycoside oleuropein. Producers demand innovative techniques to improve the performance and industrial sustainability. Consumers are interested in foods with optimal nutritional characteristics, high quality and safety, improved organoleptic characteristics, and with reduced additives. This Special Issue provides high-quality papers covering the state-of-the-art, recent progress, and perspectives related to characterization and processing of table olives. It covers a broad range of aspects, such as characterization of their chemical composition, bioavailability, advances in the processing technology, chemical and microbiological changes, optimized use of starter cultures for the improvement of the different fermentative processes, and new strategies to reduce sodium and additives for stabilizing the organoleptic properties and avoiding defects. In addition, overviews of both the main technologies used for olive fermentation, including the role of lactic acid bacteria and yeasts characterizing this process, and of the processing and storage effects on the nutritional and sensory properties of table olives, are included.

Beatriz Gandul-Rojas, Lourdes Gallardo-Guerrero
Editors

foods

MDPI

Editorial

Characterization and Processing of Table Olives: A Special Issue

Beatriz Gandul-Rojas * and Lourdes Gallardo-Guerrero

Chemistry and Biochemistry of Pigments, Food Phytochemistry Department, Instituto de la Grasa (CSIC), Campus Universitario Pablo de Olavide, Edificio 46, Ctra. Utrera km 1, 41013 Sevilla, Spain; lgallardo@ig.csic.es
* Correspondence: gandul@ig.csic.es; Tel.: +34-954-611-550

Received: 30 September 2020; Accepted: 12 October 2020; Published: 15 October 2020

Abstract: Table olives are recognized as an essential component of the Mediterranean diet, having been explicitly included in the second level of its nutritional pyramid as an aperitif or culinary ingredient, with a recommended daily consumption of one to two portions (15–30 g). Producers demand innovative techniques improving the performance and industrial sustainability, as well as the development of new products that respond efficiently to increasingly demanding consumers. The purpose of this special issue was to publish high-quality papers with the aim to cover the state-of-the-art, recent progress and perspectives related to characterization and processing of table olives. Two reviews offer an overview about the processing and storage effects on the nutritional and sensory properties of table olives, as well as the main technologies used for olive fermentation, and the role of lactic acid bacteria and yeasts characterizing this niche during the fermentation. A total of 10 research papers cover a broad range of aspects such as characterization of their chemical composition, bioavailability, advances in the processing technology, chemical and microbiological changes, optimized use of starter cultures for the improvement of the different fermentative processes, and new strategies to reduce sodium and additives to stabilize the organoleptic properties and avoid defects.

Keywords: functional food; bioaccessibility; microbiological quality; mineral nutrients; nutritional properties; predictive models; pigment composition; sensory analysis; starter cultures; user-friendly spreadsheet; volatile composition

---

The olive tree (*Olea europea* L.) is a widely distributed plant originating in the Mediterranean region. It is the most cultivated fruit tree in the world, surpassing 11 M ha. Although the olive fruits are mostly destined to obtain the highly valued olive oil, 11% of them are processed for direct consumption as table olives. This food of high nutritional value was sustenance and a source of calcium for the Mediterranean inhabitants in times of scarcity. Table olives are currently consumed as an appetizer and/or highly healthy culinary ingredients for their low sugars content, high monounsaturated fatty acids content, and additional contribution of fiber, minerals, vitamins, and bioactive components.

The olive fruit is a bitter drupe that has to be processed to transform it into an appetizing and edible food. There is a wide range of production styles, depending on the variety, ripening degree, and type of fruit (whole or split), aimed at hydrolyzing and/or diffusing to the brine the bitter oleuropein glucoside. The most widespread systems are those that use an alkaline hydrolysis or a slow acid and enzymatic hydrolysis. In addition, a fermentative process by lactic acid bacteria or yeasts is usually developed to increase palatability.

Producers demand innovative techniques improving the performance and industrial sustainability, as well as the development of new products that respond efficiently to increasingly demanding consumers. Foods with optimal nutritional characteristics, high quality, and safety, improved organoleptic characteristics, and with reduction of additives, are highly demanded. Under this framework, researchers were invited to participate in this special issue with original research papers or review articles focused on novel aspects related to table olives: characterization of their chemical composition, functional properties, and bioavailability of phytochemicals, as well as advances in the processing technology and waste treatment, including emerging techniques and optimized use of starter cultures for the improvement of the different fermentative processes. New strategies were also expected to reduce sodium and additives, to stabilize the organoleptic properties and avoid defects. Conservation methods aimed at extending the shelf life of highly valued artisanal products were also a goal. Likewise, analytical methods and prediction models for the traceability of the products and the detection of fraudulent practices related to the use of unauthorized additives were of interest. Fortunately, our proposal has had a good response, and a wide range of the above topics have been covered in this Special Issue thanks to 12 high-quality contributions.

Perpetuini et al. [1] present an overview of the main technologies used for olive fermentation and the role of lactic acid bacteria and yeasts characterizing this niche during the fermentation. The authors offer particular attention to the selection and use of microorganisms as starter cultures to speed up the process and improve the safety of table olives. In addition, they discuss the development and implementation of multifunctional starter cultures in order to obtain health-oriented table olives.

On the other hand, with the aim of giving an up-to-date overview of the processing and storage effects on the nutritional and sensory properties of table olives, Conte et al. [2] analyze the most relevant literature of the last twenty years in the review "Table Olives: An Overview on Effects of Processing on Nutritional and Sensory Quality". According to this analysis [2], the nutritional properties of table olives are mainly influenced by the processing method used, even if preharvest-factors such as irrigation and fruit ripening stage may have a certain weight. Data reveal that the nutritional value of table olives depends mostly on the balanced profile of polyunsaturated and monounsaturated fatty acids and the contents of health-promoting phenolic compounds, which are best retained in natural table olives. Studies on the use of low-salt brines and of selected starter cultures have shown the possibility of producing table olives with an improved nutritional profile. Sensory characteristics are mostly process-dependent, and a relevant contribute is achieved by starters, not only for reducing the bitterness of fruits, but also for imparting new and typical taste to table olives. Findings reported in this review confirm that table olives surely constitute an important food source for their balanced nutritional profile and unique sensory characteristics.

In the work "Benefits of the Use of Lactic Acid Bacteria Starter in Green Cracked Cypriot Table Olives Fermentation", Anagnostopoulos et al. [3] study the microbial and physicochemical changes of Cypriot green cracked table olives during spontaneous or controlled fermentation process at industrial scale. For this purpose, the authors processed Cypriot green cracked table olives directly in brine (natural olives), using three distinct methods: spontaneous fermentation, inoculation with commercial lactic acid bacteria at a 7%, or a 10% NaCl concentration. Sensory, physicochemical, and microbiological alterations were monitored at intervals, and major differences were detected across treatments. Results indicated that the predominant microorganisms in the inoculated treatments were lactic acid bacteria, while yeasts predominated in control. As a consequence, starter culture contributed to a crucial effect on olives fermentation, leading to faster acidification and lower pH, and inhibition of enterobacteria growth in a shorter period and at a significantly lower salt concentration, compared to the spontaneous fermentation. Likewise, the degradation of oleuropein was achieved faster in inoculated treatments, thus producing higher levels of hydroxytyrosol. Notably, the reduction of salt concentration, in combination with the use of starter, accented novel organoleptic characteristics in the final product, as confirmed from a sensory panel; hence, it becomes obvious that the production of Cypriot table olives at reduced NaCl levels is feasible.

The microbial starters used for table olives can be made by a few species and strains (selected starter cultures) or can consist of an indefinite number of microorganisms (natural biodiverse starter cultures). In order to select the best candidates to be used as starters, Paba et al. [4] carry out a comparative study between twenty-seven *Lactobacillus pentosus* strains, and the undefined starter for table olives from which they were isolated. Strains were characterized for their technological properties: tolerance to low temperature, high salt concentration, alkaline pH, and olive leaf extract; acidifying ability; oleuropein degradation; hydrogen peroxide and lactic acid production. Then, the authors selected two strains with appropriate technological properties, and they compared table olive fermentation in vats, with the original starter (autochthonous and undefined biodiverse starter, SIE), the selected double-strain starter (DSS), and without starter (natural fermentation, NF). Starters affected some texture profile parameters. The SIE resulted in the most effective Enterobacteriaceae reduction, acidification, and olive debittering, while the DSS batch showed the lowest antioxidant activity. Overall, the authors conclude that the best candidate strains cannot guarantee better fermentation performance than the undefined biodiverse mix from which they originate.

The presence of volatile organic compounds (VOCs) in table olives plays an unquestionable role in their sensory appeal. In the work "Lactic acid bacteria and yeast inocula modulate the volatile profile of Spanish-style green table olive fermentation", Benitez-Cabello et al [5] designed a study to support that the VOCs profile of olive fermentation may be modulated by the addition of starter culture, as suggested by different researchers. For this purpose, the authors analyzed the VOCs in brines of Manzanilla Spanish-style green table olive fermentations that were inoculated with two strain of *Lactobacillus pentosus* (LPG1 and Lp13), one of *Lactobacillus plantarum* Lpl15, the yeast *Wickerhanomyces anomalus* Y12, and a mixed culture of all them, and they applied diverse multivariate statistical techniques for studying the results. After fermentation (65 days), a total of 131 volatile compounds were found, but only 71 showed statistical differences between, at least, two fermentation processes. Results showed that inoculation with Lactobacillus strains, especially *L. pentosus* Lp13, reduced the formation of volatile compounds. On the contrary, inoculation with *W. anomalus* Y12 increased their concentrations with respect to the spontaneous process, mainly of 1-butanol, 2-phenylethyl acetate, ethanol, and 2-methyl-1-butanol. Furthermore, biplot and biclustering analyses segregated fermentations inoculated with Lp13 and Y12 from the rest of the processes. The authors point out that the use of sequential lactic acid bacteria and yeasts inocula, or their mixture, in Spanish-style green table olive fermentation, could be advisable practice for producing differentiated and high-quality products with improved aromatic profile.

The topics related to the composition of volatiles and the sensory analysis of table olives have been also addressed by Sánchez-Rodríguez et al. [6] in HydroSOStainable table olives (HydroSOS), which are produced from olive trees grown under regulated deficit irrigation (RDI) strategies. In this contribution, the authors study the volatile composition, the sensory profile and the consumer opinion and willingness to pay (at three locations) for HydroSOS table olives (cv. Manzanilla), produced from three RDI treatments and a control, and processed as Spanish-style. Volatile composition was affected by RDI, by increasing alcohols, ketones, and phenolic compounds in some treatments, while others led to a decrease in esters and the content of organic acids. Descriptive sensory analysis (10 panelists) showed an increase in green-olive flavor with a decrease in bitterness in the HydroSOS samples. Consumers, after being informed about the HydroSOS concept, preferred HydroSOS table olives to the conventional samples and were willing to pay a higher price for them. Finally, green-olive flavor, hardness, crunchiness, bitterness, sweetness, and saltiness were defined as the attributes driving consumer acceptance of HydroSOS table olives.

There is vast experience in the application of sensory analysis to green Spanish-style olives, but black olives have received scarce attention and panelists have less experience on the evaluation of this presentation. In relation to this matter, Lopez-López et al. [7], contribute to this special issue with the work entitled "Panel and panelist performance in the sensory evaluation of black ripe olives from Spanish Manzanilla and Hojiblanca cvs.". Using previously developed lexicon, ripe black olives from

Manzanilla and Hojiblanca cultivars from different origins were sensorily analyzed according to the Quantitative Descriptive Analysis (QDA). The panel (eight men and six women) was trained, and the QDA tests were performed following similar recommendations as for green olives. The data were examined while using SensoMineR v.1.07, programmed in R, which provides a diversity of easy to interpret graphical outputs. The repeatability and reproducibility of panel and panelists were good for product characterization. However, the panel performance investigation was essential in detecting details of panel work (detection of panelists with low discriminant power, those that have interpreted the scale in a different way than the whole panel, the identification of panelists who required training in several/specific descriptors, or those with low discriminant power). Besides, the study identified the descriptors of hard evaluation (skin green, vinegar, bitterness, or natural fruity/floral).

Aspects related to chemical composition, functional properties, and bioavailability of phytochemicals in table olives have also been addressed in this special issue. In this frame, Lanza et al. [8] focused their research on the study of the influence of different brining processes with iodized and non-iodized salt on mineral content, microbial biodiversity, sensory evaluation, and color change of natural fermented table olives. Iodized salt has been used in food processing to prevent iodine deficiency disorders. Then, fresh olives of Carolea and Leucocarpa cvs. were immersed in different brines prepared with two different types of salt: the PGI "Sale marino di Trapani" and the same salt enriched with 0.006% of KIO_3. PGI sea salt significantly enriches the olive flesh in macro-elements such as Na, K, and Mg, and microelements such as Fe, Mn, Cu, and Zn. Instead, Ca decreases, P remains constant, while iodine is present in trace amounts. In the olives fermented in iodized-PGI sea salt brine, the iodine content reached values of 109 μg/100 g (Carolea cv.) and 38 μg/100 g (Leucocarpa cv.). The relationships between the two varieties and the mineral composition were explained by principal component analysis (PCA) and cluster analysis (CA). Furthermore, analyzing the fermenting brines, iodine significantly reduces the microbial load, represented only by yeasts, both in Carolea cv. and in Leucocarpa cv. Candida is the most representative *genus*. The sensory and color properties weren't significantly influenced by iodized brining. Only Carolea cv. showed significative difference for b* parameter and, consequently, for C value. The authors point out that knowledge of the effects of iodized and non-iodized brining on table olives will be useful for developing new functional foods, positively influencing the composition of food products.

The research conducted by López-López et al. [9] studies, for the first time, the bioaccessibility of the mineral nutrients in table olives darkened by oxidation (ripe olives) and their contributions to the recommended daily intake (RDI), according to digestion methods (Miller's vs. Crews' protocols), digestion type (standard vs. modified, standard plus a post-digest re-extraction), and mineralization system (wet vs. ashing). The digestion protocols had significant effects on the bioaccessibility estimation of ripe olive mineral nutrients. Overall, Miller's protocol led to higher bioaccessibilities of Na, K, Ca, Mg, and Fe than the Crews' method. The modified protocols improved most of the values, and they were useful to evaluate the strength of the linkage between some elements and olive flesh components. Monovalent minerals (Na and K) were hardly bound and completely bioaccessible. In contrast, the noticeable presence of divalent (and P) elements in the final solid residue indicated that at least some of them can still be strongly linked to olive flesh even after digestion. The modified Miller's protocol, regardless of the mineralization system, led to the overall highest bioaccessibility values in ripe olives, which were: Na (96%), K (95%), Ca (20%), Mg (73%), Fe (45%), and P (60%). Their potential contributions to the RDI, based on these bioaccessibilities and 100 g olive flesh service size, were then 29, 0.5, 4, 3, 33, and 1%, respectively. This investigation has led to the proposition of the modified Miller's protocol, which includes a post-digest re-extraction, for further studies on the bioaccessibility of mineral nutrients in table olives.

Alkaline treatment is a key stage in the production of green table olives and its main aim is rapid debittering of the fruit. However, its action is complex and structural changes in the olive, and loss of bioactive components, also occur. Because chlorophylls are one of the bioactive components seriously affected, Berlanga-Del Pozo et al. [10] designed a work aimed to investigate the effect of the alkaline treatment on these pigments that are responsible for the characteristic bright green color of table olives not preserved by fermentation. Specifically, the authors investigated the effect of nine combinations of two important parameters of the alkali treatment (NaOH concentration and treatment time) on green table olives processed in the *Campo Real style* and preserved for 1 year under refrigerated conditions. They found a direct relationship between the intensity of the alkali treatment and the degree of chlorophyll degradation, with losses of more than 60% being recorded when NaOH concentration of 4% or greater were used. Oxidation with opening of the isocyclic ring was the main structural change, followed by pheophytinization and degradation to colorless products. To a lesser extent, decarbomethoxylation and dephytylation reactions were detected. An increase in NaOH from 2% to 5% reduced the treatment time from 7 to 4 h, but fostered greater formation of allomerized derivatives, and caused a significant decrease in the chlorophyll content of the olives. However, NaOH concentrations between 6% and 10% did not lead to further time reductions, which remained at 3 h, nor to a significant increase in oxidized compounds, though the proportion of isochlorin *e4*-type derivatives was modified. Chlorophyll compounds of series *b* were more prone to oxidation and degradation reactions to colorless products than those of series *a*. However, the latter showed a higher degree of pheophytinization, and, exclusively, decarbomethoxylation and dephytylation reactions.

Another important aspect in table olive processing concerns advances in conservation methods aimed at extending the shelf life of highly valued artisanal products, maintaining microbial quality, and ensuring safety. The *Clostridium sp.* is a large group of spore-forming, facultative or strictly anaerobic, gram-positive bacteria that can produce food poisoning. The table olive industry is demanding alternative formulations to respond to market demand for the reduction of acidity and salt contents in final products, while maintaining the appearance of freshness of fruits. In the work by Valero et al. [11], the authors develop logistic regression models for non-adapted and acid-adapted *Clostridium sporogenes* strains to study the influence of pH, NaCl, and incubation time on the probability of germination of their spores. They select the factor ranges so that the model could be applied to table olive processing. A *Clostridium sporogenes* cocktail was not able to germinate at pH < 5.0, although the adaptation of the strains produced an increase in the probability of germination at 5.0–5.5 pH levels and 6% NaCl concentration. At acidic pH values (5.0), the adapted strains germinated after 10 days of incubation, while those that were non-adapted required 15 days. At pH 5.75 and with 4% NaCl, germination of the adapted strains took place before 7 days, while several replicates of the non-adapted strains did not germinate after 42 days of storage. The model was validated in natural green olive brines with good results (>81.7% correct prediction cases). The information will be useful for the industry and administration to assess the safety risk in the formulation of new processing conditions in table olives and other fermented vegetables.

Finally, Bevilacqua et al. [12] also contributes to this section on microbial quality and safety assurance of the Special Issue. The purpose of their manuscript was to develop a decision support tool based on simple input parameters to assess the potential for spoilage of green olives processed by the *Spanish-style* during the post-fermentation stage. The duration of this stage is quite variable (from a month to a year) and depends on demand and market prices. If pH and NaCl are not strictly controlled, a microbial spoilage can occur due to a variety of microorganisms (*Aerobacter*, bacilli, propionibacteria, oxidative yeasts, molds, etc.). In this paper, the authors propose a user-friendly spreadsheet (Excel interface), a designated MoS (Micro-Olive-Spreadsheet), as a tool to point out spoiling phenomena in *Bella di Cerignola* olive brines. The spreadsheet was designed as a protected Excel worksheet, where users input values for the microbiological criteria and pH of brines, and the output is a visual code, much like a traffic light: three red cells indicate a spoiling event, while two red cells indicate the possibility of a spoiling event. The input values are: (a) Total Aerobic Count (TAC);

(b) Lactic Acid Bacteria (LAB); (c) yeasts; (d) staphylococci; (e) pH. TAC, LAB, yeasts, and pH are the input values for the first section (quality), while staphylococci count is the input for the second section (technological history). The worksheet can be modified by adding other indices or by setting different breakpoints; however, it is a simple tool for an effective application of hazard analysis and predictive microbiology in table olive production.

To conclude, the present special issue consists of 10 original research papers and two review articles. The research papers, focused on recent research advances related to characterization and processing of table olives, have covered microbiological and chemical changes in table olives during spontaneous or controlled fermentation employing different cultivars [3], characterization of their composition of volatiles and the sensory profile [5–7], mineral composition [8] and bioavailability [9], changes in bioactive components (chlorophylls) by processing [10], optimized use of starter cultures for the improvement of the different fermentative processes [4,5], and new strategies to reduce sodium and additives, to stabilize the organoleptic properties and avoid defects [11,12].

Author Contributions: Both authors equally contributed to organizing the Special Issue, to editorial work, and to writing this editorial. All authors have read and agreed to the published version of the manuscript.

Funding: This work was funded by Spanish Government [Projects AGL RTI2018–095415–B–I00], partially financed by European regional development funds (ERDF).

Acknowledgments: We thank the authors for their contributions to this Special Issue with manuscripts of high quality, the reviewers for their constructive scientific evaluations and the editorial staff of MDPI for their professional support throughout the editorial process.

Conflicts of Interest: The authors declare no conflict of interest.

References

1. Perpetuini, G.; Prete, R.; Garcia-Gonzalez, N.; Khairul Alam, M.; Corsetti, A. Table Olives More than a Fermented Food. *Foods* **2020**, *9*, 178. [CrossRef] [PubMed]
2. Conte, P.; Fadda, C.; Del Caro, A.; Urgeghe, P.P.; Piga, A. Table Olives: An Overview on Effects of Processing on Nutritional and Sensory Quality. *Foods* **2020**, *9*, 514. [CrossRef] [PubMed]
3. Anagnostopoulos, D.A.; Goulas, V.; Xenofontos, E.; Vouras, C.; Nikoloudakis, N.; Tsaltas, D. Benefits of the Use of Lactic Acid Bacteria Starter in Green Cracked Cypriot Table Olives Fermentation. *Foods* **2020**, *9*, 17. [CrossRef] [PubMed]
4. Paba, A.; Chessa, L.; Daga, E.; Campus, M.; Bulla, M.; Angioni, A.; Sedda, P.; Comunian, R. Do Best-Selected Strains Perform Table Olive Fermentation Better than Undefined Biodiverse Starters? A Comparative Study. *Foods* **2020**, *9*, 135. [CrossRef] [PubMed]
5. Benítez-Cabello, A.; Rodríguez-Gómez, F.; Morales, M.L.; Garrido-Fernández, A.; Jiménez-Díaz, R.; Arroyo-López, F.N. Lactic Acid Bacteria and Yeast Inocula Modulate the Volatile Profile of Spanish-Style Green Table Olive Fermentations. *Foods* **2019**, *8*, 280. [CrossRef] [PubMed]
6. Sánchez-Rodríguez, L.; Cano-Lamadrid, M.; Carbonell-Barrachina, Á.A.; Sendra, E.; Hernández, F. Volatile Composition, Sensory Profile and Consumer Acceptability of HydroSOStainable Table Olives. *Foods* **2019**, *8*, 470. [CrossRef] [PubMed]
7. López-López, A.; Sánchez-Gómez, A.H.; Montaño, A.; Cortés-Delgado, A.; Garrido-Fernández, A. Panel and Panelist Performance in the Sensory Evaluation of Black Ripe Olives from Spanish Manzanilla and Hojiblanca Cultivars. *Foods* **2019**, *8*, 562. [CrossRef] [PubMed]
8. Lanza, B.; Di Marco, S.; Simone, N.; Di Marco, C.; Gabriele, F. Table Olives Fermented in Iodized Sea Salt Brines: Nutraceutical/Sensory Properties and Microbial Biodiversity. *Foods* **2020**, *9*, 301. [CrossRef] [PubMed]
9. López-López, A.; Moreno-Baquero, J.M.; Garrido-Fernández, A. In Vitro Bioaccessibility of Ripe Table Olive Mineral Nutrients. *Foods* **2020**, *9*, 275. [CrossRef] [PubMed]
10. Berlanga-Del Pozo, M.; Gallardo-Guerrero, L.; Gandul-Rojas, B. Influence of Alkaline Treatment on Structural Modifications of Chlorophyll Pigments in NaOH—Treated Table Olives Preserved without Fermentation. *Foods* **2020**, *9*, 701. [CrossRef] [PubMed]

11. Valero, A.; Olague, E.; Medina-Pradas, E.; Garrido-Fernández, A.; Romero-Gil, V.; Cantalejo, M.J.; García-Gimeno, R.M.; Pérez-Rodríguez, F.; Posada-Izquierdo, G.D.; Arroyo-López, F.N. Influence of Acid Adaptation on the Probability of Germination of *Clostridium sporogenes* Spores Against pH, NaCl and Time. *Foods* **2020**, *9*, 127. [CrossRef] [PubMed]
12. Bevilacqua, A.; Speranza, B.; Campaniello, D.; Sinigaglia, M.; Corbo, M.R. A Preliminary Report for the Design of MoS (Micro-Olive-Spreadsheet), a User-Friendly Spreadsheet for the Evaluation of the Microbiological Quality of Spanish-Style Bella di Cerignola Olives from Apulia (Southern Italy). *Foods* **2020**, *9*, 848. [CrossRef] [PubMed]

Publisher's Note: MDPI stays neutral with regard to jurisdictional claims in published maps and institutional affiliations.

foods

MDPI

Review

Table Olives More than a Fermented Food

Giorgia Perpetuini, Roberta Prete, Natalia Garcia-Gonzalez, Mohammad Khairul Alam and Aldo Corsetti *

Faculty of BioScience and Technology for Food, Agriculture and Environment, University of Teramo, 641000 Teramo, Italy; giorgia.perpetuini@gmail.com (G.P.); rprete@unite.it (R.P.); ngarciagonzalez@unite.it (N.G.-G.); mohammadkhairul.alam@studenti.unite.it (M.K.A.)
* Correspondence: acorsetti@unite.it; Tel.: +39-086-126-6896

Received: 16 December 2019; Accepted: 7 February 2020; Published: 12 February 2020

Abstract: Table olives are one of the oldest vegetable fermented foods in the Mediterranean area. Beside their economic impact, fermented table olives represent also an important healthy food in the Mediterranean diet, because of their high content of bioactive and health-promoting compounds. However, olive fermentation is still craft-based following traditional processes, which can lead to a not fully predictable final product with the risk of spontaneous alterations. Nowadays, food industries have to face consumer demands for safe and healthy products. This review offers an overview about the main technologies used for olive fermentation and the role of lactic acid bacteria and yeasts characterizing this niche during the fermentation. Particular attention is offered to the selection and use of microorganisms as starter cultures to fasten and improve the safety of table olives. The development and implementation of multifunctional starter cultures in order to obtain heath-oriented table olives is also discussed.

Keywords: table olives; starter cultures; LAB; yeasts; fermented food; probiotic table olives; non-dairy probiotics

1. Introduction

Table olives are defined as "the sound fruit of varieties of the cultivated olive trees (*Olea europaea* L.) that are chosen for their production of olive whose volume, shape, flash-to-stone ratio, fine flesh, taste, firmness, and ease of detachment from the stone make them particularly suitable for processing; treated to remove their bitterness and preserved by natural fermentation; or by heat treatment, with or without the addition of preservatives; packed with or without covering liquid" [1]. Table olives are considered one of the oldest fermented vegetables in the Mediterranean basin and are an important element for the economy of several countries. Their production exceeded 2.9 million tons in the 2017/2018 season and the main producers are Spain, Egypt, Turkey, Algeria, Italy, Greece, and Portugal [2]. However, their production is increasing also in other countries, such as South America, Australia, and the Middle East [2]. Moreover, in 2010 they have been added in the Healthy Eating Pyramid of the Mediterranean diet (https://dietamediterranea.com/), because of their high content of bioactive compounds, dietary fibers, fatty acids, and antioxidants [3].

The olive fruit is a drupe which cannot be consumed directly from the tree because of the presence of a bitter compound called oleuropein. The bitterness can be removed by alkaline treatment, or by brining/salting, fermentation, and acidification [4]. According to the International Olive Oil Council (IOOC) [1], the main goals of olive processing are to improve their sensory characteristics and to ensure safety of consumption. The "trade standard applying to table olives" [1] describes the type of preparation of table olives; however, some traditional processes are still applied, such as the Castelvetrano system. This method is diffused in Sicily and mainly is based on the exploitation of the Nocellara del Belice variety. Only olives of more than 19 mm in diameter are used, which are placed in vessels and treated with a 1.8%–2.5% NaOH solution for one hour. After that, 5–8 kg of salt are added, and the olives are maintained in this brine for 10–15 days. A mild washing step is performed to avoid the total elimination of lye [5].

The main trade preparations are reported in Table 1.

Table 1. Olive processing methods according to the International Olive Oil Council (IOOC) [1].

Preparation Method	Process
Treated olives	It is applied to green olives, olives turning color, or black olives. Olive debittering is achieved through an alkaline treatment (lye 2.5%–3% w/v). Olives are then placed in brine (NaCl 10%–11% w/v) where the fermentation takes place and lasts 3–7 months. Fermentation is driven by lactic acid bacteria.
Natural olives	It is applied to green olives, olives turning color. or black olives. Olives are placed directly in brine. With a salt concentration of about 6%–10% (w/v). Oleuropein is removed through the enzymatic activities (mainly β-glucosidase and esterase) of indigenous microorganisms. The fermentation process can last 8–12 months and it is mainly driven by yeasts and lactic acid bacteria.
Dehydrated and/or shriveled olives	It is applied to green olives, olives turning color, or black olives. Olives are subjected or not to a mild alkaline treatment, preserved in brine, or partially dehydrated in dry salt and/or by heating.
Olives darkened by oxidation	It is applied to green olives or olives turning color. Olives are preserved in brine, fermented or not, and darkened by oxidation in an alkaline medium. They are stored in hermetically sealed containers and subjected to heat sterilization.
Specialties	Olives prepared in a different way than those above following traditional recipes.

2. Table Olives Associated Microbiota

Olive fermentation is a complex process involving a wide array of microorganisms and mainly lactic acid bacteria (LAB) (e.g. *Lactobacillus plantarum* and *Lactobacillus pentosus*) and yeasts (*Saccharomyces cerevisiae, Wickerhamomyces anomalus, Candida boidinii*, etc.) [6]. Their enzymatic activities shape the characteristics of the final products, e.g., flavor, texture, and safety [6]. Moreover, strains isolated from table olives show specific probiotic traits and are able to adhere to the fruit's epidermis, which could thus be ingested by consumers, turning olives into a carrier for these beneficial microbes [7].

The role of LAB during olive fermentation has been investigated in detail [8–15]. The majority of studies indicated that *L. plantarum* and *L. pentosus* are the main LAB isolated from table olives [10,13,15,16]. They are facultative heterofermentative; therefore, they can produce different end products, such as lactic acid, acetic acid, and carbon dioxide or only produce lactic acid depending on the environmental conditions [13]. Hurtado et al. [13] highlighted that *L. plantarum* produced a higher amount of acetic acid during olive fermentation than *L. pentosus*, suggesting the lower ability of the latter species to preserve a homofermentative metabolism under stress conditions. The main species are reported in Figure 1. LAB are the main bacteria responsible of olive debittering thanks to their enzymatic reservoir (β-glucosidase and esterase). *L. pentosus* is characterized by a strong β-glucosidase activity [11]. This enzyme catalyzes oleuropein degradation and the release of glucose and aglycone. This last compound is converted to non-bitter compounds, such as elenolic acid and hydroxytyrosol, by an esterase [17]. They also play a key role in the decrease of pH and provide microbiological stability to the final product as well as an extended shelf life. The production of lactic acid induces an acidification of brine that prevent the growth of spoilage microorganisms and pathogens [17,18].

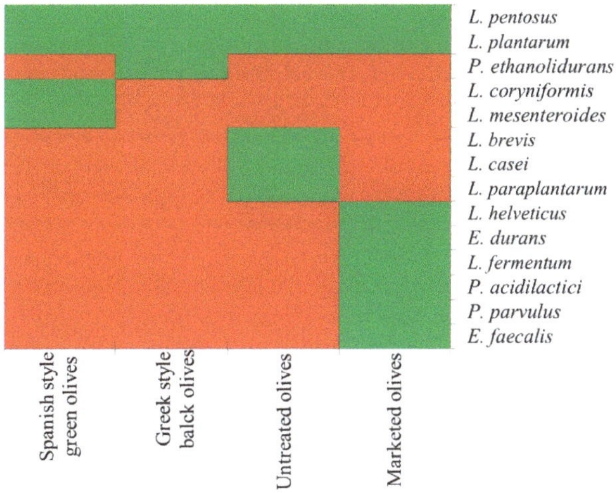

Figure 1. LAB species detected in table olives. The green color indicates the presence of the species, while red its absence. Spanish-style olives are debittered through the addition of lye. In the Greek style, olives are put directly in brine and oleuropein is removed by the enzymatic activities of indigenous microorganisms.

Yeasts can play a double role during olive fermentation; in fact, they are associated with the production of volatile compounds (e.g., alcohols, ethyl acetate, and acetaldehyde) and metabolites that improve the taste and aroma and the preservation characteristics of this fermented food. Moreover, they can enhance LAB growth by the release of nutritive compounds, either synthesizing vitamins, amino acids, and purins, or by metabolizing complex carbon sources [19–21]. Finally, they show esterase and lipase activities. The first one improves the olive taste since it is involved in the production of esters from free fatty acids, while the second one changes the free fatty acids composition of olives improving the characteristics of the final product [22]. On the other hand, yeasts may cause gas-pocket formation and softening of the olive tissue, or even package bulging, clouding of the brines, and production of off flavors and odors [20].

Microbiological studies revealed that *W. anomalus*, *S. cerevisiae*, *Pichia kluyveri*, and *Pichia membranifaciens* are the yeast mainly present in olive brine [6,20,23,24]. *S. cerevisiae* and several species of the *Pichia* genus showed antioxidant activity which protects fruits from oxidation and peroxide formation [21]. Hernandez et al. [21,25] underlined the relevance of *W. anomalus* during olive fermentation. In fact, it presents β-glucosidase activity, as well as produces anti-oxidant compounds and killer toxins against human pathogens and spoilage microorganisms.

Moreover, *D. hansenii*, *P. membranifaciens*, and *W. anomalus* showed strain-specific killer activity against spoilage yeasts [20,23,25,26].

A recent study started to study the biogeography of the microbial communities associated with Spanish-style green olive fermentations [27]. The authors studied the microbial biodiversity of 30 ten-ton fermenters of three different fermentations yards (*patios*) during the fermentation process. Some species were constant, representing the core microbiota of this area. *L. pentosus*, *Pediococcus parvulus*, *Lactobacillus collinoides/paracollinoides*, *Lactobacillus coryniformis*, *L. plantarum*, *Pichia manshurica*, and *Candida thaimueangensis* were found in every *patio*. In particular, cosmopolitan strains belonged to the following species: *L. pentosus*, *P. parvulus*, *L. collinoides/paracollinoides*, and *P. manshurica*.

3. Microbial Spoilage of Table Olives

Olive fermentation is still craft-based; therefore, it is not fully predictable, and some alterations can occur. During the first phase of Spanish fermentation, the Gram-negative bacteria prevail. This phase lasts until LAB grow up inducing a decrease in pH. If this reduction is not too fast, "gas pockets", resulting in the softening and breakage of the cuticle, can appear [28]. A high pH can also favor the development of *Clostridium* spp., which could induce a putrid or butyric fermentation, which cause the appearance of off-flavors and off-odors [28].

The softening of olive drupe is another alteration due to the development of pectinolytic yeasts (e.g., *P. manshurica*, *Pichia kudriavzevii*,, *Saccharomyces oleaginosus*, etc.), molds (*Aspergillus niger*, *Fusarium* spp., and *Penicillium* spp.) and some bacteria (*Bacillus* spp., *Aerobacter* spp., etc.) [29]. These microorganisms release degrading enzymes, which act on pectic substances and cellulose, hemicellulose, and polysaccharides, causing the loss of the structural integrity of the olive drupe [28,29].

Seville-style table olives can undergo a defect called "white spot". These spots develop between the skin and the flesh and are associated to the development of some *L. plantarum* strains [30].

Finally, when the final product is not pasteurized *Propionibacterium* can develop, producing acetic and propionic acids. This alteration is called "zapateria" and cause an increase in volatile acidity and the formation of cyclohexanecarboxylic acid [31] and the production of biogenic amines, such as cadaverine and tyramine [32].

4. Table Olives' Starter Cultures

The use of starter cultures for table olives fermentation is highly recommended [17]. An appropriate inoculum reduces the effects of spoilage microorganisms, inhibits the growth of pathogenic microorganisms, and helps to achieve a controlled process, reducing debittering time and improving the sensorial and hygienic quality of the final product [17,33–35]. Two different types of starter cultures can be applied. Natural starter cultures are made up of microorganisms that spontaneously colonize the raw materials [3]. Their composition is often not reproducible; however, they guarantee a high biodiversity, which contributes to enrich the final product with particular sensory characteristics mostly linked to the region of origin of the raw material itself [36]. On the other hand, selected starter cultures provide numerous advantages (Table 2). They are usually represented by a single strain or by a mixture of strains previously selected on the basis of specific features: A high survival capacity in the fermentation environment (low pH, high concentrations of salts, and low fermentation substrates); high acidifying activity (through organic acid production); the ability to hydrolyze phenolic compounds (such as oleuropein); as well as the possibility of producing volatile molecules and/or specific enzymatic activities that contribute positively to the development of the sensory profile of the final product [17]. Another important characteristic of a starter culture is its ability to dominate the indigenous microbiota [17]. Dominance of the starter culture would be exerted by its fast and predominant growth under fermentation conditions and/or its ability to produce antagonistic substances [37]. In addition, for commercial purpose, it is necessary that starter cultures resist the freezing or freeze-drying process [17].

Despite these advantages, the application of starter cultures for olive fermentation is still limited [6]. Some of the most important olive varieties are still processed without their addition [3].

Among LAB species, the most often proposed as starter cultures are *L. plantarum* and *L. pentosus* [15,17,38], used alone or in combination with other bacterial or yeast species (Tables 3 and 4).

Several studies were conducted to drive the fermentation processes and to improve the quality and sensory profiles of different table olive cultivars using both autochthonous and commercial oleuropeinolytic strains belonging to the *L. plantarum* group [33,34,39–44].

Different *L. pentosus* and *L. plantarum* starter cultures have been found to dominate and improve the fermentation process of green table olives in terms of processing time, microbiological quality, color stability, and aroma profile [39–41].

A strain of *Lb. pentosus* (1MO) was used as a starter to shorten the debittering process of different cultivars (cv. Itrana and Leccino) at the pilot and industrial scale [45]. The use of the selected strain *L. pentosus* (1MO) significantly improved the quality and safety aspects of the fermented table olives, allowing to successfully end the fermentation process within eight days, while more than one week or even months are usually required for biological spontaneous fermentation [46–48].

Table 2. Characteristics and advantages in the use of selected starter cultures.

Properties	Characteristics	Advantages
Safety	Safe and stable activity Standardized activity Easy to manage and reproduce	Reproducibility Controlled and stable fermentation Continuous monitoring of fermentation
Technological	Ability to colonize olives surface (i.e., biofilm formation) Low demand for nutrients	Rapid and predominant growth High adaptation ability
	Growth at different pH (high/low) Salt tolerance Ability to survive/growth at low temperatures	Dominance during the fermentation
	Biodegradation of phenolic compounds Debittering activity (i.e., oleuropeinolytic activity) High acidification activity	Reduction of fermentation time Avoided use of chemicals (microbial biotransformation)
Functional	Antimicrobial activity vs. pathogens (i.e., bacteriocins production, competitive action on nutrients) Biocontrol agents vs. spoilage microorganisms (i.e., production of killer factors)	Protection from undesirable and/or pathogenic microorganisms Improvement of final product stability and shelf-life extension
	Enzymatic activities (i.e., lipase, alkaline/acid phosphatase, β-glucosidase) Vitamins production Production of aromatic compounds	Enhancement of organoleptic, nutritional and sensory profile of the final product
Probiotic	Survival under gastrointestinal conditions (i.e., low pH, gastric and pancreatic digestion, bile salts) Ability to adhere and persist in the intestinal mucosa Modulation of host immune system Antimicrobial activity against pathogens	Ensuring product safety Quality enhancement of the final product Production of a health-promoting functional food

Recently, a starter culture made up of two *L. pentosus* strains was successfully used to debitter green table olives (cv. Itrana) [35] and was patented (Patent N0. 0001428559).

Interestingly, the use of *L. plantarum* strains as starter strains has been investigated also for the ability to positively affect the fermentation process in term of quality preservation and stability during storage. Sherhai et al. [42] found a protective effect of *L. plantarum* on fatty acid oxidation and peroxidation processes, as well as a strong antioxidant activity during the Spanish-style fermentation process. In line with that, a recent study on inoculated Nocellara Etnea table olives with six different starter cultures made up of *L. plantarum, L. pentosus*, and *L. paracasei* confirmed the dominance of *L. plantarum* during fermentation and its positive impact on table olives [34].

Furthermore, a sequential inoculation strategy has been proposed as a promising biotechnological tool to produce low salt Nocellara Etnea table olives. The authors reported on the use of a β-glucosidase-positive strain, *L. plantarum* strain, followed after 60 days by the inoculum of a *L. paracasei* probiotic strain. This strategy reduced the processing time, and positively affected the polyphenol content and sensory profile of the final product, which was characterized by a low salt concentration (5%) [43].

In recent years, several studies focused on the development of yeast starter cultures, both alone and in combination with LAB [20,23,26,49–51]. *L. plantarum* and *L. pentosus* strains have been used with excellent results in combination with an autochthonous *Wickerhamomyces anomalus* strain to accelerate the fermentation of Bella di Cerignola table olives [33]. A functional starter strain of *L. pentosus*, with and without *P. membranifaciens*, was successfully used to drive fermentations of Conservolea black olives, which allow producing a functional product with an improved sensory profile [52].

A sequential inoculation strategy (firstly yeasts, then bacteria) was developed by Tufariello et al. [53]. In particular, the authors tested different yeast species (*S. cerevisiae, D. hansenii*, and *W. anomalus*) in combination with *L. plantarum* and *Leuconostoc mesenteroides* in order to improve the sensory and organoleptic properties of table olives. Pilot-scale fermentations with the sequential inoculation of LAB and yeast strains reduced the fermentation time (from 180 to 90 days), as well as improved the organoleptic characteristics of the final product [53].

Other yeasts species, such as *Debaryomyces* spp., *Pichia* spp., and *Rhodotorula* spp., were recently investigated in order to select the appropriate strains to use in combination with LAB [3,54] (Table 2).

Bonatsou et al. [54] selected *P. guilliermondii* and *W. anomalus* among several yeast strains, isolated from black table olives, and screened for their technological and probiotic properties as promising multifunctional starters to use in real olive fermentations. The use of yeasts is also linked to their ability to favor the formation of multispecies biofilms on biotic (drupes) and abiotic (fermenter vats) surfaces [3]. Several studies showed the ability of some yeast species, such as *D. hansenii*, *Geotrichum candidum*, *P. guilliermondii*, and *W. anomalus*, to form biofilm and create a positive environment for *L. pentosus* growth [7,23,55–59].

Recently, the application of autochthonous strains has arisen to face consumers' demand for more traditional products with a unique sensory profile and peculiar organoleptic properties [60]. Autochthonous strains, being well adapted to the raw material conditions, can easily lead the fermentation process by dominating the table olives microenvironment [3,51]. However, only few studies report the application of autochthonous starter cultures [36,53,61,62]. Martorana et al. [36] used autochthonous starter cultures as a *"Pied de cuve"* to ferment Nocellara del Belice olives [36]. The application of autochthonous starter cultures could be useful for achieving IGP and PDO (Protected Designation of Origin) product specifications, linking the fermented final product to the region where it comes from [3].

Table 3. Main starter strains used for table olive fermentation.

Bacterial Starter Cultures	Cultivar	References
L. plantarum	Alorena	[40]
	Bella di Cerignola	[33,63–65]
	Carolea/Cassanese	[66]
	Conservolea	[41]
	Gordal	[40]
	Halkidiki	[67–69]
	Hojiblanca	[40,70]
	Kalamata/Chalkidikis	[62,71]
	Manzanilla	[40]
	Mele	[28]
	Nocellara del Belice/Nocellara Messinese	[66]
	Nocellara Etnea	[34]
	Picholine	[72]
	Pishomi	[42]
	Tonda di Cagliari	[39,61]
	Leccino	[44]
L. pentosus	Arbequina	[73]
	Conservolea	[41,52]
	Gordal	[55,74]
	Halkidiki	[67–69]
	Itrana	[15,35]
	Manzanilla	[40,75–79]
	Nocellara del Belice	[36,80]
	Nocellara Etnea	[34]
	Tonda di Cagliari	[39,61,81]
L. paracasei	Bella di Cerignola	[9]
L. rhamnosus	Giaraffa e Grossa di Spagna	[82]
Yeast starter cultures	**Cultivar**	**References**
N. molendini-olei/C. matritensis/C. adriatica/ C. diddensiae/W. anomalus/S. cerevisiae	Taggiasca	[83]

Table 4. Main multi-starter strains used for table olive fermentation.

Multi-starter Cultures	Cultivar	References
L. plantarum/L. pentosus	Bella di Cerignola	[65]
	Halkidiki	[67,68]
	Nocellara Etnea	[34]
L. plantarum/L. casei	Nocellara Etnea	[84]
L. plantarum/L. paracasei	Giaraffa e Grossa di Spagna	[82]
L. plantarum/L. paracasei	Nocellara Etnea	[43]
	Nocellara Etnea	[34,85]
L. plantarum/P. pentosaceus	Green olives	[70]
L. plantarum/E. faecieum	Green olives	[70]
L. paracasei/L. pentosus	Nocellara Etnea	[34]
L. pentosus/L. coryniformis	Nocellara del Belice	[12]
L. plantarum/L. paracasei/L. rhamnosus	Giaraffa e Grossa di Spagna	[82]
L. plantarum/L. paracasei/L. pentosus	Nocellara Etnea	[34]
L. plantarum/D. hansenii	Conservolea	[53,86]
L. plantarum/C. famata/C. guilliermondii	Bella di Cerignola	[64]
L. plantarum/S. cerevisiae	Leccino	[53,86]
L. plantarum/W. anomalus	Cellina di Nardò	[53,86]
L. plantarum/W. anomalus	Bella di Cerignola	[33,65]
L. plantarum/W. anomalus/L. pentosus	Bella di Cerignola	[33]
L. pentosus/P. membranifaciens	Conservolea	[52,53]
L. pentosus/C. boidinii	Manzanilla	[87]
L. mesenteroides/S. cerevisiae	Kalamata	[53,86]

5. New Trend in Olive Production: Probiotic Table Olives

The concept of functional food was born in Japan around the 1980s; in 1991, the acronym FOSHU (Foods for Specified Health Use) was coined. Nowadays, the accepted definition is the one recognized by the European Union Food Information Council (EUFIC), based on which functional foods are defined as "foods similar in appearance to conventional foods that are consumed as part of a normal diet, and have demonstrated physiological benefits and/or the capacity to reduce the risk of chronic disease beyond their basic nutritional functions" [88]. Probiotics and prebiotics represent the most-used strategies for the production of functional foods [89–94]. Probiotics are defined as "live microorganisms which, when administered in adequate amounts, as part of a food or a supplement, confer a health benefit on the host" [95]. Generally, probiotics are bacteria isolated from human sources, mostly from the gastrointestinal tract [96], and mainly belong to *Bifidobacterium* and *Lactobacillus* genera [88]. Indeed, it has been recently showed that also naturally occurring food-associated microbes can reach the gut as viable cells, interact with the human host, and potentially provide benefits to gut health [97]. In this context, a diet may represent not only a source of nutrients to the body, but can be also a vehicle of exogenous microorganisms with positive effects on human health [98,99].

Table olives represent a wide reservoir of putative beneficial microbes. Thus, several studies have been conducted to assess the probiotic effects of strains isolated from different fermented olives cultivars and/or already used as starter cultures, belonging to the most widely spread species *L. plantarum* and *L. pentosus*, as well as to species less frequently used, such as *L. paracasei*, *L. casei*, and *L. paraplantarum* [43,85,100–106]. Some studies revealed that some LAB strains isolated from table olives were able to adhere to porcine jejune epithelial cells IPEC-J2 and produced antimicrobial compounds able to inhibit *Helicobacter pylori*, *Propionibacterium* spp., and *Clostridium perfringens* [10,70,107–109]. Probiotic potential, based on the ability to outcompete foodborne pathogens for cell adhesion, was also characterized in several *L. pentosus* isolated from different table olive cultivars (i.e., Nocellara del Belice and Aloreña green table olives) [103,110]. Strains isolated from both cultivars showed the ability to adhere to human intestinal epithelial Caco-2 [110] and vaginal cells [103], as well as the ability to auto-aggregate and co-aggregate with pathogenic bacteria, to ferment some prebiotics, and to in vivo exert protective effects in *Caenorhabditis elegans* [103,110]. Beside antimicrobial activity, different strains of *L. pentosus* and *L. plantarum* isolated from table olives stimulated the release of pro-inflammatory (IL-6) and anti-inflammatory (IL-10) interleukins on macrophages, suppressed the secretion of IL-8, and showed anti-proliferative activity on the HT-29 cell line [111].

Table olives of different cultivars have already been validated as a promising carrier for delivering different probiotics strains into the human GI tract [112] (patent application EP2005/0104138 [9,113]. Table olives can be considered an ideal matrix for the survival of probiotics due to the nutrients released by the fruits and the fact that drupes are coated with a hydrophobic epicuticular wax that promote microbial adhesion [6,7,52,56,112–116].

The probiotic *L. paracasei* strain LMGP22043 was able to colonize the human gut, positively influencing fecal bacteria and biochemical parameters [113]. Lavermicocca et al. [112] used table olives as carrier for the probiotic *L. paracasei* strain IMPC2.1. The strain was recovered in human feces after fermented olive intake, confirming the possibility to use table olives as carrier of probiotics into the human gastrointestinal tract [112]. An autochthonous potential probiotic *L. pentosus* strain [23,75,103] showed to be able to survive for 200 days in packed olives, confirming the possibility to incorporate probiotic strains and thus produce functional table olives [76].

The genetic basis of LAB strains adhesion on olive surfaces is still in its infancy. Perpetuini et al. [115] revealed that the sessile state represented the prevailing *L. pentosus* life-style during table olive fermentation and that the three genes *eno*A1, *gpi* and *oba*C were necessary in *L. pentosus* to form an organized biofilm on the olive skin. The first two genes encoded for cytosolic enzymes involved in the glycolysis pathway and in the adhesion to some specific components of olive skin, while *oba*C for a putative fatty acid binding protein of the DegV family, which could bind some lipids of the epicuticular wax. More recently, Pérez Montoro et al. [116] analyzed the adhesion to mucin of *L. pentosus* strains isolated from Aloreña green table olives. They revealed the presence of four moonlighting proteins over-produced in adhesive strains, which were not produced in non-adhesive strains. These proteins were involved in the glycolytic pathway (phosphoglycerate mutase and glucosamine-6-phosphate deaminase), stress response (small heat shock protein), and transcription (transcription elongation factor GreA). A new in silico approach confirmed that moonlighting proteins are involved in the adhesion to both the extracellular matrix (i.e., olive surface) and host cells, as well as in host immunomodulation [117]. Due to the importance of the genetic background on health-promoting traits, Calero-Delgado et al. [118] recently published the draft genome sequences of five *L. pentosus* strains isolated from biofilms on the skin of green table olives. In particular, most of the strains evaluated harbored two copies of the *luxS* gene, involved in the production of the universal bacterial communicator autoinductor-2. Genes encoding for bacteriocin, exopolysaccharide, and MucBP proteins, which could play an important role in microbe-eukaryote cell adhesion, were also found [118]. The main feature of these studied strains was their ability to adhere to the surface of olives during fermentation, forming biofilms, and turning table olives into carriers of beneficial microorganisms to consumers [114,115,119].

Recently, different studies have been focusing on the yeast microbiota associated with table olives fermentations in order to find potential probiotic candidates to be used as starter cultures [23,26,54,100–106]. *Saccharomyces boulardii* represent the only yeast with claimed probiotic effects [120]. Evidences of other yeast species showing probiotic features, mainly associated with table olive microbiota, such as *D. hansenii, T. delbrueckii, K. lactis, and S. cerevisiae,* are emerging [121–124].

Different *Torulaspora delbrueckii* and *Debaryomyces hansenii* strains have been found to survive in the presence of high bile salt concentrations and low pH values, as well as to have antimicrobial activity against foodborne pathogens [26]. Furthermore, Silva et al. [125] found some *P. membranifaciens* and *Candida oleophila* strains within a native yeast population of Portuguese olives to be promising candidates as multifunctional starter cultures, by having both technological (oleuropeinolytic activity) and beneficial potential (vitamins production, mycogenic, and antimicrobial activities).

In this context two important issues to be considered are the assessment of technological factors influencing the survival of probiotic starter cultures and the starter effect on olives' sensory profile.

Rodríguez-Gómez et al. [77] evaluated the effects of inoculation strategies on the survival of *L. pentosus* TOMC-LAB2—a potential probiotic strain when used as a starter culture in large-scale fermentations of green Spanish-style olives. They proposed an inoculation immediately after brining to reduce the presence of initial natural microbiota, the re-inoculation to replace the possible initial died starter and an early processing in the season when starter survival is higher. Concerning the second aspect, a recent study analyzed the organoleptic characteristics of traditional, spontaneously fermented green table olives and green table olives inoculated with *L. pentosus* TOMC-LAB2. Consumers perceived them similarly, only saltiness had a marked adverse effect [78].

Probiotics are generally carried through dairy products. However, the increased incidence of lactose intolerance, concerns over cholesterol, and the wide spread of new lifestyles (vegans and vegetarians) drove new researches toward non-dairy probiotic foods, such as fruits and vegetables, which are rich in vitamins, minerals, carbohydrates, fibers, and antioxidant compounds [126,127]. Recently, it has been shown that vegetable-derived products (i.e., fruits, fruits juices, cereals, and legumes) can act as carriers for positive microbes because of their intrinsic structure; thus, microorganisms can colonize pores, lesions, lenticels, and irregularities present on the surface [119]. Moreover, vegetables are also rich in prebiotic compounds, which protect probiotic microorganisms from the harsh GI tract conditions and are a source of nutrients that positively influences bacterial survival [128,129]. Actually, vegetable-based probiotic foods are available on the market. However, further studies are necessary to better understand the viability of selected strains in the human GI tract and their interactions with human microbiota. In vivo studies are required to assess if carried bacteria and the food matrix have a positive impact on human health. In this case, health claims could be proposed.

6. Conclusions

Table olives have a great impact on the economy of several countries. According to Bonatsou et al. [6], olives are considered in the food industry as the "food of the future". Despite the many advances made, table olives are still produced according to ancient and local recipes, refusing the addition of starter cultures. Olive industries will face several challenges in the next future, including crop management, olive quality, production methods, and health issues. The application of starter cultures represents the main biotechnological challenge/innovation in this field. In this review the main criteria used for starter cultures selection are reported. LAB and yeasts are the main microbial groups studied and several strains have been characterized in order to develop new starter cultures. The use of autochthonous starter cultures is gaining attention since they offer several advantages in terms of adaptability to stressful niches and characterization of the final product, offering a link with the product origin. Another interesting aspect is the characterization of probiotic strains. This issue is the main research trend in this field since it responds to consumer demand for health-oriented products. The potential addition of probiotics in table olive fermentation on one hand give rise to new questions to be solved in terms of cost-effectiveness and acceptance by consumers, but on the other hand can improve the entire production process by positively affecting the aroma and sensory profile, product shelf-life, and by providing additional health-promoting properties to the consumers. Moreover, the development of probiotic table olives could have a positive economic impact, since this product is produced also in less developed countries.

In our opinion, further studies are necessary to isolate and characterize more strains from different table olive cultivars in order to prepare autochthonous starter culture collections and produce healthy products with enhanced sensory characteristics. Additional researches are also needed to implement fermentation strategies to favor the survival and dominance of starter strains and develop new starters by combining LAB and yeasts, to mimic the natural microbiota of olives. Moreover, concerning probiotic strains, further validation in *in vivo* trials with more complex animal or human systems should be performed to gain a deeper understanding of their potential health-promoting features for humans. Finally, further studies should develop new approaches for the treatment of wastewater produced by table olive industries in order to have healthy eco-friendly products.

Author Contributions: A.C. conceived the idea; G.P., R.P. and A.C. drafted the paper; N.G.-G. and M.K.A. prepared the tables and figure; All authors have read and agreed to the published version of the manuscript.

Funding: This work is part of the project "Tracciabilità, certificazione e tutela della qualità dell'olio di oliva e delle olive da tavola Azione 4d" supported by a grant from UNAPROL (Reg. CE No. 867/2008 Misura 4). It has received financial support from the European Union's Horizon 2020 research and innovation program under the Marie Skłodowska-Curie grant agreement 713714 ESR 07 to Natalia Garcia-Gonzalez and from "PON Ricerca e Innovazione 2014-20," azione I.1 "Dottorati innovativi con caratterizzazione industriale," A.Y. 2018-19, XXXIV Cycle, for the PhD project grant of Mohammad Khairul Alam.

Conflicts of Interest: The authors declare no conflict of interest.

References

1. IOOC, International Olive Oil Council. *Trade Standard Applying to Table Olives*; International Olive Oil Council: Madrid, Spain, 2004.
2. IOOC, International Olive Oil Council. World Table Olive Figures. 2019. Available online: http://www.internationaloliveoil.org/estaticos/view/132-world-table-olive-figures (accessed on 20 January 2019).
3. Campus, M.; Degirmencioglu, N.; Comunian, R. Technologies and Trends to Improve Table Olive Quality and Safety. *Front. Microbiol.* **2018**, *9*, 617. [CrossRef] [PubMed]
4. Garrido-Fernández, A.; Fernández-Díez, M.J.; Adams, R.M. *Table Olives: Production and Processing*, 1st ed.; Chapman and Hall: London, UK, 1997.
5. Rejano, L.; Montaño, A.; Casado, F.J.; Sánchez, A.H.; de Castro, A. Table olives: Varieties and variations. In *Olives and Olive Oil in Health and Disease Prevention*; Preedy, V.R., Watson, R.R., Eds.; Oxford Academic Press: Oxford, UK, 2010; pp. 5–15.
6. Bonatsou, S.; Tassou, C.C.; Panagou, E.Z.; Nychas, G.J.E. Table olive fermentation using starter cultures with multifunctional potential. *Microorganisms* **2017**, *5*, 30. [CrossRef] [PubMed]
7. Benítez-Cabello, A.; Romero-Gil, V.; Rodríguez-Gómez, F.; Garrido-Fernández, A.; Jiménez-Díaz, R.; Arroyo-López, F.N. Evaluation and identification of poly-microbial biofilms on natural green Gordal table olives. *Antonie Van Leeuwenhoek* **2015**, *108*, 597–610. [CrossRef] [PubMed]
8. Randazzo, C.L.; Restuccia, C.; Romano, A.D.; Caggia, C. *Lactobacillus casei*, dominant species in naturally fermented Sicilian green olives. *Int. J. Microbiol.* **2004**, *90*, 9–14. [CrossRef]
9. De Bellis, P.; Valerio, F.; Sisto, A.; Lonigro, S.L.; Lavermicocca, P. Probiotic table olives: Microbial populations adhering on olive surface in fermentation sets inoculated with the probiotic strain *Lactobacillus paracasei* IMPC2.1 in an industrial plant. *Int. J. Food Microbiol.* **2010**, *140*, 6–13. [CrossRef]
10. Bevilacqua, A.; Altieri, C.; Corbo, M.R.; Sinigaglia, M.; Ouoba, L.I.I. Characterization of lactic acid bacteria isolated from Italian Bella di Cerignola table olives: Selection of potential multifunctional starter cultures. *J. Food Sci.* **2010**, *75*, 536–544. [CrossRef]
11. Franzetti, A.; Gandolfi, I.; Gaspari, E.; Ambrosini, R.; Bestetti, G. Seasonal variability of bacteria in fine and coarse urban air particulate matter. *Appl. Microbiol. Biotechnol.* **2011**, *90*, 745–753. [CrossRef]
12. Aponte, M.; Blaiotta, G.; La Croce, F.; Mazzaglia, A.; Farina, V.; Settanni, L. Use of selected autochthonous lactic acid bacteria for Spanish style table olive fermentation. *Food Microbiol.* **2012**, *30*, 8–16. [CrossRef]
13. Hurtado, A.; Reguant, C.; Esteve-Zarzoso, B.; Bordons, A.; Rozès, N. Microbial population dynamics during the processing of Aberquina table olives. *Food Res. Int.* **2008**, *41*, 738–744. [CrossRef]
14. Doulgeraki, A.I.; Hondrodimou, O.; Iliopoulos, V.; Panagou, E.Z. Lactic acid bacteria and yeast heterogeneity during aerobic and modified atmosphere packaging storage of natural black Conservolea olives in polyethylene pouches. *Food Control* **2012**, *26*, 49–57. [CrossRef]
15. Tofalo, R.; Perpetuini, G.; Schirone, M.; Ciarrocchi, A.; Fasoli, G.; Suzzi, G.; Corsetti, A. *Lactobacillus pentosus* dominates spontaneous fermentation of Italian table olives. *LWT Food Sci. Technol.* **2014**, *57*, 710–717. [CrossRef]
16. Campaniello, D.; Bevilacqua, A.; D'Amato, D.; Corbo, M.R.; Altieri, C.; Sinigaglia, M. Microbial characterization of table olives processed according to Spanish and natural styles. *Food Technol. Biotechnol.* **2005**, *43*, 289–294.
17. Corsetti, A.; Perpetuini, G.; Schirone, M.; Tofalo, R.; Suzzi, G. Application of starter cultures to table olive fermentation: An overview on the experimental studies. *Front. Microbiol.* **2012**, *3*, 248. [CrossRef]

18. Caggia, C.; Randazzo, C.L.; Di Salvo, M.; Romeo, F.V.; Giudici, P. Occurrence of *Listeria monocytogenes* in green table olives. *J. Food Prot.* **2004**, *67*, 2189–2194. [CrossRef]
19. Alves, M.; Gonçalves, T.; Quintas, C. Microbial quality and yeast population dynamics in cracked green table olives' fermentations. *Food Control* **2012**, *23*, 363–368. [CrossRef]
20. Arroyo-López, F.N.; Querol, A.; Bautista-Gallego, J.; Garrido-Fernández, A. Role of yeasts in table olive production. *Int. J. Food Microbiol.* **2008**, *128*, 189–196. [CrossRef]
21. Hernández, A.; Martín, A.; Aranda, E.; Pérez-Nevado, F.; Córdoba, M.G. Identification and characterization of yeast isolated from the elaboration of seasoned green table olives. *Food Microbiol.* **2007**, *24*, 346–351. [CrossRef]
22. Bautista-Gallego, J.; Rodriguez-Gomez, F.; Barrio, E.; Querol, A.; Garrido- Fernandez, A.; Arroyo-López, F.N. Exploring the yeast biodiversity of green table olive industrial fermentations for technological applications. *Int. J. Food Microbiol.* **2011**, *147*, 89–96. [CrossRef] [PubMed]
23. Arroyo-López, F.N.; Romero-Gil, V.; Bautista-Gallego, J.; Rodríguez-Gómez, F.; Jiménez-Díaz, R.; García-García, P.; Querol, A.; Garrido-Fernández, A. Potential benefits of the application of yeast starters in table olive processing. *Front. Microbiol.* **2012**, *5*, 34. [CrossRef]
24. Tofalo, R.; Perpetuini, G.; Schirone, M.; Suzzi, G.; Corsetti, A. Yeast biota associated to naturally fermented table olives from different Italian cultivars. *Int. J. Food Microbiol.* **2013**, *161*, 203–208. [CrossRef]
25. Hernández, A.; Martín, A.; Córdoba, M.G.; Benito, M.J.; Aranda, E.; Pérez-Nevado, F. Determination of killer activity in yeasts isolated from the elaboration of seasoned green table olives. *Int. J. Food Microbiol.* **2008**, *121*, 178–188. [CrossRef]
26. Psani, M.; Kotzekidou, P. Technological characteristics of yeaststrains and their potential as starter adjuncts in Greek-style black olive fermentation. *World J. Microbiol. Biotechnol.* **2006**, *22*, 1329–1336. [CrossRef]
27. Lucena-Padrós, H.; Ruiz-Barba, J.L. Microbial biogeography of Spanish-style green olive fermentations in the province of Seville, Spain. *Food Microbiol.* **2019**, *82*, 259–268. [CrossRef] [PubMed]
28. Lanza, B. Abnormal fermentations in table-olive processing: Microbial origin and sensory evaluation. *Front. Microbiol.* **2013**, *4*, 91. [CrossRef] [PubMed]
29. Golomb, B.L.; Morales, V.; Jung, A.; Yau, B.; Boundy-Mills, K.L.; Marco, M.L. Effects of pectinolytic yeast on the microbial composition and spoilage of olive fermentations. *Food Microbiol.* **2013**, *33*, 97–106. [CrossRef]
30. Kailis, S.; Harris, D. *Producing Table Olives*; Landlinks Press: Collingwood, VIC, Australia, 2007; p. 328.
31. Montano, A.; de Castro, A.; Rejano, L.; Brenes, M. 4- hydroxycyclohexanecarboxylic acid as a substrate for cyclohexane carboxylic acid production during the "Zapatera" spoilage of Spanish-style green table olives. *J. Food Prot.* **1996**, *59*, 657–662. [CrossRef]
32. Garcia, P.G.; Barranco, C.R.; Quintana, M.C.; Fernandez, A.G. Biogenic amine formation and "zapatera" spoilage of fermented green olives: Effect of storage temperature and debittering process. *J. Food Prot.* **2004**, *67*, 117–123. [CrossRef] [PubMed]
33. De Angelis, M.; Campanella, D.; Cosmai, L.; Summo, C.; Rizzello, C.G.; Caponio, F. Microbiota and metabolome of un-started and started Greek-type fermentation of *Bella di Cerignola* table olives. *Food Microbiol.* **2015**, *52*, 18–30. [CrossRef]
34. Randazzo, C.L.; Todaro, A.; Pino, A.; Corona, O.; Caggia, C. Microbiota and metabolome during controlled and spontaneous fermentation of Nocellara Etnea table olives. *Food Microbiol.* **2017**, *65*, 136–148. [CrossRef]
35. Perpetuini, G.; Caruso, G.; Urbani, S.; Schirone, M.; Esposto, S.; Ciarrocchi, A.; Prete, R.; Garcia-Gonzalez, N.; Battistelli, N.; Gucci, R.; et al. Changes in polyphenolic concentrations of table olives (cv. Itrana) produced under different irrigation regimes during spontaneous or inoculated fermentation. *Front. Microbiol.* **2018**, *9*, 1287. [CrossRef]
36. Martorana, A.; Alfonzo, A.; Settanni, L.; Corona, O.; La Croce, F.; Caruso, T.; Moschetti, G.; Francesca, N. An innovative method to produce green table olives based on "pied de cuve" technology. *Food Microbiol.* **2015**, *50*, 126–140. [CrossRef] [PubMed]
37. Marugg, J.D. Bacteriocins, their role in developing natural products. *Food Biotechnol.* **1991**, *5*, 305–312. [CrossRef]
38. Hurtado, A.; Reguant, C.; Bordons, A.; Rozès, N. Lactic acid bacteria from fermented table olives. *Food Microbiol.* **2012**, *31*, 1–8. [CrossRef] [PubMed]

39. Comunian, R.; Ferrocino, I.; Paba, A.; Daga, E.; Campus, M.; Di Salvo, R.; Cauli, E.; Piras, F.; Zurru, E.; Cocolin, L. Evolution of microbiota during spontaneous and inoculated Tonda di Cagliari table olives fermentation and impact on sensory characteristics. *LWT Food Sci. Technol.* **2017**, *84*, 64–72. [CrossRef]
40. Ramírez, E.; Medina, E.; García, P.; Brenes, M.; Romero, C. Optimization of the natural debittering of table olives. *LWT Food Sci. Technol.* **2017**, *77*, 308–313. [CrossRef]
41. Chranioti, C.; Kotzekidou, P.; Gerasopoulos, D. Effect of starter cultures on fermentation of naturally and alkali-treated cv. Conservolea green olives. *LWT Food Sci. Technol.* **2018**, *89*, 403–408. [CrossRef]
42. Sherahi, M.H.A.; Shahidi, F.; Yazdi, F.T.; Hashemi, S.M.B. Effect of *Lactobacillus plantarum* on olive and olive oil quality during fermentation process. *LWT Food Sci. Technol.* **2018**, *89*, 572–580. [CrossRef]
43. Pino, A.; Vaccalluzzo, A.; Solieri, L.; Romeo, F.V.; Todaro, A.; Caggia, C.; Arroyo-López, F.N.; Bautista-Gallego, J.; Randazzo, C.L. Effect of Sequential Inoculum of Beta-Glucosidase Positive and Probiotic Strains on Brine Fermentation to Obtain Low Salt Sicilian Table Olives. *Front. Microbiol.* **2019**, *10*, 174. [CrossRef]
44. Caponio, F.; Difonzo, G.; Calasso, M.; Cosmai, L.; De Angelis, M. Effects of olive leaf extract addition on fermentative and oxidative processes of table olives and their nutritional properties. *Food Res. Int.* **2019**, *116*, 1306–1317. [CrossRef]
45. Servili, M.; Settanni, L.; Veneziani, G.; Esposto, S.; Massitti, O.; Taticchi, A.; Urbani, S.; Montedoro, G.F.; Corsetti, A. The use of *Lactobacillus pentosus* 1MO to shorten the debittering process time of black table olives (*Cv.* Itrana and Leccino): A pilot-scale application. *J. Agric. Food Chem.* **2006**, *54*, 3869–3875. [CrossRef]
46. Tassou, C.C.; Panagou, E.Z.; Katsaboxakis, K.Z. Microbiological and physicochemical changes of naturally black olives fermented at different temperatures and NaCl levels in the brines. *Food Microbiol.* **2002**, *19*, 605–615. [CrossRef]
47. Ciafardini, G.; Marsilio, A.; Lanza, B.; Pozzi, N. Hydrolysis of oleuropein by *Lactobacillus plantarum* strains associated with olive fermentation. *Appl. Environ. Microbiol.* **1994**, *60*, 4142–4147. [CrossRef] [PubMed]
48. Vega Leal-Sánchez, M.; Ruiz-Barba, J.L.; Sánchez Gómez, A.H.; Rejano, L.; Jiménez-Dıíaz, R.; Garrido Fernández, A. Fermentation profile and optimization of green olive fermentation using *Lactobacillus plantarum* LPCO10 as a starter culture. *Food Microbiol.* **2003**, *20*, 421–430. [CrossRef]
49. Nisiotou, A.A.; Chorianopoulos, N.; Nychas, G.J.E.; Panagou, E.Z. Yeast heterogeneity during spontaneous fermentation of black Conservolea olives in different brine solutions. *J. Appl. Microbiol.* **2010**, *108*, 396–405. [CrossRef]
50. Bevilacqua, A.; Beneduce, L.; Sinigaglia, M.; Corbo, M.R. Selection of yeasts as starter cultures for table olives. *J. Food Sci.* **2013**, *78*, 742–751. [CrossRef]
51. Bevilacqua, A.; De Stefano, F.; Augello, S.; Pignatiello, S.; Sinigaglia, M.; Corbo, M.R. Biotechnological innovations for table olives. *Int. J. Food Sci. Nutr.* **2015**, *66*, 127–131. [CrossRef]
52. Grounta, A.; Doulgeraki, A.I.; Nychas, G.J.E.; Panagou, E.Z. Biofilm formation on Conservolea natural Black olives during single and combined inoculation with a functional *Lactobacillus pentosus* starter culture. *Food Microbiol.* **2016**, *56*, 35–44. [CrossRef]
53. Tufariello, M.; Durante, M.; Ramires, F.A.; Grieco, F.; Tommasi, L.; Perbellini, E.; Falco, V.; Tasioula-Margari, M.; Logrieco, A.F.; Mita, G.; et al. New process for production of fermented black table olives using selected autochthonous microbial resources. *Front. Microbiol.* **2015**, *6*, 1007. [CrossRef]
54. Bonatsou, S.; Benítez, A.; Rodríguez-Gómez, F.; Panagou, E.Z.; Arroyo-López, F.N. Selection of yeasts with multifunctional features for application as starters in natural black table olive processing. *Food Microbiol.* **2015**, *46*, 66–73. [CrossRef]
55. Domínguez-Manzano, J.; Olmo-Ruiz, C.; Bautista-Gallego, J.; Arroyo-López, F.N.; Garrido Fernández, A.; Jiménez-Díaz, R. Biofilm formation on abiotic and biotic surfaces during Spanish style green table olive fermentation. *Int. J. Food Microbiol.* **2012**, *157*, 230–238. [CrossRef]
56. Grounta, A.; Panagou, E.Z. Mono and dual species biofilm formation between *Lactobacillus pentosus* and *Pichia membranifaciens* on the surface of black olives under different sterile brine conditions. *Ann. Microbiol.* **2014**, *64*, 1757–1767. [CrossRef]
57. Grounta, A.; Doulgeraki, A.I.; Panagou, E.Z. Quantification and characterization of microbial biofilm community attached on the surface of fermentation vessels used in green table olive processing. *Int. J. Food Microbiol.* **2015**, *203*, 41–48. [CrossRef] [PubMed]

58. León-Romero, Á.; Domínguez-Manzano, J.; Garrido-Fernández, A.; Arroyo-López, F.N.; Jiménez Díaz, R. Formation of *in vitro* mixed-species biofilms by *Lactobacillus pentosus* and yeasts isolated from Spanish-style green table olive fermentations. *Appl. Environ. Microbiol.* **2016**, *82*, 689–695. [CrossRef] [PubMed]
59. Porru, C.; Rodríguez-Gómez, F.; Benítez-Cabello, A.; Jiménez-Díaz, R.; Zara, G.; Budroni, M.; Mannazzu, I.; Arroyo-Lopez, F.N. Genotyping, identification and multifunctional features of yeasts associated to Bosana naturally black table olive fermentations. *Food Microbiol.* **2018**, *69*, 33–42. [CrossRef]
60. Medina, E.; Ruiz-Bellido, M.A.; Romero-Gil, V.; Rodríguez-Gómez, F.; Montes-Borrego, M.; Landa, B.B.; Arroyo-López, F.N. Assessment of the bacterial community in directly brined Aloreña de Málaga table olive fermentations by metagenetic analysis. *Int. J. Food Microbiol.* **2016**, *236*, 47–55. [CrossRef]
61. Campus, M.; Sedda, P.; Cauli, E.; Piras, F.; Comunian, R.; Paba, A.; Daga, E.; Schirru, S.; Angioni, A.; Zurru, R.; et al. Evaluation of a single strain starter culture, a selected inoculum enrichment, and natural microflora in the processing of Tonda di Cagliari natural table olives: Impact on chemical, microbiological, sensory and texture quality. *LWT Food Sci. Technol.* **2015**, *64*, 671–677. [CrossRef]
62. Tataridou, M.; Kotzekidou, P. Fermentation of table olives by oleuropeinolytic starter culture in reduced salt brines and inactivation of *Escherichia coli* O157:H7 and *Listeria monocytogenes*. *Int. J. Food Microbiol.* **2015**, *208*, 122–130. [CrossRef]
63. Perricone, M.; Bevilacqua, A.; Corbo, M.R.; Sinigaglia, M. Use of *Lactobacillus plantarum* and glucose to control the fermentation of "Bella di Cerignola" Table Olives, a traditional variety of Apulian region (Southern Italy). *J. Food Sci.* **2010**, *75*, 430–436. [CrossRef]
64. Perricone, M.; Corbo, M.R.; Sinigaglia, M.; Bevilacqua, A. Use of starter cultures in olives: A not-correct use could cause a delay of performances. *Food Nutr. Sci.* **2013**, *4*, 721–726. [CrossRef]
65. Cosmai, L.; Campanella, D.; De Angelis, M.; Summo, C.; Paradiso, V.M.; Pasqualone, A.; Caponio, F. Use of starter cultures for table olives fermentation as possibility to improve the quality of thermally stabilized olive-based paste. *LWT Food Sci. Technol.* **2018**, *90*, 381–388. [CrossRef]
66. Benincasa, C.; Muccilli, S.; Amenta, M.; Perri, E.; Romeo, F.V. Phenolic trend and hygienic quality of green table olives fermented with *Lactobacillus plantarum* starter culture. *Food Chem.* **2015**, *186*, 271–276. [CrossRef]
67. Argyri, A.A.; Nisiotou, A.A.; Mallouchos, A.; Panagou, E.Z.; Tassou, C.C. Performance of two potential probiotic *Lactobacillus* strains from the olive microbiota as starters in the fermentation of heat shocked green olives. *Int. J. Food Microbiol.* **2014**, *171*, 68–76. [CrossRef]
68. Blana, V.A.; Grounta, A.; Tassou, C.C.; Nychas, G.J.E.; Panagou, E.Z. Inoculated fermentation of green olives with potential probiotic *Lactobacillus pentosus* and *Lactobacillus plantarum* starter cultures isolated from industrially fermented olives. *Food Microbiol.* **2014**, *38*, 208–218. [CrossRef]
69. Blana, V.A.; Polymeneas, N.; Tassou, C.C.; Panagou, E.Z. Survival of potential probiotic lactic acid bacteria on fermented green table olives during packaging in polyethylene pouches at 4 and 20 °C. *Food Microbiol.* **2016**, *53*, 71–75. [CrossRef]
70. Ruiz-Barba, J.L.; Caballero-Guerrero, B.; Maldonado-Barragán, A.; Jiménez-Díaz, R. Coculture with specific bacteria enhances survival of *Lactobacillus plantarum* NC8, an autoinducer-regulated bacteriocin producer, in olive fermentations. *Food Microbiol.* **2010**, *27*, 413–417. [CrossRef] [PubMed]
71. Kaltsa, A.; Papaliaga, D.; Papaioannou, E.; Kotzekidou, P. Characteristics of oleuropeinolytic strains of *Lactobacillus plantarum* group and influence on phenolic compounds in table olives elaborated under reduced salt conditions. *Food Microbiol.* **2015**, *48*, 58–62. [CrossRef]
72. Lamzira, Z.; Asehraou, A.; Brito, D.; Oliveira, M.; Faid, M.; Peres, C. Bloater spoilage of greenolives. *Food Technol. Biotechnol.* **2005**, *43*, 373–377. [CrossRef]
73. Hurtado, A.; Reguant, C.; Bordons, A.; Rozès, N. Evaluation of a single and combined inoculation of a *Lactobacillus pentosus* starter for processing cv. Arbequina natural green olives. *Food Microbiol.* **2010**, *27*, 731–740. [CrossRef]
74. Bautista-Gallego, J.; Arroyo-López, F.N.; Romero-Gil, V.; Rodríguez-Gómez, F.; Garcia-Garcia, P.; Garrido-Fernández, A. Chloride salt mixtures affect Gordal cv. green Spanish-style table olive fermentation. *Food Microbiol.* **2011**, *28*, 1316–1325. [CrossRef]
75. Rodríguez-Gómez, F.; Bautista-Gallego, J.; Arroyo-López, F.N.; Romero-Gil, V.; Jiménez-Díaz, R.; Garrido-Fernández, A.; Garcia-Garcia, P. Table olive fermentation with multifunctional *Lactobacillus pentosus* strains. *Food Control* **2013**, *34*, 96–105. [CrossRef]

76. Rodríguez-Gómez, F.; Romero-Gil, V.; Bautista-Gallego, J.; García-García, P.; Garrido-Fernández, A.; Arroyo-López, F.N. Production of potential probiotic Spanish-style green table olives at pilot plant scale using multifunctional starters. *Food Microbiol.* **2014**, *44*, 278–287. [CrossRef] [PubMed]
77. Rodríguez-Gómez, F.; Romero-Gil, V.; Arroyo-López, F.N.; Roldán-Reyes, J.C.; Torres-Gallardo, R.; Bautista-Gallego, J.; Garcia-Garcia, P.; Garrido-Fernandez, A. Assessing the challenges in the application of potential probiotic lactic acid bacteria in the large-scale fermentation of Spanish-style table olives. *Front. Microbiol.* **2017**, *8*, 915. [CrossRef]
78. López-López, A.; Moreno-Baquero, J.M.; Rodríguez-Gómez, F.; García-García, P.; Garrido-Fernández, A. Sensory Assessment by Consumers of Traditional and Potentially Probiotic Green Spanish-Style Table Olives. *Front. Nutr.* **2018**, *5*, 53. [CrossRef] [PubMed]
79. de Castro, A.; Sánchez, A.H.; Cortés-Delgado, A.; López-López, A.; Montaño, A. Effect of Spanish-style processing steps and inoculation with *Lactobacillus pentosus* starter culture on the volatile composition of cv. Manzanilla green olives. *Food Chem.* **2019**, *271*, 543–549. [CrossRef] [PubMed]
80. Martorana, A.; Alfonzo, A.; Gaglio, R.; Settanni, L.; Corona, O.; La Croce, F.; Vagnoli, P.; Caruso, T.; Moschetti, N.; Francesca, N. Evaluation of different conditions to enhance the performances of *Lactobacillus pentosus* OM13 during industrial production of Spanish-style table olives. *Food Microbiol.* **2017**, *61*, 150–158. [CrossRef]
81. Campus, M.; Cauli, E.; Scano, E.; Piras, F.; Comunian, R.; Paba, A.; Daga, E.; Di Salvo, R.; Sedda, P.; Angioni, A.; et al. Towards controlled fermentation of table olives: Lab starter driven process in an automatic pilot processing plant. *Food Bioprocess Technol.* **2017**, *10*, 1063–1073. [CrossRef]
82. Randazzo, C.L.; Todaro, A.; Pino, A.; Pitino, I.; Corona, O.; Mazzaglia, A.; Caggia, C. Giarraffa and Grossa di Spagna naturally fermented table olives: effect of starter and probiotic cultures on chemical, microbiological and sensory traits. *Food Res. Int.* **2014**, *62*, 1154–1164. [CrossRef]
83. Ciafardini, G.; Zullo, B.A. Use of selected yeast starter cultures in industrial-scale processing of brined Taggiasca black table olives. *Food Microbiol.* **2019**, *84*, 103250. [CrossRef]
84. Randazzo, C.L.; Fava, G.; Tomaselli, F.; Romeo, F.V.; Pennino, G.; Vitello, E.; Caggia, C. Effect of kaolin and copper based products and of starter cultures on green table olive fermentation. *Food Microbiol.* **2011**, *28*, 910–919. [CrossRef]
85. Pino, A.; De Angelis, M.; Todaro, A.; Van Hoorde, K.; Randazzo, C.L.; Caggia, C. Fermentation of Nocellara Etnea Table Olives by Functional Starter Cultures at Different Low Salt Concentrations. *Front. Microbiol.* **2018**, *9*, 1125. [CrossRef]
86. Durante, M.; Tufariello, M.; Tommasi, L.; Lenucci, M.S.; Bleve, G.; Mita, G. Evaluation of bioactive compounds in black table olives fermented with selected microbial starters. *J. Sci. Food Agric.* **2017**, *98*, 96–103. [CrossRef] [PubMed]
87. Zhu, Y.; González-Ortiz, G.; Benítez-Cabello, A.; Calero-Delgado, B.; Jiménez-Díaz, R.; Martín-Orúe, S.M. The use of starter cultures in the table olive fermentation can modulate antiadhesive properties of brine exopolysaccharides against enterotoxigenic *Escherichia coli*. *Food Funct.* **2019**, *10*, 3738–3747. [CrossRef]
88. Shah, P.N. Functional cultures and health benefits. *Int. Dairy J.* **2007**, *17*, 1262–1277. [CrossRef]
89. Fuller, R. Probiotics in man and animal. *J. Appl. Bacteriol.* **1989**, *66*, 365–378. [PubMed]
90. Fuller, R.; Gibson, G.R. Probiotics and prebiotics: microflora management for improved gut health. *Clin. Microbiol. Infect.* **1998**, *4*, 477–488. [CrossRef]
91. Gorbach, S.L. Probiotics in the third millennium. *Dig. Liver Dis.* **2002**, *34*, S2–S7. [CrossRef]
92. Guarner, F.; Malagelada, J.R. Gut flora in health and disease. *Lancet* **2003**, *361*, 512–519. [CrossRef]
93. Holzapfel, W.H.; Schillinger, U. Introduction to pre- and probiotics. *Food Res. Int.* **2002**, *35*, 109–116. [CrossRef]
94. Nagpal, R.; Kumar, A.; Kumar, M.; Behare, P.V.; Jain, S.; Yadav, H. Probiotics, their health benefits and applications for developing healthier foods: A review. *FEMS Microbiol. Lett.* **2002**, *334*, 1–15. [CrossRef]
95. FAO. *Probiotics in Food: Health and Nutritional Properties and Guidelines for Evaluation*; FAO: Rome, Italy, 2006; p. 85.
96. Haller, D.; Colbus, H.; Ganzle, M.G.; Scherenbacher, P.; Bode, C.; Hammes, W.P. Metabolic and functional properties of lactic acid bacteria in the gastro-intestinal ecosystem: A comparative in vitro study between bacteria of intestinal and fermented food origin. *Syst. Appl. Microbiol.* **2001**, *24*, 218–226. [CrossRef]

97. David, L.A.; Maurice, C.F.; Carmody, R.N.; Gootenberg, D.B.; Button, J.E.; Wolfe, B.E.; Ling, A.V.; Varma, Y.; Fischbach, M.A.; Biddinger, S.B.; et al. Diet rapidly and reproducibly alters the human gut microbiome. *Nature* **2014**, *505*, 559–563. [CrossRef] [PubMed]
98. Sekirov, I.; Russell, S.L.; Antunes, L.C.; Finlay, B.B. Gut microbiota in health and disease. *Physiol. Rev.* **2010**, *90*, 859–904. [CrossRef] [PubMed]
99. O'Hara, A.M.; O'Regan, P.; Fanning, A.; O'Mahony, C.; Macsharry, J.; Lyons, A.; Bienenstock, J.; O'Mahony, L.; Shanahan, F. Functional modulation of human intestinal epithelial cell responses by *Bifidobacterium infantis* and *Lactobacillus salivarius*. *Immunology* **2006**, *118*, 202–215. [CrossRef] [PubMed]
100. Argyri, A.A.; Zoumpopoulou, G.; Karatzas, K.A.G.; Tsakalidou, E.; Nychas, G.J.E.; Panagou, E.Z.; Tassou, C.C. Selection of potential probiotic lactic acid bacteria from fermented olives by in vitro tests. *Food Microbiol.* **2013**, *33*, 282–291. [CrossRef]
101. Bautista-Gallego, J.; Arroyo-López, F.N.; Rantsiou, K.; Jiménez-Díaz, R.; Garrido-Fernández, A.; Cocolin, L. Screening of lactic acid bacteria isolated from fermented table olives with probiotic potential. *Food Res. Int.* **2013**, *50*, 135–142. [CrossRef]
102. Botta, C.; Langerholc, T.; Cencic, A.; Cocolin, L. In vitro selection and characterization of new probiotic candidates from table olive microbiota. *PLoS ONE* **2014**, *9*, e94457. [CrossRef] [PubMed]
103. Pérez Montoro, B.; Benomar, N.; Lavilla Lerma, L.; Castillo Gutiérrez, S.; Gálvez, A.; Abriouel, H. Fermented Aloreña table olives as a source of potential probiotic *Lactobacillus pentosus* strains. *Front. Microbiol.* **2016**, *7*, 1583. [CrossRef]
104. Prete, R.; Tofalo, R.; Federici, E.; Ciarrocchi, A.; Cenci, G.; Corsetti, A. Food-associated *Lactobacillus plantarum* and yeasts inhibit the genotoxic effect of 4-Nitroquinoline-1-Oxide. *Front. Microbiol.* **2017**, *8*, 2349. [CrossRef]
105. Garcia-Gonzalez, N.; Prete, R.; Battista, N.; Corsetti, A. Adhesion Properties of Food-Associated *Lactobacillus plantarum* Strains on Human Intestinal Epithelial Cells and Modulation of IL-8 Release. *Front. Microbiol.* **2018**, *9*, 2392. [CrossRef]
106. Prete, R.; Long, S.L.; Gallardo, A.L.; Gahan, C.G.; Corsetti, A.; Joyce, S.A. Beneficial bile acid metabolism from *Lactobacillus plantarum* of food origin. *Sci. Rep.* **2020**, *10*, 1165. [CrossRef]
107. Delgado, A.; Brito, D.; Peres, C.; Noe-Arroyo, F.; Garrido-Fernández, A. Bacteriocin production by *Lactobacillus pentosus* B96 can be expressed as a function of temperature and NaCl concentration. *Food Microbiol.* **2005**, *22*, 521–528. [CrossRef]
108. Kacem, M.; Karam, N.E. Microbiological study of naturally fermented Algerian green olives: Isolation and identification of lactic acid bacteria and yeasts along with the effects of brine solutions obtained at the end of olive fermentation on *Lactobacillus plantarum* growth. *Grasas y Aceites* **2006**, *57*, 292–300. [CrossRef]
109. Mokhbi, A.; Kaid-Harche, M.; Lamri, K.; Rezki, M.; Kacem, M. Selection of *Lactobacillus plantarum* strains for their use as starter culturesin Algerian olive fermentations. *Grasas y Aceites* **2009**, *60*, 82–89. [CrossRef]
110. Guantario, B.; Zinno, P.; Schifano, E.; Roselli, M.; Perozzi, G.; Palleschi, C.; Uccelletti, D.; Devirgiliis, C. In vitro and in vivo selection of potentially probiotic lactobacilli from Nocellara del Belice table olives. *Front. Microbiol.* **2018**, *9*, 595. [CrossRef] [PubMed]
111. Benítez-Cabello, A.; Torres-Maravilla, E.; Bermúdez-Humarán, L.; Langella, P.; Martín, R.; Jiménez-Díaz, R.; Arroyo-López, F.N. Probiotic properties of *Lactobacillus* strains isolated from table olive biofilms. *Probiotics Antimicrob. Proteins* **2019**. [CrossRef] [PubMed]
112. Lavermicocca, P.; Valerio, F.; Lonigro, S.L.; de Angelis, M.; Morelli, L.; Callegari, M.L. Study of adhesion and survival of lactobacilli and bifidobacteria on table olives with the aim of formulating a new probiotic food. *Appl. Environ. Microbiol.* **2005**, *71*, 4233–4240. [CrossRef] [PubMed]
113. Valerio, F.; de Candia, S.; Lonigro, S.L.; Russo, F.; Riezzo, G.; Orlando, A.; De Bellis, P.; Sisto, A.; Lavermicocca, P. Role of the probiotic strain *Lactobacillus paracasei* LMGP22043 carried by artichokes in influencing faecal bacteria and biochemical parameters in human subjects. *J. Appl. Microbiol.* **2011**, *111*, 155–164. [CrossRef]
114. Argyri, A.A.; Nisiotou, A.A.; Pramateftaki, P.; Doulgeraki, A.I.; Panagou, E.Z.; Tassou, C.C. Preservation of green table olives fermented with lactic acid bacteria with probiotic potential under modified atmosphere packaging. *LWT Food Sci. Technol.* **2015**, *62*, 783–790. [CrossRef]
115. Perpetuini, G.; Pham-Hoang, B.N.; Scornec, H.; Tofalo, R.; Schirone, M.; Suzzi, G.; Cavin, J.F.; Waché, Y.; Corsetti, A.; Licandro-Seraut, H. In *Lactobacillus pentosus*, the olive brine adaptation genes are required for biofilm formation. *Int. J. Food Microbiol.* **2016**, *216*, 104–109. [CrossRef]

116. Pérez Montoro, B.; Benomar, N.; Caballero Gómez, N.; Ennahar, S.; Horvatovich, P.; Knapp, C.W.; Alonso, E.; Gálvez, A.; Abriouel, H. Proteomic analysis of *Lactobacillus pentosus* for the identification of potential markers of adhesion and other probiotic features. *Food Res. Int.* **2018**, *111*, 58–66. [CrossRef] [PubMed]
117. Abriouel, H.; Pérez Montoro, B.; Casimiro-Soriguer, C.S.; Pérez Pulido, A.J.; Knapp, C.W.; Caballero Gómez, N.; Castillo-Gutiérrez, S.; Estudillo-Martínez, M.D.; Gálvez, A.; Benomar, N. Insight into potential probiotic markers predicted in *Lactobacillus pentosus* MP-10 genome sequence. *Front. Microbiol.* **2017**, *8*, 891. [CrossRef] [PubMed]
118. Calero-Delgado, B.; Pérez-Pulido, A.J.; Benítez-Cabello, A.; Martín-Platero, A.M.; Casimiro-Soriguer, C.S.; Martínez-Bueno, M.; Arroyo-López, F.N.; Jiménez Díaz, R. Multiple genome sequences of *Lactobacillus pentosus* strains isolated from biofilms on the skin of fermented green table olives. *Microbiol. Resour. Announc.* **2019**, *8*, e01546-18. [CrossRef] [PubMed]
119. Martins, E.M.F.; Ramos, A.M.; Vanzela, E.S.L.; Stringheta, P.C.; Pinto, C.L.O.; Martins, J.M. Products of vegetable origin: A new alternative for the consumption of probiotic bacteria. *Food Res. Int.* **2013**, *51*, 764–770. [CrossRef]
120. Sazawal, S.; Hiremath, G.; Dhingra, U.; Malik, P.; Deb, S.; Black, R.E. Efficacy of probiotics in prevention of acute diarrhoea: A meta-analysis of masked, randomised, placebo-controlled trials. *Lancet Infect. Dis.* **2006**, *6*, 374–382. [CrossRef]
121. Pennacchia, C.; Blaiotta, G.; Pepe, O.; Villani, F. Isolation of *Saccharomyces cerevisiae* strains from different food matrices and their preliminary selection for a potential use as probiotics. *J. Appl. Microbiol.* **2008**, *105*, 1919–1928. [CrossRef]
122. Kourelis, A.; Kotzamanidis, C.; Litopoulou-Tzanetaki, E.; Scouras, Z.G.; Tzanetakis, N.; Yiangou, M. Preliminary probiotic selection of dairy and human yeast strains. *J. Biol. Res.* **2010**, *13*, 93–104.
123. Moslehi-Jenabian, S.; Lindegaard Pedersen, L.; Jespersen, L. Beneficial effects of probiotic and food borne yeasts on human health. *Nutrients* **2010**, *2*, 449–473. [CrossRef]
124. Etienne-Mesmin, L.; Livrelli, V.; Privat, M.; Denis, S.; Cardot, J.M.; Alric, M.; Blanquet-Diot, S. Effect of a new probiotic *Saccharomyces cerevisiae* strain on survival of *Escherichia coli* O157:H7 in a dynamic gastrointestinal model. *Appl. Environ. Microbiol.* **2011**, *77*, 1127–1131. [CrossRef]
125. Silva, T.; Reto, M.; Sol, M.; Peito, A.; Peres, C.M.; Peres, C.; Malcata, F.X. Characterization of yeasts from Portuguese brined olives, with a focus on their potentially probiotic behaviour. *LWT Food Sci. Technol.* **2011**, *44*, 1349–1354. [CrossRef]
126. Granato, D.; Branco, G.F.; Nazzaro, F.; Cruz, A.G.; Faria, J.A.F. Functional foods and non-dairy probiotic food development: Trends, concepts, and products. *Compr. Rev. Food Sci. Food Saf.* **2010**, *9*, 292–302. [CrossRef]
127. Vijaya Kumar, B.; Vijayendra, S.V.N.; Reddy, O.V.S. Trends in dairy and non-dairy probiotic products—A review. *J. Food Sci. Technol.* **2015**, *52*, 6112–6124. [CrossRef] [PubMed]
128. Ranadheera, S.; Baines, S.K.; Adams, M.C. Importance of food in probiotic efficacy. *Food Res. Int.* **2010**, *43*, 1–7. [CrossRef]
129. Lamsal, B.P.; Faubion, J. The beneficial use of cereal and cereal components in probiotic foods. *Food Rev. Int.* **2009**, *25*, 103–114. [CrossRef]

foods

Review

Table Olives: An Overview on Effects of Processing on Nutritional and Sensory Quality

Paola Conte, Costantino Fadda, Alessandra Del Caro, Pietro Paolo Urgeghe and Antonio Piga *

Dipartimento di Agraria, Università degli Studi di Sassari, Viale Italia 39/A, 07100 Sassari, Italy; pconte@uniss.it (P.C.); cfadda@uniss.it (C.F.); delcaro@uniss.it (A.D.C.); paolou@uniss.it (P.P.U.)
* Correspondence: pigaa@uniss.it; Tel.: +39-(0)79-229272

Received: 19 March 2020; Accepted: 13 April 2020; Published: 20 April 2020

Abstract: Table olives are a pickled food product obtained by a partial/total debittering and subsequent fermentation of drupes. Their peculiar sensory properties have led to a their widespread use, especially in Europe, as an appetizer or an ingredient for culinary use. The most relevant literature of the last twenty years has been analyzed in this review with the aim of giving an up-to-date overview of the processing and storage effects on the nutritional and sensory properties of table olives. Analysis of the literature has revealed that the nutritional properties of table olives are mainly influenced by the processing method used, even if preharvest-factors such as irrigation and fruit ripening stage may have a certain weight. Data revealed that the nutritional value of table olives depends mostly on the balanced profile of polyunsaturated and monounsaturated fatty acids and the contents of health-promoting phenolic compounds, which are best retained in natural table olives. Studies on the use of low salt brines and of selected starter cultures have shown the possibility of producing table olives with an improved nutritional profile. Sensory characteristics are mostly process-dependent, and a relevant contribute is achieved by starters, not only for reducing the bitterness of fruits, but also for imparting new and typical taste to table olives. Findings reported in this review confirm, in conclusion, that table olives surely constitute an important food source for their balanced nutritional profile and unique sensory characteristics.

Keywords: composition; nutritional properties; polyphenols; sensory analysis; table olives

1. Introduction

The olive (*Olea europaea* L.) originates in the Mediterranean countries; it can be found in the wild form in the Middle East and it is widely distributed around the world, especially in the Mediterranean region, where about 96% of the world's production of olives occurs [1]. It grows in form of an evergreen tree, and the first domestic cultivation dates to the Minoan period (3500–1500 BC) in Crete [2]. The fruits are mainly used to produce oil and table olives, a widely consumed food of the Mediterranean countries. The World Catalogue of Olive Cultivars [3] reports about 2500 olive varieties, but only 10% of them can be considered commercial, and their selected use (oil, table or both) is determined by different parameters. Table olives, in fact, are prepared from varieties low in oil content, medium to large in size and appropriate in shape, with flesh-to-pit ratios higher than 4, green to black skin and appropriate texture (depending on the skin color). The main table olive varieties used in the five major producing countries are Gordal, Manzanilla and Hojiblanca for Spain; Aggezi Shami, Hamed and Toffahi for Egypt; Gemlik, Memecik and Memely for Turkey; Konservolia, Chalkidiki and Kalamon for Greece; Azeraj and Sigoise for Algeria. The International Olive Oil Council has estimated for the 2017/2018 crop year that Egypt, with 655.000 tons, will be for the first time the world leading country for table olive production. The olive trees produce drupes that are each constituted by a thin epidermis and a soft mesocarp surrounding a stone containing the seed [4]. The epidermis (1.5–3% of the total weight) has a protective function against external attacks and it is mainly constituted of cellulose and cutin [5,6]. Olive mesocarp represents 70–90% of the weight. The stone accounts for the 10–30%, while the seed is about 1–3% of the whole fruit, and it is made up mainly of lipids [7]. Olives fruits have a round to ovoid shape, and their weight ranges from 0.5 to 20g, with a major frequency in the weight class of 3–10 g. Additionally, they are characterized by a strong bitter taste that decreases with fruit ripening, during which the peel color changes from green to light-yellow, purple-red and purple-black. The principal components of olives are water (60–75%), lipids (10–25%), reducing sugars (2–5%) and phenolic substances (1–3%) [4,8]. Olives, moreover, have good amounts of tocopherols, carotenoids [9] and minerals [10]. Among the cited components, olives are very rich in polyphenols, which are important for the sensory properties of olives, and may have various health promoting activities [11]. Polyphenols in olives belong to the following five different classes [12,13]: acids (caffeic, gallic, syringic); alcohols (tyrosol, hydroxytyrosol); flavonoids (luteolin-7-glucoside, cyanidin-3-glucoside); secoiridoids, such as the bitter oleuropein that diminishes during maturation, demethyloleuropein and the dialdehydic form of elenolic acid linked to tyrosol and hydroxytyrosol—whose amount in contrast, increases with fruit maturation; and lignans (1-acetoxypinoresinol, pinoresinol). The International Olive Oil Council (IOC) [14] has recently reported on the importance of table olives in an every-day diet, as this specialty is the most consumed fermented food in Europe and accounts for a worldwide production of close to 3 million tons. Some authors recommend daily consumption of a serving size [15,16].

Tables olive processing involves the removal of the bitter taste, and in most cases the subsequent fermentation that imparts to the fruits a well-defined sensory profile, while avoiding the growth of pathogenic bacteria and giving proper stability [4]. Unit operations involved during processing and storage, on the other hand, may have important effects on the nutritional and sensory characteristics of fresh olives, and this review has the aim of giving an up-to-date overview of how the processing and storage of table olives may affect the nutritional and sensory characteristics of this pickled food.

2. How Processing Influences the Nutritional Properties of Table Olives

According to IOC [17] "Table olives are the product prepared from the sound fruits of varieties of the cultivated olive tree that are chosen for their production of olives whose volume, shape, flesh-to-stone ratio, fine flesh, taste, firmness and ease of detachment from the stone make them particularly suitable for processing; treated to remove its bitterness and preserved by natural fermentation, or by heat treatment with or without the addition of preservatives; packed with or without covering liquid". Classification of table olives could be made on the basis of the ripening stage at harvest (green, turning color and black), trade preparations (treated, natural, darkened by oxidation, dehydrated and/or shriveled, specialties) and styles (whole, pitted, stuffed, salad and other). According to the trade preparations, about 80% of the world's production is covered by three commercial processing methods: treated green olives (or Spanish style green olives); olives darkened by oxidation (ripe olives) (Californian style); natural (mainly black) olives (Greek style) (Figure 1). Processing, in any case, promotes a quantitative and qualitative evolution in the phenolic compounds of table olives, thereby changing their sensory and health properties [18].

In the following pages, we will review the effects of the trade preparation methods and styles, and the influence of the microbial starter on the table olives' nutritional quality.

The following databases were used for the bibliographic research: Web of Science (2000–2020), Scopus (2000–2020) and Food Science and Technology Abstracts (2000–2020). Some papers deal with either nutritional or sensory topic; thus, we discuss them in both sections. Results are summarized in Table 1.

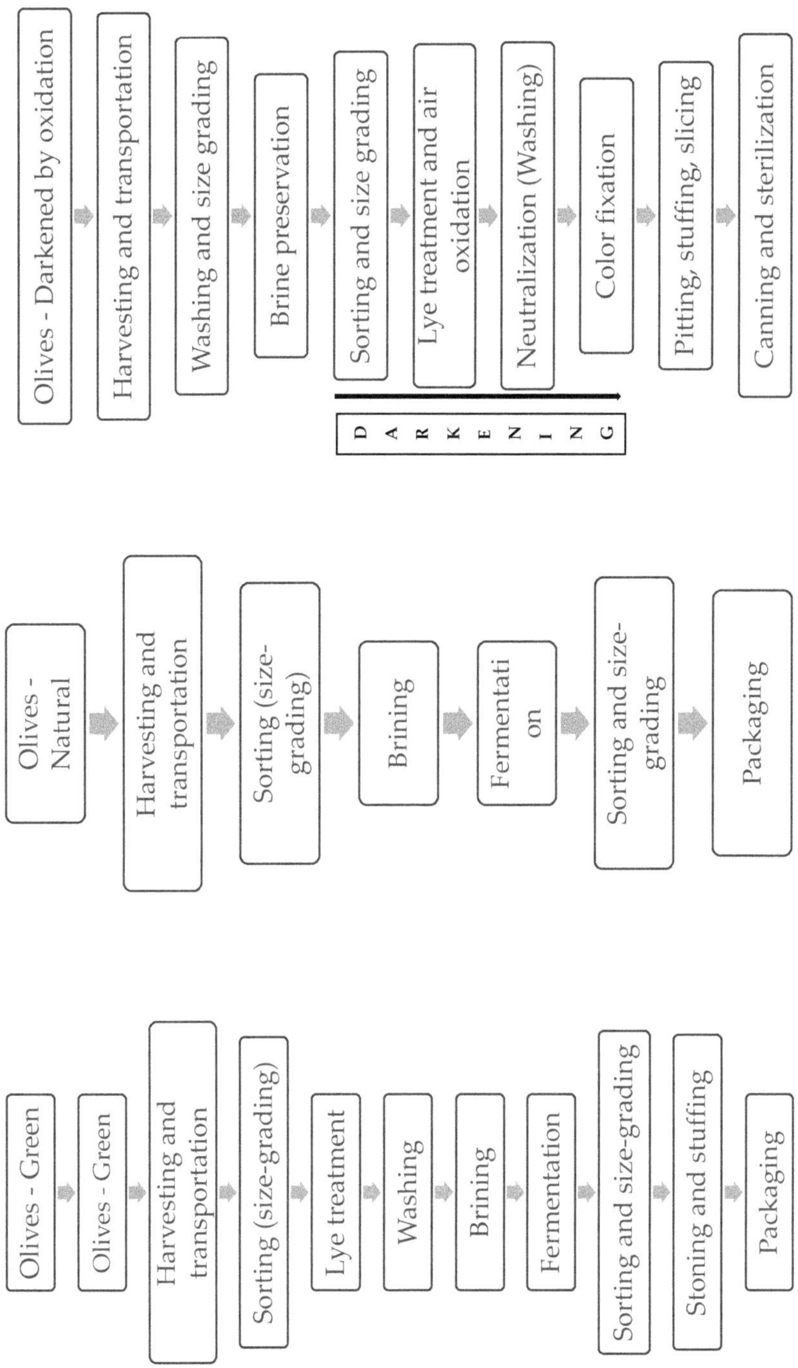

Figure 1. Flow sheets of processing of treated green, natural and darkened by oxidation olives.

2.1. Trade Preparations

2.1.1. Treated Green Olives or "Spanish Style"

Basically, lactic acid bacteria (LAB) ferment brined olives, which have been previously debittered through a chemical hydrolysis of oleuropein by a lye treatment in a 1.5–4.5% (w/v) NaOH solution until 2/3 of the mesocarp is interested; after that, olives are drained and washed with water. The alkali treatment speeds up the fermentation, as it increases the skin permeability and the efflux of fermentable compounds and nutrients in the sodium chloride (NaCl) brine. Diffused sugars are converted into lactic acid by LAB, which predominate after the first days of processing. The final product, which is obtained after 30–60 days from brining, has a pH of 3.8–4 and 5–6% NaCl, and it is shelf stable in its final pack. The use of sorbic acid or application of pasteurization have been reported to extend the shelf life [12]. Nutritional losses from the fresh olives result both from the alkali treatment and from leakage of soluble constituents from olive mesocarp to brine. Sakouhi et al. [19] studied how ripening and processing may change the contents of α-tocopherol and fatty acids (FA) of three Tunisian varieties (Meski, Picholine and Sayali). They harvested fruits while green but also at cherry and black stages of ripening, and showed that α-tocopherol increased during ripening and decreased after fermentation, especially when black olives were used. Data on fatty acids revealed that for the three cultivars, the ratio of polyunsaturated (PUFA) to saturated fatty acids (SFA) was lower and irregular at the cherry stage, with respect to the other two ripening stages, but this ratio increased after processing in Meski and Picholine olives at a value higher than 1.5, which is usually associated with health-promoting capacity [20]. Lanza et al. [21] focused on the nutritional properties of Spanish style Italian "Intosso d'Abruzzo" fermented olives. The authors showed that these olives may be considered a food with a high nutritional potential for their low and balanced fat profile, especially for the high rate between monounsaturated fatty acids (MUFA) and SFA; the appreciable amounts of polyphenols, α-tocopherol, minerals and fiber; an adequate content of essential amino acids; and a normal NaCl level. López-López et al. [22] verified whether the Spanish-style could affect the FA and triacylglycerol (TAG) composition of Manzanilla ad Hojiblanca olive cultivars. The authors used principal component analysis (PCA) to analyze data, and found that FA, TAG and nutritional fat subclasses were influenced mainly by cultivar and to a very low extent by processing, and that the ratio PUFA/SFA was slightly lower than the 0.5 that is recommended for prevention of coronary hearth diseases. Cano-Lamadrid et al. [23] evaluated the influences of three different irrigation regimes, from normal to moderate stress, on the fatty acid composition of green Manzanilla olives. Results showed that olives grown under moderate irrigation stress had the highest content of PUFA, and in particular, linoleic acid. The same research group [24] carried out a study by applying a similar irrigation experimental plan and evaluated its influence on the phenolic profile of Spanish style Manzanilla olives. Results evidenced that a moderate level of irrigation increased the amounts of some polyphenols in table olives, especially those of compounds with health promoting activities, such oleuropein and oleoside diglucoside. Mastralexi et al. [25] followed the evolution of hydrophilic and hydrophobic antioxidants of the protected denomination of origin (PDO) "Prasines Elies Chalkidikis" olives prepared at industrial scale and following a storage period of 12 months. The authors found that NaOH debittering and subsequent washing reduced the total polyphenol content by at least 2/3. Oleuropein was completely removed by the alkali treatment, and only hydroxytyrosol, tyrosol and oleoside-11-methyl ester were found at the end of washing in significantly higher concentrations with respect to fresh olives. The hydrophobic nutrients α-tocopherol and squalene were not affected by processing. The ensuing storage in brine led to a decrease of squalene and phenols; the latter, however, were high enough to permit to use the health claim on olive oil polyphenols [26]; however, the authors highlighted some concerns about an overly high final salt content.

Results of the papers above-discussed reveal that treated green olives may have an important nutritional profile, regarding the adequate FA content, and the appropriate PUFA/SFA ratio and α-tocopherol minerals. Polyphenols, on the other hand, despite undergoing a severe loss due to the lye treatment, remain at a good level.

2.1.2. Natural Olives

This trade preparation is performed by harvesting olives at the three ripening stages and then fermenting them directly in brine. Aids may be used to further preserve olives. Primarily, fruits at the black stage are used, and this preparation is known as "Greek style". The fermentation may be achieved in an 8–10% NaCl brine in anaerobic or aerobic conditions. In the last case, a modification of the fermenter is obtained by bubbling air through a central column. The regulation of NaCl in brine drives the type of fermentation, because when NaCl is higher than 8% yeast (Y) predominates, while an NaCl concentration of 3–6% may promote the LAB growth in turning or black olives. The anaerobic fermentation requires a long time, from 8 to 12 months, to solubilize oleuropein in the brine, while the aerobic system significantly reduces the process time and limits gas-pocket spoilage and shriveling of fruits [27]. The obtained olives may be packed directly in brine and sold, or they may also be submitted to pasteurization or even preserved with the addition of sorbic acid at 0.5 % to the packing brine. The nutritional loss is mainly caused by leakage of soluble compounds into the brine during fermentation and storage. Boskou et al. [28] worked on five different commercial samples of black Greek-style fermented olives: he analyzed the polyphenolic pool. He identified 13 different polyphenols; the main ones were hydroxytyrosol, oleanolic acid and tyrosol. Data obtained on the different samples evidenced, regardless of the cultivar and preparation, an appropriate amount of polyphenols for covering the requested daily intake, which could be satisfied by 5 or 10–12 olives in North European countries or Greece, respectively. Similar results were obtained by Pires-Cabral et al. [29] on fermented table olives belonging to three Portuguese cultivars (Cobrançosa, Galega and Maçanilha Algarvia). Results showed an important dietary fiber and polyphenol content and high amounts of PUFA in all samples, with particular emphasis on the Maçanilha cultivar, which was able to provide 13.1% of the recommended daily intake of PUFA. In a further paper of Pires-Cabral et al. [30] the authors studied the nutritional properties of a Portuguese olive cultivar fermented in a reduced NaCl brine (4% NaCl + 4% KCl). The authors highlighted the use of this technology in halving the Na content of olives and in increasing by six and four-times, K and Ca, respectively, in comparison to samples fermented in a conventional brine (8% NaCl). D'Antuono et al. [31] revealed by liquid chromatography–mass spectrometry/mass spectrometry (LC-MS/MS), for the first time in Greek-style processed olives, the presence of three nutritionally important polyphenols—hydroxytyrosol acetate (HTAc), caffeoyl-6′-secologanoside (SEC) and comselogoside (COM); the first was previously found only in the olive oil of table olives [32], and the other two in air-dried olives [33]. The authors also showed the good bioaccessibility of these phenolic compounds and evidenced that these table olives can be considered a functional food. In a more recent paper of Fernández-Poyatos et al. [34], a complete characterization of the polyphenolic and inorganic fractions of Cornezuelo natural processed table olives was carried out. The authors identified thirty phenolic compounds, the most representative being oleuropeine and comselogoside isomers, and a high amount of Ca. The authors submitted the polyphenolic extract to a simulated gastrointestinal digestion and found that, although almost 50% of it has been digested, an important residual in vitro antioxidant activity remained. In a study of Rodríguez et al. [35], thirty-two commercial samples obtained with ten cultivars and different styles were analyzed for phenol composition, with emphasis on the nutritional important compound 3,4-dihydroxyphenylglycol (DHPG) [36]. Data obtained on natural-style olives processed at the black stage revealed a high concentration of DHPG; thus, these samples have an interesting nutritional potential.

Direct brine fermentation confirms, thus, the disadvantage, with respect to treated olives' trade preparation, of longer processing times, but it results in olives that contain health-promoting polyphenols.

Table 1. Effect of trade preparation and processing style on nutritional quality of table olives.

Compound Class	Trade Preparation/Processing Style/Starters (LAB*, Y)	Olive Cultivar	Olive Ripening Stage	Nutritional Results Related to the Compound Class (Results Related to Other Compounds)	References
Polyphenols	Treated	Manzanilla	Green	A moderate level of irrigation increased oleuropein and oleoside diglucoside in table olives	Sánchez-Rodríguez et al. [24]
		Prasines Elies Chalkidikis	Green	Hydroxytyrosol, tyrosol and oleoside-11-methyl ester at higher concentrations in treated olives with respect to fresh ones (decrease of squalene in brine stored olives)	Mastralexi et al. [25]
	Natural	Various	Black	Greek-style olives are a good source of polyphenols	Boskou et al. [28]
		Bella di Cerignola	Black	Presence of hydroxytyrosol acetate, caffeoyl-6′-secologanoside and comselogoside with good bioaccessibility	D'Antuono et al. [31]
		Cornezuelo	Green	Prevalence of oleuropeine and comselogoside isomers, polyphenols retain a good antioxidant activity after digestion (high Ca content)	Fernández-Poyatos et al. [34]
		Eight different Greek cultivars	Black	High concentration of the nutritional important compound 3,4-dihydroxyphenylglycol	Rodríguez et al. [35]
	Dried	Thassos	Black	Storage at 20 °c of dried olives allowed the best retention of polyphenols	Mantzouridou et al. [33]
		Majatica di Ferrandina	Black	High content of biophenols	Lanza et al. [37]
	Other (Water debittering)	Megaritiki	Black	New compounds were detected, two are unique for the species (rengyoxide and cleroindicin C) and one for table olives (haleridone)	Mousori et al. [38]
	Other (High Pressure Processing)	Different Turkish cultivars	Black	Increase of phenolics following the HPP treatment	Tokuşoğlu et al. [39]
	Mixed (Natural and drying with salt)	Meski, Chemlali, Besbessi, Tounsi	Black	Olives of Tunisian market have an important content of polyphenols	Ben Othman et al. [40]
	Mixed (Water dipping and Spanish Style)	Istrska belica, Storta	Green	Debittering with water dipping resulted in higher biophenol content in olives	Valenčič et al. [41]
	Mixed (Five different processing styles)	Nine different Greek cultivars	Black	The highest oleuropein content was found on dry-salted Throuba Thassos olives	Zoidou et al. [42]
	Mixed (Darkened by oxidation and drying with salt)	Manzanilla, Mission, Throuba Thassos	Black	Confirms results of Zoidou et al. [42]	Melliou et al. [43]

Table 1. *Cont.*

Compound Class	Trade Preparation/Processing Style/Starters (LAB*, Y)	Olive Cultivar	Olive Ripening Stage	Nutritional Results Related to the Compound Class (Results Related to Other Compounds)	References
	LAB (Natural)	Frantoio, Carolea, Coratina, Leccino	Black	LAB produced tissue skin degradation with consequent higher polyphenol leakage and reduced debittering time	Servili et al. [44]
		Kalamata, Chalkidikis	Black, green	LAB fermented olives had a significantly higher content in phenols, especially hydroxytyrosol and tyrosol	Tataridou et al. [45]
	LAB + Y (Natural)	Taggiasca	Black	Polyphenol loss from flesh to brines depend on process temperature and not on starter use	Pistarino et al. [46]
	Y + LAB (Natural)	Cellina di Nardò, Conservolea, Kalamàta, Leccino	Black	Olives are rich in polyphenolic compounds (rich in MUFA, polyphenols, tocopherols and triterpenic acids)	Durante et al. [47]
		Bella di Cerignola, Termite di Bitetto, Cellina di Nardò	Black	The use of starters allowed to obtain olives with high content of tyrosol and hydroxytyrosol that were up to eight times higher with respect to the virgin olive oils obtained by the same olives	D'Antuono et al. [48]
	Y (Natural)	Picual, Manzanilla, Kalamàta	Black, green	The Y allowed an increase in hydroxytyrosol, tyrosol and verbascoside on the olives, with respect to the control sample, thus improving the nutritional value	Tufariello et al. [49]
Fatty acids	Treated	Meski, Picholine, Sayali	Black, cherry, green	Decrease of FA and increase of the PUFA/SFA ratio after processing (α-tocopherol decreased after fermentation, mainly in black olives)	Sakouhi et al. [19]
		Intosso d'Abruzzo	Green	Optimal MUFA/SFA ratio (appreciable amounts of polyphenols, α-tocopherol, minerals and fibre)	Lanza et al. [21]
		Manzanilla, Hojiblanca		PUFA/SFA ratio lower than 0.5	Lopez-Lopez et al. [22]
		Manzanilla	Green	High PUFA content in olives grown under moderate irrigation regime	Cano-Lamadrid et al. [23]
	Natural	Maçanilha, Cobrançosa, Galega	Black	A serving size of Maçanilha olives provide 13.1% of the recommended daily intake of PUFA (important content of dietary fiber and polyphenols)	Pires-Cabral et al. [29]
		Cellina di Nardò, Conservolea, Kalamàta, Leccino	Black	High content of MUFA	Durante et al. [47]
	Oven-dried	Ferrandina	Black	PUFA/SFA ratio higher than 0.5 (dried fruits contained appreciable amounts of phenols and tocopherols)	Lanza et al. [37]

Table 1. *Cont.*

Compound Class	Trade Preparation/Processing Style/Starters (LAB*, Y)	Olive Cultivar	Olive Ripening Stage	Nutritional Results Related to the Compound Class (Results Related to Other Compounds)	References
	Other (Alcaparras)	Not reported	Green	Oleic acid content up to 81% (lower content of Vitamin E if compared to olives prepared with other styles)	Sousa et al. [50]
		Cobrançosa, Madural, Negrinha de Freixo, Santulhana, Verdeal, Transmontana	Green	Content of some fatty acids and of SFA, MUFA and PUFA permitted a statistical discrimination among cultivars	Malheiro et al. [51]
	Mixed (Treated, natural, darkened by oxidation)	Alorena, Arbequiña, Cacereña, Carrasqueña, Gordal, Hojiblanca, Manzanilla, Verdial	Green, black	The fat profile was useful to discriminate olive cultivars	López-López et al. [52]
Minerals	Cracked	Aloreña	Green	Packing brines with combinations of $CaCl_2$, KCl and NaCl resulted in significant reduction of flesh Na content, with respect to the traditional packed product	Moreno-Baquero et al. [53]
		Maçanilha Algarvia	Green	Brines at 4% NaCl + 4% KCl gave olives with increased K and reduced Na contents (lower fat, similar dietary fiber, phenolic compounds and Ca content with respect to the control brine (8% NaCl)	Saúde et al. [54]
	Mixed (Treated, water dip, scratched plus $CaCl_2$ dipping and reduced salt brines)	Domat	Green	Reduced salt content in scratched olives processed in low-salt brines	Savas et al. [55]
Triterpenic acids	Mixed (Treated, natural, darkened by oxidation)	Seventeen different cultivars	Black, green	Natural-style olives have the highest content of triterpenic acids, with respect to the other trade preparations and to virgin olive oil	Romero et al. [56]
Fibres	Mixed (Natural, darkened by oxidation, dried)	Douro, Hojiblanca, Cassanese, Conservolia, Taggiasca, Thasos	Black, green	High content of fibers in all samples	Jiménez et al. [57]

*LAB= lactic acid bacteria; Y=yeast; FA= fatty acids; MFA=monounsaturated fatty acids; PF=polyunsaturated fatty acids; SFA=saturated fatty acids.

2.1.3. Dehydrated and/or Shriveled Olives

According to IOC [17], this trade preparation is carried out on "green olives, olives turning color or black olives that have undergone or not to mild alkaline treatment, preserved in brine or partially dehydrated in dry salt and/or by heating or by any other technological process".

Drying is one of the oldest unit operations for food stabilization, and it is based on constitutive water removal under water activity (a_W) values below the threshold for microbial growth. Drying of foods like olives is often carried out using cabinet drying equipment. Nutritional loss is expected due to thermal damage. Mantzouridou et al. [33] evaluated the influences of mild drying conditions and storage for 6 months at 4 or 20 °C in an air, nitrogen or vacuum atmosphere on the phenol composition of intermediate moisture olives (a_W =0.89). After drying, the authors found a significant decrease of the single polyphenols, up to 73%, that continued during storage. They also evidenced that the best combination in reducing such a loss during storage was keeping olives under vacuum at 20 °C, which assured the highest contents of nutritionally important polyphenols, such as oleuropeine and hydroxytyrosol. Other important results were the high contents of eleanolic acid and elanolic acid glicoside, which are hydrolytic oleuropein derivatives and may be considered bioactive compounds. The authors concluded that, despite this loss, the olives maintain a sufficient polyphenol content to assure a proper shelf life. In another paper, Lanza et al. [37] evaluated the nutritional properties of oven-dried Ferrandina table olives. Although the authors found a low protein content, they observed an important contribution from some essential amino acids. They also revealed that the fat content was high, but with a balanced composition of PUFA. Moreover, the dried fruits contained appreciable amounts of phenols and tocopherols.

The expected nutritional loss due to the thermal treatment has been demonstrated by the above-cited papers, which, however, highlighted adequate contents, in the finished products, of some specific compounds, such as essential amino acids, fats, polyphenols and tocopherols.

2.1.4. Other Processing Methods and Stabilization Treatments

The "alcaparras" are a Portuguese table olive specialty. Alcaparras are prepared with olives harvested at the green or yellow-green stage. Fruits are cut with a hammer to separate pulp from stone and then halved; thus, they may be classified as "stoned halved olives" [17]. After that, halved fruits are dipped in water several times over a week, in order to remove oleuropein by diffusion, and this results in a significant loss of all polyphenols [58]. Stabilization of olives is carried out by placing them in brine. Sousa et al. [50] did a complete nutritional characterization of thirty stoned alcaparras table olive samples along three production seasons. The caloric value of processed olives is lower with respect to the majority of other commercial samples; moreover, they have a higher content of oleic acid and a lower content of α-tocopherol, if compared to olives prepared in other styles, although the authors evidenced that a serving size may provide a moderate contribution to the daily intake of tocopherols. Similar results were obtained by a further study of the same research group that investigated the effects of the influence of cultivar on main nutritional quality of alcaparras olives [51].

Another diffused style in the major producing countries is the "cracked olives style," which is when the "whole olives [are] subjected to a process whereby the flesh is opened without breaking the stone (pit), which remains whole and intact inside the fruit" [17]. Moreno-Baquero et al. [53] substituted up to 50% of NaCl in packing brines with combinations of $CaCl_2$, KCl and NaCl, and checked for the influence on the mineral nutrients of cracked Aloreña olives, one of the three processing styles of this fruit (a complete description will be reported in Section 3.1.4). The authors found that the reduction of NaCl resulted in significant reduction of flesh Na content, with respect to the traditional packed product; moreover, the contents of K and Ca increased. An important contribution to the knowledge of polyphenol changes during table olive processing was given by Mousori et al. [38], who used nuclear magnetic resonance (NMR) to detect new compounds and relative metabolites from Megaritiki table olives and wastewaters. The authors detected, for the first time, compounds that are unique for the species (rengyoxide and cleroindicin C) and for table olives (haleridone), and found four lactones derived from oleuropein hydrolysis. Another promising non-chemical debittering unit operation has been proposed by Habibi et al. [59], who checked the influence of ultrasound on the nutritional content (protein, ash and fat) of natural fermented table olives. Ultrasound-assisted debittering (UAD) was carried out both in water and in brine, and two unassisted controls were considered. The UAD significantly decreased the debittering time and left unchanged all the nutritional parameters, with respect to the controls, except for ashes that increased in UAD samples. Saúde et al. [54] tested the effect of brine NaCl replacement with $CaCl_2$ and/or KCl on nutritional properties of cracked Maçanilha Algarvia table olives. The authors evidenced that the combination of 4%NaCl+4%KCl—that was the only one sensorially accepted—resulted in samples with lower fat contents; similar dietary fiber contents, phenolic compound contents and Ca contents with respect to the control (8% NaCl); and increased K and reduced Na, thereby improving the nutritional quality of the obtained reduced salt olives.

With the aim of substitute traditional thermal stabilization technologies, the use of high hydrostatic pressure (HHP) was proposed on Greek-style Turkish table olives [39]. HHP assures both microbial elimination and no heat damage; thus, nutritional characteristics may be maintained [60]. The authors showed that the HHP treatment, aside from stabilizing the product, resulted in an increase up to 2.1–2.5-fold of total phenolics, and in particular, of hydroxytirosol, probably because the HPP treatment allows a higher extraction rate. The conclusion was that this unit operation can be proposed as an alternative to the traditional heat treatments.

Results obtained using new unit operations such as UAD and HPP or reducing brine salt content have, thus, proved to be beneficial to improving the nutritional power of table olives.

2.1.5. Comparison among Different Trade Preparations and Styles

Several papers focused comparisons of the three main trade preparations—treated olives, natural olives and olives darkened by oxidation—on the nutritional quality of olives, and on other styles.

In the very comprehensive paper of Romero et al. [32], the effects of the above-cited main trade preparations and of cultivars on the single polyphenolic compounds extracted from aqueous and lipid phases of table olives were studied. Concerning the aqueous phase, the authors showed the highest amounts of polyphenols in turning color olives, as they are submitted only to two dilution operations, while ripe oxidized olives had the lowest content. The advanced degree of ripeness of black olives processed with the Greek-style resulted in the highest anthocyanin content. The analysis of the lipidic phase, therein carried out for the first time, gave very important knowledge, as the authors evidenced a unique phenolic profile that is different from that of raw fruits. In fact, aglycons of oleuropein and ligustroside were absent in table olives, which, instead, presented for the first time the compound catechol. The authors concluded that table olives are a rich source of antioxidants, in some cases even more than virgin olive oils.

In three papers of López-López et al. [52,61,62], an extensive analysis of FA composition, provitamin A carotenoids and Vitamin B was carried out on 67 commercial samples of table olives prepared according to the three above-cited preparations and with different cultivars and styles. Data presented by the authors showed which preparation-styles had the better nutritional profiles. In particular, the authors showed that there is a large variability in carotenoids content and that this is mainly due to the cultivar used [61], while they demonstrated that is possible to discriminate cultivars and commercial preparations by statistically analyzing the fat profiles with a discriminant analysis [52] and that the best trade preparation to maintain the Vitamin B content is the natural style, followed by the treated olive style [62]. Romero et al. [56] made for the first time an important study on triterpenic acids—which have been reported to have an anti-cancer activity [63]—of seventeen different olive cultivars processed according to the Spanish, Greek and Californian styles. The authors evidenced that placing olives directly in brine resulted in a very high content of triterpenic acids, with respect to the alkaline treatment of Spanish style. Additionally, the authors found that natural black olives have a much higher content of these bioactive compounds than olive oil, thereby concluding that table olives should be nutritionally reevaluated. Jiménez et al. [57] tested the effects of cultivar, processing type (darkening by oxidation, brine fermentation, or drying by oven or salt), and the ensuing storage on the fat and dietary fiber of six table olive cultivars, with emphasis on some properties of dietary fiber. The authors reported that obtained olives had a high content of fiber, but also that the water holding capacity of the alcohol insoluble residue is like that reported for other vegetables. Ben Othman et al. [40] studied the total polyphenol contents, the single polyphenols and the antioxidant capacities of four Tunisian table olives (Meski, Chemlali, Besbessi and Tounsi); one of them was harvested at four different ripening stages and processed with the natural style or dry-salting. The authors detected 14 different phenolic compounds, mostly hydroxytyrosol and tyrosol, while oleuropein was not detected. Results obtained for total phenol content and antioxidant activity values encouraged the authors to conclude that studied samples had an important amount of antioxidant compounds. Valenčič et al. [41] compared the effects of two processing methods, the traditional regional and modified Spanish style, and of storage (60 and 180 days), on the phenol contents of two Slovenian table olives. The traditional method involves debittering olives in water for 10 days followed by fermentation in brines at increasing NaCl concentration. The authors found a significantly higher biophenol content in the olives processed with the traditional method, and this resulted in the inhibition of LAB growth. Zoidou et al. [42] made a very comprehensive study on several commercial samples of Greek table olives (nine different cultivars and five processing-styles) to find which of them possessed the highest concentration of the nutritionally important phenolics oleuropein and hydroxytyrosol. The authors found that the dry-salting process of the Throuba Thassos olives allows obtaining olives with a very high oleuropein content.

Lanza et al. [64] studied the effects of two modified Greek methods of preparation on the nutritional quality of each of two Italian table olive cultivars, Itrana and Oliva bianca di Itri. The first method implies a first water immersion step, followed, after 15–45 days, by NaCl addition to obtain an 8% brine. The second one was carried out by placing olives directly in brine prepared with a double-salting procedure that consists of adding half of the NaCl at the beginning and the other half after 15 days of brining. The authors found that the best method was the double-salting, as both olive cultivars have an appreciable amount of fiber and polyphenols, and Itrana cultivar has a PUFA/SFA ratio of about 0.4–0.5; that is the value recommended by nutritional guidelines [65]. Savas et al. [55] used different debittering methods in relation to the nutritional properties of Turkish Domat cultivar table olives. The methods were the following: lye at 1% NaOH, immersion in tap water and scratching followed by $CaCl_2$ immersion and different brining replacements with reduced-salt brines. The best preparation was the scratching method in low-salt brines, as it resulted in table olives with reduced salt contents. Melliou et al. [43] set up an advanced HPLC-MS fast method to detect single polyphenols in darkened by oxidation (Manzanilla) and dry-salted (Mission and Throuba Thassos) table olives. The authors confirmed the results of Zoidou et al. [42], which showed that dry salting is the best processing style for retaining polyphenols in table olives.

Comparisons of the main trade preparations and other styles seem to confirm that the best nutritional table olive profile could be achieved by placing olives directly in brine.

2.2. *Influence of Starters*

The use of LAB or Y starter cultures could notably improve the fermentation of table olives, as the process may be shortened due to a rapid decrease of the bitter oleuropein trough hydrolysis mediated by the microbial enzymatic activity. These starter cultures should grow rapidly and predominate starting from refrigerated temperatures; they have a homofermentative metabolism and tolerate NaCl and glucosidated polyphenols and inhibit foodborne pathogens [66]. The above-cited characteristics are rarely found contemporarily in starter cultures, and this results in the scarce success of commercially available starter cultures, because the strains have not been adequately adapted for this use [67]. More success could surely come from starter cultures isolated from the olive fruits and brine. Recent studies highlighted the potential of starter cultures to improve the nutritional profiles of table olives. Servili et al. [44] evidenced the role of a *Lactobacillus pentosus* strain in polyphenol release from flesh to brine. The authors used appropriate scanning electron microscopy (SEM) to follow the microstructural changes of cells and tissues during the brining process, and found that skin cuticle tissues of LAB inoculated olives were totally altered, while normally fermented olive tissues were intact. The authors hypothesize that the skin degradation resulted in an increased permeability and diffusion of polyphenols from flesh to brine, thereby reducing the debittering time. Opposite results were found by Pistarino et al. [46], who showed that polyphenol loss from olives was enhanced by the process temperature and not by the use of LAB or LAB plus Y. Tataridou et al. [45] studied the effect of indigenous strains of *Lactobacillus plantarum* able to hydrolyze oleuropein on the phenol profile of green and black table olives placed in brines with low NaCl content. The authors found that the starter had an inhibitory effect on pathogen growth and that the obtained olives were significantly richer in phenols, especially hydroxytyrosol and tyrosol, and had lower NaCl content, with respect to an industrial product, thereby having a better nutritional profile. Durante et al. [47] used different starter cultures made up of a mixture of Y and LAB to assess changes in carotenoids, phenolics, triterpenic acids, vitamins, fatty acid profiles and antioxidant activity on black table olives from Italian and Greek cultivars prepared in the natural style. The authors, although they did not make any comparison with samples fermented with spontaneous microbiota, found that the obtained table olives were rich in monounsaturated fatty acids (MUFA), polyphenols, tocopherols and triterpenic acids, so that they may provide health benefits. D'Antuono et al. [48] evaluated the content of single polyphenols of natural table olives obtained with autochthonous LAB and Y starters and compared them with samples obtained by the market. The authors found that LAB+Y olives had quite always significantly higher polyphenol content than commercial samples, but no valid explanation was reported for this result. In particular, they found high contents of tyrosol and hydroxytyrosol that were up to eight times higher with respect to the virgin olive oils obtained by the same olives. Tufariello et al. [49] compared the effects of a previously selected *Saccharomyces cerevisiae* strain with a commercial preparation of the same Y and with a control without Y in terms of the nutritional properties of three olive cultivars prepared with the natural style at the black stage. The starters allowed a more rapid debittering process and permitted an increase in hydroxytyrosol, tyrosol and verbascoside on the olives, with respect to the control sample, thereby improving the nutritional value.

The above-cited papers highlight the importance of autochthonous starters on table olive fermentation and leave the field open to additional research directed toward finding appropriate commercial starter cultures, at least for the main trade preparations.

3. How Processing Influences the Sensorial Quality

The conversion of fresh, inedible olive fruits to edible table olives involves, mainly, fruit debittering and the development of sensory characteristics (odor, taste and texture) that are unique to this food specialty. Researchers have given importance to this topic both to find the effects of unit operations and starters on sensory characteristics of table olives and to find appropriate sensorial profiles of selected trade preparations. Sensory analysis by trained assessors is generally carried out by quantitative descriptive analysis (QDA), unless differently indicated, using internationally recognized standards. Results will be summarized in Table 2.

Table 2. Effect of trade preparation and processing style on sensory quality of table olives.

Trade Preparation/Processing Style/Starters (LAB*, Y)	Olive Cultivar	Test Used	Descriptors	Main Results	References
Treated	Not reported	QDA	Acidity, bitterness, color, saltiness, intensity and persistency of nasal aroma	Color, firmness, acidity and saltiness best characterized the olive	González et al. [68]
	Nocellara messinese	QDA	Appearance, color, odor, flavor, texture, overall	Olives treated with CO_2 are more acidic that control	Marsilio et al. [69]
	Çelebi, Domat, Kaba, Ayvalık	QDA Preference	Appearance, aroma, flavor, texture	Cultivars were sensorially different	Yilmaz et al. [70]
	Gordal	QDA	Abnormal fermentation type, cooking effect, earthy, metallic, musty, rancid, soapy, winey-vinegary; acidity, bitterness, saltiness; crunchiness, fibrousnesses, hardness	Saltiness was significantly related to NaCl and KCl levels; bitterness, hardness, fibrousness, and crunchiness were related to the $CaCl_2$ percentage	Moreno-Baquero et al. [71]
	Manzanilla	QDA	As previous	Decrease in saltiness and increase in bitterness at increasing Ca amounts in the pulp. Ca content highly correlated with some kinaesthetic and taste attributes	López-López et al. [72]
	Manzanilla	QDA Acceptability	Color and size; aftertaste, bitter, green olive flavor, salt, sour, sweet; crunchiness, fibrousness, hardness, pit removal	Olives grown under soft stress conditions were preferred and rated as the best for the more important descriptors	Cano-Lamadrid et al. [73]
	Gordal, Manzanilla, Hojiblanca	QDA	Acetic acid, grass, green fruit, hay, lactic acid, lupin, ripe fruit, musty, winery; alcohol, bitter, salty, sour; astringent, piquant, pungent	Development of a lexicon for the sensory characteristics of Spanish-style olives	López-López et al. [74]
	Manzanilla and Hojiblanca	QDA	A total of 33 descriptors (see paper)	A certain number of the descriptors attributes fit sample discrimination	López-López et al. [75]
	Manzanilla	QDA Acceptability	As in López-López et al. [72]	Increase of the green olive flavor and decrease of bitter taste in olives subjected to deficit of irrigation. Consumer preference for the same samples.	Rodríguez et al. [76]

Table 2. *Cont.*

Trade Preparation/Processing Style/Starters (LAB*, Y)	Olive Cultivar	Test Used	Descriptors	Main Results	References
Natural	Not reported	QDA	Abnormal fermentation, cooking effect, musty, rancid	Data analysis gave a good discrimination between unacceptable, acceptable and marginal samples and evidenced that olives could be discriminated by an electronic nose developed in the study	Panagou et al. [77]
	Brandofino, Castriciana, Manzanilla, Nocellara del Belice, Passalunara		Brightness, intensity of the green color; odor of green olives, off odor; crispness, easy peeling, juiciness; acid, bitter, salt, sweet; astringent; green olive flavor, off flavor; overall	Sensory data were affected mainly by cultivar and the overall assessment was below the imposed threshold of acceptability after 150 days of fermentation	Aponte et al. [78]
	Tonda di Cagliari	Preference		Assessors preferred olives obtained with the lowest salt concentration for the lower salt and bitter taste	Fadda et al. [79]
	Itrana	QDA	Butyric fermentation, putrid fermentation; acid, bitter, salty; crunchiness, fibrousness, hardness	All sample were rated as "Extra or Fancy", or as "First, 1st, Choice or Select". The analysis was able to separate in different areas the defected and un-defected samples	Lanza and Amoruso [80]
Darkened by oxidation	Not reported	QDA Degree of liking	A total of 34 descriptors (see paper) of appearance, aroma, flavor, taste and texture	The QDA showed that country of origin well separated samples for showed that aroma and flavor, while appearance and texture were the descriptors that best discriminated the olive products. The American consumers expressed an important score of acceptability for samples produced in California	Lee et al. [81]
	Cacereña, Gordal, Hojiblanca, Manzanilla	QDA	Brightness, skin defects, surface color; acid, bitter, salty; abnormal fermentation, other defects; crunchiness, fibrousnesses, hardness, pit release, skin strength; metallic taste, soap taste, typical flavur	The sensory analysis found significant changes only for surface color of whole olives. The classification of 'extra' was attributed to almost all samples	García-García et al. [82]

Table 2. *Cont.*

Trade Preparation/Processing Style/Starters (LAB*, Y)	Olive Cultivar	Test Used	Descriptors	Main Results	References
	Hojiblanca, Manzanilla	QDA	Alcohol, artificial fruity/floral, briny, cheesy, earthy/soil-like, fishy/ocean-like, natural fruity/floral, nutty, oak barrel, sautéed mushroom, vinegary	Cultivars were sensorially discriminated only for the briny descriptor. Analysis of data accurately predicted the nutty flavor and permitted the identification of the aroma compounds volatiles that highly contributed to the attributes of olives processed at the black stage	Sanchez et al. [83]
Dried (hot air or salt)	Ascolana Tenera	Preference		The highest preference was expressed for the least bitter olives, that were also judged saltier, with respect to the other samples	Gambella et al. [84]
	Various (see paper)	Preference		Assessors preferred the salted olives as salt had a masking effect on bitterness	Piga et al. [85]
	Gemlik	QDA	Black, black-brown, brown; bitterness, off flavor, rancidity, saltiness; softness, pit-flesh detachment; overall eating quality	MAP and vacuum-packaged olives as well as those stored at 4 °C obtained the best scores	Değirmencioğlu et al. [86]
Other (Cured, fresh green, traditional)	Aloreña de Málaga	QDA	Descriptors were developed in the work	The panel developed nine specific descriptors: odour (fruity, green, seasoning, lactic), aroma (fruit, seasoning), basic tastes (acid, bitter), texture (crunchy)	Galán-Soldevilla and Ruiz Perez-Cacho [87]
	Aloreña de Málaga	QDA	Acidic, bitterness, crunchiness; hardness, salty; appreciation of defects, darkening, overall acceptability	Olives subjected to a hot water dipping maintained a better green color, with respect to the control	Rodríguez-Gómez et al. [88]
Other (Fresh)	Aloreña de Málaga	QDA	Descriptors were developed in the work	Assessors selected 15 descriptors for aroma, basic, odor, aroma, trigeminal and texture attributes. The processing style significantly influenced fruit odor, bitter taste, firmness and odor, while each style resulted in differences for all the descriptors	Galán-Soldevilla et al. [89]
	Aloreña de Málaga	QDA	Acidity, bitterness, saltiness; color; crispness, firmness, fibrousness; odor	Olives treated with 0.075 $ZnCl_2$ obtained higher scores for acidic taste, color, odor, saltiness	Bautista-Gallego et al. [90]; Bautista-Gallego et al. [91]

Table 2. *Cont.*

Trade Preparation/Processing Style/Starters (LAB*, Y)	Olive Cultivar	Test Used	Descriptors	Main Results	References
Other (Traditional)	Aloreña de Málaga	QDA Acceptability	Acidic, bitter, salty; crunchiness, hardness, appreciation of external damages and any kind of defects, browning	The highest acceptance was obtained by olives with a shelf life from 6 to 42 days, while a drastic decrease in sensorial quality was found at 131 days	Romero-Gil et al. [92]
Other (Pitted, reduction to a paste)	Taggiasca	QDA	Abnormal fermentation, cooking effects, musty, rancid, other defects present; acid, bitter, salty	Assessors rated the rancidity defect with a defect predominant perceived <3, which is the threshold for the extra category, for paste olives up to 18 months storage, while for pitted olives this limit was overcome after 12 months	Lanza et al. [93]
Other (Cracked)	Maçanilha	Acceptability		Assessors gave the highest acceptability to olives brined with HCl and with the mixture of citric and lactic acid	Alves et al. [94]
Other (Alcaparras)	Cobrançosa, Negrinha de Freixo		Bitter, pungent, salty, sweet	Data of sensory analysis were corelated with those obtained with an electronic tongue and revealed that this device is effective in monitoring the changes in bitter, pungent and sweet intensities	Rodrigues et al. [95]
LAB (Natural)	Ascolana Tenera	QDA	Acid/sour, bitter; color; odor; crispness, firmness	The LAB olives were more appreciated than non-inoculated ones, for their less bitter taste, a higher odor intensity, and good textural attributes	Marsilio et al. [96]
	Nocellara Etnea	QDA	Acid, bitter, salty, sweet; crunchiness, fibrousness, hardness	Panelists judged olives treated and fermented by LAB the best for acidic and salty tastes and for gave the highest scores for acidity, crunchiness and saltiness	Randazzo et al. [97]
	Giarraffa, Grossa di Spagna	QDA	Bright, green color; green olive aroma, off odor; crisp, easy stone, juicy; acid, bitter, salt, sweet; astringent; green olive flavor, off flavor; overall	Results evidenced that the sensory characteristics were cultivar dependent	Randazzo et al. [98]

Table 2. *Cont.*

Trade Preparation/Processing Style/Starters (LAB*, Y)	Olive Cultivar	Test Used	Descriptors	Main Results	References
	Tonda di Cagliari	QDA	Bitterness	Samples obtained with LAB were debittered at the end of processing, while control olives needed 12 months	Campus et al. [99]
	Tonda di Cagliari	QDA	Acidity, bitterness, saltiness; crunchiness, fibrousness, freestone, hardness	The use of *L. pentosus* resulted in olives with a sensory profile very close to the natural-style samples, naturally fermented ones, with respect to *L. plantarum*	Communian et al. [100]
	Nocellara del Belice	QDA	Green olive aroma; crunchiness; acid, bitter, complexity, salty, sweet; off odor, off flavor	The use of pied de cuve resulted in olives with the highest scores of sensory complexities and with the absence of off-odors and off flavors	Martorana et al. [101]
	Nocellara del Belice	QDA	Green color intensity; green olive aroma, off odors; crispness, easy stone detachment; astringent, bitter, complexity, juicy, salt, sour, sweet; off flavors	Mechanically harvested and LAB fermented olives were sensorially like the manually harvested olives	Martorana et al. [102]
	Tonda di Cagliari	QDA	Acetic, acid, bitter, fruity, mushroom, saltiness, silage; astringent, crunchiness, fibrousness, fleshy, freestone, hardness, juiciness	Samples fermented in an automated pilot plant obtained the same bitterness of commercial sample after 90 days, while control olives had a significantly higher bitter taste after 180 days	Campus et al. [103]
	Nocellara Etnea	QDA	Cooking effect, earthy, metallic, musty, rancid, soapy, winey-vinegary; acidity, bitterness, saltiness; crunchiness, fibrousness, hardness	LAB fermented olives obtained the significantly highest overall acceptability score and the sample brined with the 5% NaCl obtained the best appreciation	Pino et al. [104]
	Nocellara Etnea	QDA	Green color, bright; green olive aroma, off odor; green olive flavor, off flavor; acid, bitter, salty, sweet; crunchiness, easy stone separation, juiciness; astringent	Significant differences in bitterness, bright, crunchiness, green color, green olive aroma and juiciness for LAB samples, control olives had the highest bitterness value	Randazzo et al. [105]

Table 2. *Cont.*

Trade Preparation/Processing Style/Starters (LAB*, Y)	Olive Cultivar	Test Used	Descriptors	Main Results	References
	Aitana, Caiazzana, Nocellara del Belice	QDA	Acid, bitter, salty; crunchiness, fibrousness, hardness; abnormal fermentation, other defects	All the tested cultivars had good sensory characteristics, and the highest scores for flesh consistency and crunchiness was obtained by Nocellara del Belice olives	Romeo et al. [106]
	Nocellara Etnea	QDA	See Pino et al. [104]	The control olives obtained the highest scores for acidity, the highest bitter taste was scored in the olives without LAB. The samples at 5% and 8% NaCl added with the LAB received the highest overall acceptability	Pino et al. [107]
LAB + Y (Natural)	Bella di Cerignola	QDA	Crunchiness; acid, bitter, salty, sweet; olive flavor, off flavor	The assessors ranked better the olives fermented with starters, with respect to the control that obtained the lowest values for crunchiness and olive flavor and the best evaluations for acid, bitter and off flavor	De Angelis et al. [108]
	Conservolea, Kalamata	QDA	Acidity, bitterness, saltiness; odor; hardness; overall	Kalamàta olives obtained the best scores for aroma and overall acceptability when the Y+LAB and MIX inoculations were used, while Conservolea olives showed the same results when LAB+Y were inoculated	Chytiri et al. [109]
Y (Natural)	Taggiasca		Acid, bitter, salty; crunchiness, fibrousness, hardness	The best combination may be obtained with the use of Y on acidified brines at the highest NaCl concentration	Ciafardini et al. [110]
LAB (treated)	Nocellara del Belice	QDA	Bright, green color; green olive aroma, off odor; crisp, easy stone, juicy; acid, bitter, salt, sweet; astringent; green olive flavor, off flavor; overall.	Data highlighted that *L. pentosus* improved the sensory characteristics of olives, with respect to control samples	Aponte et al. [111]

*LAB= lactic acid bacteria; Y=yeast.

3.1. Trade Preparations

3.1.1. Treated Green Olives or "Spanish Style"

This trade preparation produces table olives in which the bitter taste is absent, while salty and acidic taste and other flavors derived from fermentation are present.

González et al. [68] tried to find a correlation between sensory and objective results with the aim to find the best match between parameters. The QDA considered the descriptors acidity, bitterness, color, firmness, saltiness and intensity and persistency of nasal aroma. Several direct and inverse correlations were found between sensory descriptors and instrumental data, such as that between instrumental and subjective color and polyphenol content and fruit color; the best descriptors that characterized the table olives were color, firmness, acidity and saltiness. Marsilio et al. [69] did a sensory study (using Nocellara messinese olives at the green stage) on the influence of alkali neutralization with CO_2, in comparison with traditional washing with water. The eight-member trained panel rated the appearances, colors, flavors (acid, bitter, salty), odors and textures (crispness and firmness) of processed olives [112]. The assessors judged olives treated with CO_2 as more acidic than the control, while no differences were found for texture, although care should be taken to reduce the increase of the buffering potential of brines that can result in inadequate pH lowering of brines. Yilmaz et al. [70] carried out a sensory evaluation of different table olives and investigated the consumer preferences. The six-member panel used the descriptors appearance, aroma, flavor and texture of commercial green table olives of four Turkish cultivars [113]. A total of 50 people carried out the consumer test by using a scale from 0 to 9. Results evidenced that the sensory differences were cultivar dependent, and that, for consumer preference, the most important factor in willingness to buy was the mouth feeling. The effects of brines obtained with different NaCl concentrations on gustatory and kinesthetic sensations of treated green table olives were tested by Moreno-Baquero et al. [71]. The authors used fifteen different brines made up of NaCl (4–10%), KCl (0–4%) and $CaCl_2$ (0–6%). A panel of nine trained assessors evaluated negative, gustatory and kinesthetic attributes [114]. Multivariate statistical analysis (MSA) was used to correlate the initial brine concentrations with sensory attributes. Saltiness was significantly related to NaCl and KCl levels, while bitterness, hardness, fibrousness and crunchiness were in relation to the $CaCl_2$ percentage. The authors concluded that the models developed in the work can be useful in the production of particular table olives. Villegas Vergara et al. [115] proposed two different brine acidification methods—the first with CO_2 gas, and the second by mixing LAB with lactic and hydrochloric acids—and evaluated their influence on the sensory properties of olives (cv. Conservolea). A ten-member panel carried out the sensory analysis [116]. The authors found that the acidification step is useful in helping the fermentation process and it has no effect on the sensory profile of olives. Bautista-Gallego et al. [117] used fermented Manzanilla olives to evaluate the influence of the addition of zinc chloride ($ZnCl_2$ at 0.00%, 0.25%, 0.50%, 0.75% and 1.00%) to brine on increasing the olives' shelf life and improving their sensory properties. A panel of twelve trained members used two protocols [81]. In the first one, the ranking test, 0.00 $ZnCl_2$ was used as the control and panelists ranked the other samples by dissimilarity to the standard (1 more similar, 5 less similar). In the second protocol, the A–Not A, judges were asked to decide if samples were the same (sure or not sure) or different (sure or not sure). The two tests did not give significant differences between the control and the olives added with $ZnCl_2$, thereby suggesting that this salt does not affect the sensory characteristics of the samples studied. The influence of the substitution of NaCl with KCl and $CaCl_2$ on the sensory profile of Manzanilla olives was studied by López-López et al. [72]; they used 16 brines with different salt concentration ranges (40–100 g/L of NaCl, 0–60 g/L of Kcl and 0–60 g/L of $CaCl_2$). Nine experienced assessors determined negative sensations and used the descriptors for taste and kinesthetic attributes. [116]. Data were statistically treated with partial least square analysis (PLS) and principal component analysis (PCA). The assessors found a decrease in saltiness and an increase in bitterness at increasing Ca amounts in the olive pulp. Data of Ca contents were highly correlated by PLS both with some kinesthetic (hardness, fibrousness, crunchiness) and taste attributes (bitterness and saltiness); PLS used Ca, K and Na pulp content to estimate sensory characteristics of samples. The influences of three different irrigation regimes, from normal to moderate stress, on sensory properties of green fermented Manzanilla olives, were evaluated by Cano-Lamadrid et al. [73]. Eight trained panelists evaluated attributes related to main sensory attributes of flavor and texture [116]. A consumer acceptability test with a nine-point scale was also carried out by 100 assessors. Olives grown under soft stress conditions were rated as the best for the more important descriptors, and they were preferred among Spanish consumers. Results confirmed those obtained in a previous work [23]. López-López et al. [74] developed a sensory profile for the main Spanish table olive cultivars (Gordal, Manzanilla, Hojiblanca) cultivated in seven different areas. A total of 15 panelists used a set of descriptors for aroma, taste and mouthfeel [116]. PCA and hierarchical clustering analysis (HCA) were useful to visualize the panel capacity and characterization of samples and their discrimination. The study allowed them to develop a vocabulary for the sensory characteristics of treated green olives from diverse cultivars and production areas. PCA analysis, moreover, permitted them to find correlations among sensory attributes and sample discrimination. A similar paper has been published by the same research group [75] to sensorially describe Manzanilla and Hojiblanca olives processed at the black stage using a list of descriptors able to characterize the product according to varieties, place of growth and duration of shelf life [116]. A total of 14 panelists used a set of descriptors for visual appearance, aroma, flavor, taste and texture. Data were analyzed by MSA. Results indicated the existence of a certain number of attributes that fit the sample discrimination, such as skin sheen, skin red, flesh yellow and others. A relevant effect on the sensory profile was found for the previously cited variables. Sánchez-Rodríguez et al. [76] recently studied the effect of cultivation under regular deficit irrigation (RDI) on sensory quality of fermented Manzanilla table olives. RDI was applied as moderate to severe grade and compared to fully irrigated control trees. Sensory analysis was carried out by 10 trained panelists, who developed an adequate lexicon, or by a consumer acceptance test with 100 consumers [116]. The QDA analysis evidenced an increment of the green olive flavor and a drop of bitter taste in the RDI olives. The customers, who were informed about the irrigation strategy used, preferred the RDI samples, with respect to control, and declared it to be favorable to pay more for these olives. The authors also found that the descriptors driving the consumer acceptance of RDI olives were both gustative, such as bitterness and saltiness, and kinesthetics, such as hardness. The work of Mastralexi et al. [25], also cited in Section 3.1.1., studied the effect of Spanish style processing and a storage period of twelve months on the sensory characteristics of the protected denomination of origin (PDO) "Prasines Elies Chalkidikis" olives prepared at industrial scale. An accredited panel used the attributes related to defects (abnormal fermentation and other defects), taste (acid, bitter and salty) and kinesthetics (crunchiness, fibrousness, and hardness) to evaluate the olives [116]. The sensory panel considered the stored olives as quite satisfactory for texture descriptors and that they could be graded as "extra".

Research for sensory characterization of treated olives is, thus, at an important level and is to highlight studies directed at developing a vocabulary for descriptors.

3.1.2. Natural Olives

This trade preparation produces olives with a residual bitter taste, and acidic and salty taste and other flavors derived from the microbial fermentation.

Piga et al. [118] evaluated the responses of three Sardinian table olives (Bosana, Manna and Sivigliana sarda) in terms of sensory acceptability after natural fermentation carried out in the Greek-style. Ten untrained laboratory persons performed an informal tasting at 50 days of fermentation and wrote on the presence of off flavors, consistency and crispness, and expressed their preferences. No off flavors were detected by assessors, which found all the cultivars with a balanced taste and satisfactory consistency. The assessors, moreover, considered all the olives excellent and ready to eat after 150 days of brining, preferring the Bosana olives for their best consistency and crispness. The same research group proposed some technological corrections to avoid the main technological problems related to the processing of green natural olives and to improve their sensory properties [119]. The authors controlled and periodically adjusted the following process parameters during the fermentation: brine NaCl concentration, pH, temperature of fermentation and brine level in the fermenters. The same sensory protocol described in [118] was used, and the attribute saltiness was also expressed. The assessors did not detect negative tastes or odors; they judged as excellent the fermented olives after 210 days of brining; and preferred samples obtained with NaCl at 4% for the more intense salty taste. Kanavouras et al. [120] focused their work on the influences of different brines on sensory descriptors of black fermented olives. The authors tested three different brines: a traditional brine with NaCl at 16%; a NaCl-free brine buffered at pH 4.7 with CH_3COOH (0.05 M) and $Ca(OH)_2$ (0.025 M); and a 12.8% NaCl brine buffered at pH 4.3 with CH_3COOH (0.05 M) and $Ca(OH)_2$ (0.025 M). A total of 39 untrained assessors evaluated the fermented samples for appearance and taste, on a 9-point scale, whereas a 3-point scale and a preference was used for the intensity of salt, vinegar, pungency, level of fermentation and unpleasant characteristics. Assessors preferred the olives fermented in the NaCl brine buffered at pH 4.3, because the samples showed a more pungent, fermented and mildly vinegary taste, had a sufficient salty taste and had a low level of unpleasant flavor, while samples processed with a simple NaCl brine were judged as the worst. The authors concluded that consumers seemed to prefer olives with lower NaCl content. An electronic nose was developed by Panagou et al. [77] to sensorially discriminate fermented green table olives on the basis of their aroma compounds. A 15-member panel classified the volatile profiles of the olives as unacceptable, acceptable and marginal, while a specific electronic nose generated a chemical map of the aromatic compounds of fermented samples. All obtained data were analyzed by MSA and artificial neural networks (ANN). The MSA analysis gave a good discrimination between unacceptable, acceptable and marginal samples, while the ANN use resulted in a good performance in discriminating the three classes, as only in two cases of the 66 samples were there misclassifications. The authors suggested that the developed device may be proposed for quality discrimination of green table olives as it had several advantages, such as the low price and the rapidity of analysis. The influences of fruit ripeness (green, turning color and black) and salt concentration (5% and 10% of NaCl) on the sensory properties of Arbequina table olives were evaluated by Hurtado et al. [121]. A 16-member panel judged the olives after fermentation and storage in acidified brine for 45 days according to UNE [122] for color, taste, texture and flesh stone. They also rated any sensory diversity between a sample processed at lab with a 10% NaCl brine and a commercial sample. Results indicated that panelists preferred the olives with a green color and that it was not possible to distinguish commercial samples from laboratory-scale processed olives. Aponte et al. [78] sensorially characterized five naturally fermented table olive cultivars picked at the green stage. Ten judges used a descriptive method [123] with fifteen descriptors for aspect, color, odor and off odor, flavor and off flavor, taste and kinesthetics sensations. Sensory data were affected mainly by cultivar, and the overall assessment was below the imposed threshold of acceptability after 150 days of fermentation. The authors suggested modifying the unit operations to ameliorate olive quality. The use of two NaCl brine concentrations (4% and 7%) was tested in order to see the influence on sensory properties of fermented green olives [79]. Thirty untrained judges expressed their preferences of samples at the end of fermentation by using a paired preference test [124]. The assessors gave positive judgements on both samples but preferred the olives obtained with the lowest NaCl concentration for their lower saltiness and bitterness. Lanza and Amoruso [80] evaluated the sensory characteristics of fermented Itrana table olives, obtained according to two styles differing from the ripening stage at harvest, the green one (Oliva Bianca di Itri) and black one (Oliva di Gaeta). A total of 8–10 panelists used IOOC standards [116] to check for gustatory, kinesthetic and negative sensations of samples. MSA discriminated between samples with or without defects. Assessors graded all samples as "Extra or Fancy," or as "First, 1st, Choice or Select". The MSA was able to separate in different areas the defected and un-defected samples and that "Extra or Fancy" olives with a defect higher than 1.0 were judged closer to samples with defects.

Analysis of literature on sensory properties of natural olives evidenced the lack of studies dealing with the development of a common lexicon that could be proposed for sensorially describing these table olives.

3.1.3. Olives Darkened by Oxidation or Californian-Style

This trade preparation produces olives with no bitter taste, and sensory characteristics of pickled olives if they are fermented before processing.

Lee et al. [81] examined the sensory properties and the preferences of California consumers of sliced black olives produced in several countries from USA (California), Europe (Portugal and Spain) and Africa (Egypt and Morocco). A panel of eight judges selected thirty-four descriptors for aroma, appearance, flavor, taste, texture, and mouthfeel [125], while a consumer test was carried out by 104 consumers that assessed the level of preference of the 20 sliced samples on a nine-point hedonic scale. According to QDA, country of origin well separated samples for aroma and flavor, while appearance and texture were the descriptors that best discriminated the olive products. Californian samples had no flavor defects, while olives produced in other countries revealed gassy, metallic, rancidity and soapy/medicinal defects. The American consumers expressed an important score of acceptability for samples produced in California, probably for their familiarity with the product. The study also revealed that consumers acceptance was driven mainly by the flavor characteristics. García-García et al. [82] evaluated changes in sensory parameters of packed pitted and whole black olives of four Spanish cultivars during a three-year period of storage in simulated marketing conditions. A panel of eight trained people described sensorially the olives using the descriptors of external appearance, odor/flavor and texture on just packed olives and samples stored for 6, 12, 24 and 36 months at ambient temperature [114]. The sensory analysis found significant changes only for surface color of whole olives. The classification of 'extra' was attributed to almost all samples. Recently Sanchez et al. [83] sensorially characterized black olives (Manzanilla and Hojiblanca) by comparing the aroma profile with volatile compounds. Fourteen trained panelists assessed eleven odor attributes [81]. Volatiles were extracted with the headspace solid-phase microextraction (HS-SPME) and analyzed by gas chromatography-mass spectrometry (GC-MS). Cultivars were sensorially discriminated only for the briny descriptor. MSA with PLS regression accurately predicted the nutty flavor and permitted the identification of the aroma compounds that highly contributed to the attributes of olives processed at the black stage.

3.1.4. Other Processing Methods and Stabilization Treatments

Gambella et al. [84] studied the influence of different pre-treatments before cabinet drying on sensory properties of green table olives. Olives were subjected to the following pre-treatments: piercing with a steel brush (A), dipping in water at 50 °C for 10 min (B), piercing plus dipping (C), piercing and dipping in a 10% NaCl brine at 50 °C for 10 min (D), untreated as control. Five untrained personnel expressed the intensity of the bitter taste with a three-points scale: 1 = no bitterness, 2 = acceptable bitterness and 3 = unacceptable bitterness. A preference was also given. The sensory test revealed that bitterness was almost absent in D olives and quite strong in olives of the groups A and B. The highest preference was expressed for the least bitter olives (D), that were also judged saltier, with respect to the other groups. In another paper of Piga et al. [85] a preference was expressed on cabinet dried olives belonging to fourteen cultivars. Whole fruits were pre-treated as follows before drying: blanching in 2% NaCl brine at 90 °C for 2 min and room cooling—"blanched olives"; 2 min water blanching at 90 °C plus salting in barrels for 3 days—"salted olives"; skin piercing with a steel brush—"pierced olives". Ten untrained assessors gave a preference judgement. The sensory analysis revealed that all the olives were appreciated, even if the assessors preferred the salted olives as salt had a masking effect on bitterness. Değirmencioğlu et al. [86] tried to prolong the shelf life of dry-salted Gemlik olives by means of modified atmosphere packaging (MAP) and vacuum sealing. Olives were stored for 7 months at 4 or 20 °C and air packaged olives served as control. A total of 32 untrained assessors evaluated the attributes (color, taste, texture and flesh stone, and overall eating quality) using a nine-hedonic scale [126–128]. Assessors rated better MAP and vacuum-packaged olives as they obtained better ratings for rancidity and softness than the control. The sensory profile was not affected by storage temperature, but olives held at 4 °C were rated with the best scores. Pradas et al. [129] proposed the use of HHP (400 MPa and 800 MPa for 5 and 10 min), as an alternative to heat treatment, to sensorially improve "Cornezuelo" dressed olives, a Spanish table olive specialty prepared with the use of some condiments (garlic, fennel, salt and thyme). Sensory analysis was carried out after packaging and at 120, 186, 218, 280 and 335 days of storage by a panel of 6–8 members that are experts in sensory evaluation of table olives, and who scored the olives for odors, flavor defects and overall sensory quality. Only olives treated at 400 MPa for 5 min fulfilled the market requirements after 335 days of storage, as revealed by the sensory analysis. Galán-Soldevilla and Ruiz Perez-Cacho [87] developed a 52 h training method for the PDO Aceituna Aloreña de Málaga quality certification panel. This PDO can be proposed in 3 different styles: cured, that are directly brined for 90 days and then seasoned and packaged; fresh green, that are cracked before brining for 3 days, and after that, seasoned and packaged or stored at low temperature; traditional, that are cracked, brined for 20 days and then consumed or seasoned and packaged [130]. The paper described all the stages involved in the sensory analysis, from recruiting (15 members) to basic and specific training. The panel developed nine specific descriptors for odor (fruity, green, seasoning and lactic), aroma (fruit and seasoning), basic tastes (acid and bitter) and texture (crunchy). The panel also characterized this PDO for its fruity and seasoning odor and aroma, bitter taste and crunchy texture. In a further paper Galán-Soldevilla et al. [89] identified the sensory descriptors that may appropriately distinguish the different styles of this specialty by using nine trained members that selected 15 descriptors for aroma, basic, odor, texture and trigeminal attributes. The evaluated samples were taken by commercial packages. The results showed that the processing style significantly influenced only bitter taste, firmness and odor, while each style resulted in differences for all the descriptors. The PDO fresh green Aceituna Aloreña de Málaga olives were also used to study the effect of addition to brine of zinc chloride ($ZnCl_2$ at 0.000%, 0.050%, 0.075% and 0.100%) in increasing their shelf life and improving their sensory properties [90,91]. In a first paper [90] an 18-member panel was used for rating the descriptors of acidity, bitterness, color, crispness, firmness, fibrousness, odor and saltiness [131]. After three months of storage the $ZnCl_2$ led, in general, to a better control of microbial spoilage, with respect to the control olives, and olives treated with 0.075 $ZnCl_2$ obtained higher scores for acidic taste, color, odor and saltiness. In the second paper [91] the authors confirmed the results of the first one. The work of Malheiro et al. [51], that has been previously cited in Section 3.1.4, investigated the effect of cultivar (Cobrançosa, Madural, Negrinha de Freixo, Santulhana and Verdeal Transmontana) on the sensory characteristics of fermented alcaparras olives. Thirty-tree untrained assessors gave a preference using a nine-point hedonic scale, evaluated aroma, consistency and flavor and rated globally the samples. The consumer test showed a preference for the cultivars Verdeal Transmontana and Negrinha de Freixo, the former for the attributes firm, fleshy and fruity, and the latter for the aroma, while all parameters of the cv. Madural were scored negatively.

Lanza et al. [93] evaluated the sensory properties of two products derived from processing of Taggiasca olives. Olives were pitted or reduced to a paste, pasteurized inside glass containers filled with extra-virgin olive then stored at room temperature for 18 months. A trained panel used official methods [116] to evaluate the attributes: negative sensations and gustatory sensations. Classification (extra, first, second, not be sold) was done considering the median of defect predominant perceived (DPP). Tasters rated the rancidity defect with a DPP ≤ 3, which is the threshold for the extra category, for paste olives with up to 18 months of storage, while for pitted olives, this limit was overcome after 12 months. Alves et al. [94] evaluated the possibility of extending the shelf life of cracked green Macanilha olives, which are an appreciated table olive specialty of Southern Portugal, with the addition to packing brines of acids (citric, hydrochloric and lactic) and preservatives acids (sorbic and benzoic). Both categories of chemicals were used one at time, but a combination of lactic and citric acid was also used. An acceptability test was performed after 158 days of packaging by 20 assessors that were regular consumers of table olives. Assessors gave the highest acceptability to olives brined with hydrochloric acid and with the mixture of the other two acids, which was, thus, suggested as an acceptable strategy for the shelf life extension of this table olive specialty. Tokuşoğlu et al. [39], also cited in Section 3.1.4., investigated the effectiveness of UAD on the sensory characteristics of table olives. Twelve semi-trained panelists evaluated the bitterness of the olives with scores from 1 (unacceptable) to 5 (acceptable) at 7, 14 and 21 days of processing. The UAD operation resulted in olives with a significantly lower bitter taste, with respect to control, thereby highlighting the beneficial effect of the UAD in improving the debittering process. Rodríguez-Gómez et al. [88] used a hot water dipping treatment (5 min at 60 °C) on DPO Aloreña de Málaga olives before brining, to enhance the sensory properties of the fermented olives. Fourteen expert members evaluated the heat treated and non-heat-treated olives for the descriptors acidic, bitterness, crunchiness, hardness and salty [114]. The authors revealed a beneficial effect of the mild heat treatment, as treated olives maintained a better green color, with respect to control, and improved the stability of the samples without imparting them negative sensory attributes. The paper cited in Section 3.1.4 [60] also tested the effect of brine NaCl replacement with $CaCl_2$ and/or KCl on sensory characteristics of cracked Maçanilha Algarvia table olives. A sensory panel of fourteen trained judges used the descriptors of acidity, appearance, aroma, bitterness, firmness, flavor and saltiness on a seven-point scale. An overall sensorial evaluation was also carried out. Results of the sensory test evidenced that the olives fermented with 8% NaCl and 4% NaCl + 4% KCl brines obtained the highest scores for flavor and overall attributes, while samples processed with other brine combinations (4% NaCl + 4% $CaCl_2$ −4% KCl +4% $CaCl_2$ and 2.7% NaCl + 2.7% KCl + 2.7% $CaCl_2$) were rated as unacceptable. Another study, on DPO Aloreña de Málaga olives processed with the traditional style, was carried out by Romero-Gil et al. [92], who determined the shelf life of this olive preparation from a sensory point of view. Olives were packaged in appropriate containers filled with a brine containing acids and preservatives and sensorially checked at 0, 6, 20, 42, 74, and 131 days. A consumer panel consisting of 35 members used specific descriptors related to Aloreña de Málaga fruits and to IOOC method [116] and expressed a global evaluation of the olive quality and acceptability by using yes (olives good for purchasing) or not (poor quality). Data were analyzed by MSA and showed the highest acceptance for olives with a shelf life from 6 to 42 days, while a drastic decrease in sensorial quality was found at 131 days, as the willingness-to-buy attribute was reduced to 50%. Rodrigues et al. [95] developed an electronic tongue for monitoring the debittering of previously described alcaparras olives. Data obtained by the electronic tongue were correlated with the sensory descriptors of bitter, pungent, salty and sweet recorded by 8 trained judges following the IOOC official regulations [116]. Data obtained were analyzed with multivariate statistical techniques and evidenced that the electronic tongue is effective at evaluating changes in bitter, pungent and sweet intensities and may be proposed as a tool with different useful characteristics, such as rapidity of analysis and low environmental impact. The authors also stated the possibility to use the electronic tongue for tasting purposes.

3.1.5. Comparison among Different Trade Preparations and Styles

Only four papers were found dealing with this topic.

Panagou et al. [132] studied the sensorial characteristics of retail table olives. Sixty-nine different samples processed as green treated olives, black natural olives and other samples were considered. A ten-member panel evaluated the olives for the specific attributes of acidic taste, bitterness, crispness, odor, saltiness and overall eating quality on a 1–10 scale [127]. Assessors evidenced that the green olives were not different, and that they were principally characterized by a sufficient acidity, an adequate bitterness and a satisfactory crispness and odor; more variability was found for black olives that showed more remaining bitterness, with respect to green ones adequate acidity and odor, and high NaCl content. A higher residual bitter taste was found on dry-salted olives, with respect to other samples and were perceived as too much salty. The previously cited paper of Valenčič et al. [41] studied the sensory profile at different processing times (60 and 180 days) of Slovenian table olives prepared according to a traditional regional and modified Spanish style. Nine trained assessors evaluated the sensory characteristics of olives [113] and used the descriptors for saltiness, bitterness, sourness, hardness and fibrousness. The authors found that the intensity of bitterness, fibrousness, hardness and sourness of both cultivars were higher in the traditional technology, with respect to Spanish style samples, which were judged not suitable to be classified as Slovenian table olives. Lanza et al. [64] evaluated the sensory properties of Itrana table olives fermented for 8 and 12 months at the green (Oliva bianca di Itri) or black stage (Oliva di Gaeta), according to the place of harvest, the maturity stage and the preparations styles reported in Section 3.1.5 by the same paper. The IOC method [116] was used for sensory analysis with the evaluation of negative, gustatory and kinesthetic sensations. The DPP was also used as reported by Lanza et al. [93]. Assessors did not find defects for green and black olives processed with the double-salting methods that were, thus, rated as the best. Lanza and Amoruso [133] recently monitored on a regular basis the capacity of every assessor and of the entire panel that applied criteria and procedure of the official IOOC method for table olives [116]. Olives were sensorially evaluated by 8 expert assessors. Univariate and MSA of data were applied. Results indicated that the panel well agrees for hardness, while a case-to-case analysis was needed for other attributes.

3.2. *Influence of Starters*

The fermentation process of table olives is aimed at stabilizing the product by reducing the pH and changing positively the olive sensory properties. Brine fermentation with indigenous microbiota, although it is widely used, could be responsible for spoilage and pathogen microorganism growth in the first phases of fermentation. The use of starter cultures made up of LAB, Y or their mix may help in preventing the cited problems and producing high quality products. For these reason, extensive research has been carried out in the last 20 years, mostly for natural olives, which need the development of specific LAB starters that can grow in the presence of specific polyphenol inhibitors derived from the fruit flesh. We will review the literature starting from papers dealing with natural olives.

3.2.1. Natural Olives

Marsilio et al. [96] evaluated the sensory quality of Ascolana tenera olives that were subjected to different pre-harvest irrigation regimes and that were fermented with a LAB. The irrigation regime consisted of a rainfed control, two regimes with water depth of 33% and 66% of the estimated crop evapotranspiration (ETc) from the seed hardening stage and a four one with 66% of ETc during the whole season. Olives were fermented with a LAB made with *Lactobacillus plantarum* strain or with indigenous microbiota. An eight-member panel made a QDA analysis and assessed the olives after seven months of fermentation for color, odor, acid/sour, bitter, firmness and crispness [112]. The control olives showed overly high bitter taste, firmness and sourness, and thereby were judged as not marketable. The LAB samples were more appreciated than non-inoculated ones, as they were found to have less bitter taste, a higher odor intensity and good textural attributes. In another study olives with a pre-harvest treatment with copper-based products and kaolin were processed using two selected LAB strains (*L. casei* T19 and *L. plantarum* UT2.1) and sensorially evaluated at the end of fermentation [97]. Un-treated olives and olives fermented with indigenous population were used as control. Eight trained panelists used an official method to assess the descriptors of taste and texture on a 1–10 scale [113]. Panelists judged the treated olives and the not treated ones fermented by *L. plantarum* the best for acidic and salty tastes and for crunchiness. Additionally, there was not always a correspondence between sensory and chemical data. Randazzo et al. [98] tested LAB and probiotic LAB strains on sensory characteristics of green olives. The olives were fermented as follows: spontaneous fermentation (control), inoculation with probiotic *L. rhamnosus* H25, inoculation with commercial probiotic *L. rhamnosus* GG, inoculation with *L. plantarum* GC3 and *L. paracasei* BS21 and inoculation with *L. plantarum* GC3 plus *L. paracasei* BS21 plus *L. rhamnosus* H25. A panel of eleven trained judges used 15 descriptors to evaluate the ready-to-eat olives. Results evidenced that sensory characteristics were cultivar dependent. De Angelis et al. [108] used an omics procedure to study the capacity of LAB and Y starters to enhance the sensory properties of black Bella di Cerignola table olives. The authors used four combinations, a commercial *L. plantarum* strain (S), the same S plus the autochthonous Y *Wickerhamomyces anomalus* DiSSPA73 (SY), the autochthonous *L. plantarum* DiSSPA1A7 and *Lactobacillus pentosus* DiSSPA7 (SYL), while the fourth fermentation was carried out with the indigenous microbiota and served as control. A panel of eight trained judges used seven descriptors to evaluate the olives (crunchiness, bitter, acid, sweet, salty, flavor and off flavor) [116]. The panelists ranked better the olives fermented with starters, especially the SYL, with respect to the control that obtained the lowest values for crunchiness and olive flavor and the best evaluations for the descriptors acid, bitter and off flavor. All the started olives were judged ready for consumption after 90 days of fermentation. Campus et al. [99] compared the effect of fermentation for 156 days driven by a single strain of *L. plantarum* (SSL) and by a mix of *L. pentosus* strains (SIE), which were isolated from previous successful fermentations, with a control carried out with indigenous microbiota (NF). Sensory analysis of the fermented green olives was performed by eight trained assessors [134] that were calibrated for the "bitterness" descriptor with a standard of reference and commercial olives and olive pastes [116]. The assessors found that the two samples obtained with LAB were debittered at the end of processing, while control olives needed 12 months, thus, according to the authors, the use of these starters may be suggested to reduce fermentation times and production costs, and to limit spoilage risk, improve the process control and standardize the product. The same group of authors integrated the above-cited work with a study aimed at focusing on dynamics of microbial growth and at developing a wider set of sensory descriptors [100]. Seven trained assessors sensorially evaluated the olives using the descriptors of acidity, bitterness, crunchiness, fibrousness, freestone, hardness and saltiness [116]. The sensory analysis on the ready-to-eat olives showed that the SIE starter resulted in olives with a sensory profile very close to natural-style samples, with respect to SSL. Martorana et al. [101] proposed an innovative approach based on the wine-technology of "pied de cuve" to sensorially enhance green table olives. The preparation of pied de cuve involved a table olive fermentation of 10 days with both indigenous microbiota (control) or the autochthonous LAB strain *L. pentosus* OM13. These pre-fermented brines were used to carry out the experimental fermentations and compared with the two controls, one made with spontaneous fermentation, the other with use *L. pentosus* OM13. Sensory analysis was performed after 200 days by 12 trained assessors that used nine descriptors for odor (green olive aroma), rheological characteristics (crunchiness), taste (sweet, acid, bitter, salty and complexity), off odor and off flavor [123]. Data analyzed by MSA revealed that the use of spontaneously fermented pied de cuve resulted in olives with the highest scores of sensory complexities and with the absence of any off odors and off flavors. The effects of mechanical harvesting on sensory quality of olives has been investigated for the first time by Martorana et al. [102]. The autochthonous LAB *strain L. pentosus* OM13 was used for fermentation, while uninoculated olives were used as control. Manually harvested drupes fermented either with the LAB strain and with an indigenous microbiota served as controls. Twelve judges used 15 descriptors for the analysis of external aspect, odor, taste and off flavors [123]. MSA of data evidenced that the mechanically harvested and LAB fermented olives were sensorially similar to the manually harvested olives, thereby suggesting that mechanical harvesting and fermentation with LAB starter could substitute the manual harvesting for table olive processing. Campus et al. [103] developed an automated pilot plant (CF), in which a LAB starter was used, and compared the sensory attributes of the obtained fermented olives with samples processed with spontaneous fermentation (NF). The pilot plant was equipped with a control of: temperature, brine, internal pressure of reactor, flow rate of the circulation pump, pH, dissolved CO_2 and brine concentration. Eight trained assessors performed the QDA [116,134] with thirteen descriptors. Assessors evaluated the bitterness until it was reached a determined commercial bitter level, based on a retail sample. CF samples obtained the same level of bitterness of commercial sample after only 90 days, while NF olives had a significantly higher bitter taste than commercial sample after 180 days. Authors are encouraged to suggest this approach as the reduction of fermentation times may be considered as environmentally friendly. Pino et al. [104] evaluated the differences in sensory properties of green table olives fermented in brines with different NaCl contents (4%, 5%, 6% and 8%) and fermented with LAB starters (*L. plantarum* UT2.1 and *L. paracasei* N24) or indigenous microbiota. A ten-trained panel rated olives for negative sensations and used the descriptors for gustatory and kinesthetic sensations [116]. Panelists gave also an overall acceptability score. Panelists rated the LAB fermented olives with a significantly high overall acceptability score and the sample brined with the 5% NaCl obtained the best appreciation. The authors conclude that the formulation of LAB fermented olives with reduced NaCl is healthier and could be suggested to avoid risks in people suffering from hypertension. Randazzo et al. [105] used six LAB starters constituted by L. *plantarum*, *L. paracasei* and *L. pentosus*, alone or in combination, to process for 180 days Nocellara Etnea olives and investigated their effects on the olive sensory quality. A non-inoculated control was used as reference. A trained panel of 12 judges used 15 descriptors to evaluate the olives [73]. Results indicated significant differences in bitterness, bright, crunchiness, green color, green olive aroma and juiciness for LAB samples, while the control olives had the highest bitterness value. The authors found that fermentation with a combination of *L. plantarum* + *L. paracasei* resulted in olives with the lowest bitter taste. The LAB processed olives, moreover, obtained the best value of overall quality. Romeo et al. [106] studied the differences in the sensory properties of fermented Aitana and Caiazzana black olives and Nocellara del Belice turning color olives. Olives were processed with a commercial LAB (*L. plantarum* Lyoflora V3, Sacco) or with indigenous microbiota. A panel of 12 trained tasters used eight descriptors to evaluate the fermented olives [116]. Panelists found that all the tested cultivars had good sensory characteristics, and gave the highest scores for flesh consistency and crunchiness to Nocellara del Belice olives. Sensory differences among cultivars were mainly explained by the descriptors of acid, bitter and hardness. Pino et al. [107] used a sequential inoculum procedure with LAB starters with the aim to evaluate the sensory characteristics of green Nocellara etnea olives brined at 5% and 8% NaCl for 120 days. The inoculation procedure consisted of using at the beginning the α-glucosidase positive strain *L. plantarum* F3.3 and adding after 60 days the potential probiotic *L. paracasei* N24 strain. A control test with no starter was also considered. Ten trained panelists described and rated the olives for attributes linked to gustatory, kinesthetic and negative sensations, and also assigned an overall quality score [116]. The panelists did not perceive negative sensation and did not detect significative differences for crunchiness, fibrousness or hardness among samples. The control olives obtained the highest scores for acidity, while the highest bitter taste was scored in the samples obtained without *L. plantarum* addition. The samples at 5% and 8% NaCl added with the *L. plantarum* strain received the highest overall acceptability. Another research group proposed the sequential inoculation approach by using LAB and Y starters, as an alternative to natural fermentation, to improve the sensory properties of Conservolea and Kalamata olives [109]. The experimental plan considered the use of LAB followed by yeast (LY), the opposite (YL), the use of Y and LAB (MIX) and an indigenous fermentation (Sp). The LAB were *Leuconostoc mesenteroides* K T5-1 and *L. plantarum* A 135–5; the Y were *S. cerevisiae* and *Debaryomyces hansenii*. Fifty-one untrained and seven trained assessors carried out acceptability and descriptive tests on olives after 105 days of fermentation. Panelists scored olives for acidity, bitterness, hardness, odor and saltiness, and gave an overall score. Kalamàta olives obtained the best scores for aroma and overall acceptability when the Y+LAB and MIX inoculations were used, while Conservolea olives showed the same results when LY were inoculated. The use of only Y starters has been recently proposed by Ciafardini and Zullo [110] with the aim to study their effects on the sensory properties of black Taggiasca olives fermented with acidified brined with 8 and 12% NaCl solutions. The Y species *Candida adriatica* 1985, *C. diddensiae* 2011, *Cyteromyces matritensis* 2005, *Nakazawaea molendini-olei* 2004, *S. cerevisiae* 2046 and *Wickerhamomyces anomalus* 1960 fermented the brines for 120 days along with an inoculated control. A panel of eight judges used a QDA to describe olives for gustatory and kinesthetic attributes. The panelists detected significant differences for bitterness, saltiness and hardness among samples [116]. The bitter taste was significantly lower in olives fermented with *C. diddensiae* 2011, *C. adriatica* 1985, and *W. anomalus* 1960. The salt flavor was higher in 12% NaCl processed olives, while no defects were detected in the samples. The authors suggest that the best combination in terms of sensory quality, may be obtained with the use of Y on acidified brines at the highest NaCl concentration.

3.2.2. Treated Green Olives or Spanish Style

Aponte et al. [111] successfully attempted to obtain a more predictable fermentation by using autochthonous LAB cultures. Olives were harvested from irrigated and not irrigated fields and a preliminary LAB isolation led to the isolation of 88 different strains. The authors used *L. pentosus* OM13, alone or in combination with a *L. coryniformis* strain, to enhance the quality of olives, while a non-inoculated fermentation was used as control. Sensory analysis was done after 60 and 120 days of brining by a panel of 10 trained judges who used fifteen descriptors for aspect, flavor, odor, tactile in mouth, texture and overall judgement [123]. A nine-point scale was used for ratings. Data highlighted that *L. pentosus* improved the sensory characteristics of olives, with respect to control samples. In the work already cited in Section 3.2 by Tataridou et al. [45], it was verified the efficacy of the autochthonous oleuropeinolytic strains of *L. plantarum* on sensory properties of the fermented olives, and compared to fermentation with spontaneous microbiota. A nine-member panel evaluated the olives for the descriptors color, odor, flavor (acid, bitter, salty), firmness and crispness. A global assessment score was also expressed. The panelists did not find any statistical difference between control and LAB fermented olives for bitterness and saltiness. The same LAB *L. pentosus* OM13 was used by Martorana et al. [135] both to enhance the fermentation of treated green olives and to study its effect on sensory quality of processed olives, in comparison with two controls (one fermented by spontaneous microbiota and another one with the addition of the studied strain). To improve the growth potential of the LAB culture the following procedures were applied: addition of lactic acid to bring brine pH at 7.0 (IOP1); lactic acid and a nutrient adjuvant (IOP2); the same as IOP2, but brine acclimatization for 12 of the LAB strain before inoculation (IOP3). Twelve judges carried out the sensory analysis after 195 days of processing. A descriptive method [123], including 16 descriptors, was used. A MSA analysis of data revealed that the IOP2 and IOP3 were very close regarding the positive characteristics of complexity (odor and taste), green olive aroma and overall acceptability, while the control olives showed the negative descriptors of bitter, astringent taste and off odors. A *L. pentosus* strain (LP99) isolated in brine of Manzanilla was also recently tested by de Castro et al. [136] and compared with a spontaneous fermentation control. In this case the authors sensorially analyzed the brines to check if differences in concentration of 4-ethyl phenol, which causes off odors, between LAB fermented and control olives resulted in perceivable differences in odor. To that end, 18 trained panelists performed a triangle test. The authors found that, despite the higher concentration of 4-ethyl phenol in inoculated olives, with respect to the control, panelists did not find sensory differences between the two theses, probably because its concentration was below the odor threshold.

3.2.3. Comparison among Different Trade Preparations and Styles

The work of Marsilio et al. [137] investigated the effect of the LAB strain *L. plantarum* (LAB B1–2001) on the sensory characteristics of Greek style-olives (GSP-i) in comparison to a control fermented with indigenous microbiota (GSP-s) and olives processed with the Spanish-style (SSP). A trained panel of 17 members evaluated the fermented samples for the descriptors of odor, bitterness, firmness and crispness [43]. Sensory analysis showed that SSP olives were less bitter, crisp and firm than both GSP samples. GSP-s obtained the higher bitter and lower odor scores, with respect to GSP-i. MSA of data by PCA well discriminated GSP-s from GSP-i olives.

4. Conclusions

The increasing consumer demand for foods with high contents of phytochemicals has stimulated the industry and research to develop new products that meet this requirement, or to study more deeply the existing ones. Table olives fall surely into the second category, and are one of the basic foods in the human diet, especially in Mediterranean countries. Their balanced fatty acids and the presence of important amounts of polyphenols and fibers and the contemporary sensory peculiarities of the very high number of preparations may further improve their use in the future. For these reasons, during the last two decades, researchers have been focusing their studies on the effects of pre-harvest, cultivar and processing factors on the nutritional and sensory properties of table olives. The review has pointed out the preeminent role of trade preparations and processing styles mainly on polyphenols and lipids. Fermentation with the natural style has been confirmed as the best preparation for maintaining the highest content of polyphenols and tryacilglicerols, while no comparative study comparing the effects of the main trade preparations on fat compounds has been reported. Despite the number of papers discussing this topic, there is a real need to focus future studies on the *in vivo* effects of these supposed nutritional claims; thus, it would be advisable to carry out multidisciplinary studies that compare technological aspects with health benefits. Moreover, more effort should be made to study new debittering technologies that are able at the same time to reduce the process time and maintain the nutritional and sensory quality of olives while assuring more sustainability from an economical and environmental point of view.

The review has also revealed the increasing interest over the last two decades of researchers in describing table olives sensorially. This topic has been deeply studied by using internationally recognized sensorial procedures, and the data obtained, treated with rigorous statistical approaches, gave important knowledge for the discrimination and quality evaluations of the different trade preparations. A further goal of sensory analysis could be that of developing a unique, world-wide accepted test for each trade preparation, as in the case of virgin olive oils.

Finally, the importance of using starters to reduce processing times and improve the overall quality of olives has been thoroughly reviewed, and the need to find suitable commercial cultures in the future has emerged.

Author Contributions: P.C.: Investigation, Validation, Writing-Review and editing. C.F.: Investigation, Writing—Review and editing. A.D.C.: Investigation, Validation; Writing—Review and editing. P.P.U.: Investigation, Writing—Review and editing. A.P.: Conceptualization, Supervision, Writing—original draft, review and editing. All authors have read and agreed to the published version of the manuscript.

Funding: This research received no external funding.

Conflicts of Interest: The authors declare no conflicts of interest.

References

1. Food and Agriculture Organization of the United Nations Corporate Statistical Database (FAOSTAT). Food and Agriculture Data. 2017. Available online: http://www.fao.org/faostat/en/#data (accessed on 20 January 2020).
2. Kiritsakis, A. *Olive Oil from the Tree to the Table*, 2nd ed.; Food & Nutrition Press: Trumbull, CT, USA, 1998; pp. 53–95, 119–226.
3. International Olive Oil Council (IOC). *The World Catalogue of Olive Varieties—Olive Germplasm, Cultivars and World-Wide Collections*; International Olive Oil Council: Madrid, Spain, 2013.
4. Garrido Fernández, A.; Fernández Diez, M.J.; Adams, M.R. *Table Olives: Production and Processing*; Chapman and Hall: London, UK, 1997; p. 461.
5. Frega, N.; Lercker, G. La composizione dei lipidi della drupa di olivo durante maturazione. *Agrochimica* **1985**, *29*, 300–308.
6. Bianchi, G.; Murelli, C.; Vlahov, G. Surface waxes from olive fruits. *Phytochemistry* **1992**, *31*, 3503–3506. [CrossRef]
7. Rodríguez, G.; Lama, A.; Rodríguez, R.; Jiménez, A.; Guillén, R.; Fernández-Bolaños, J. Olive stone an attractive source of bioactive and valuable compounds. *Bioresour. Technol.* **2008**, *99*, 5261–5269. [CrossRef] [PubMed]
8. Montaño, A.; Sánchez, A.H.; López-López, A.; De Castro, A.; Rejano, L. Chemical Composition of Fermented Green Olives: Acidity, Salt, Moisture, Fat, Protein, Ash, Fiber, Sugar, and Polyphenol. In *Olives and Olive Oil in Health and Disease Prevention*; Preedy, V.R., Watson, R.R., Eds.; Elsevier: Oxford, UK, 2010; pp. 291–297.

9. Evangelou, E.; Kiritsakis, K.; Sakellaropoulos, N.; Kiritsakis, A. Table olives production, postharvest processing, and nutritional qualities. In *Handbook of Vegetables and Vegetable Processing*, 2nd ed.; Siddiq, M., Uebersax, M.A., Eds.; John Wiley & Sons Ltd: Chichester, UK, 2018; Volume 2, pp. 727–744.
10. Guo, Z.; Jia, X.; Zheng, Z.; Lu, X.; Zheng, Y.; Zheng, B.; Xiao, J. Chemical composition and nutritional function of olive (*Olea europaea* L.): A review. *Phytochem. Rev.* **2018**, *17*, 1091–1110. [CrossRef]
11. Johnson, R.L.; Mitchell, A.E. Reducing Phenolics Related to Bitterness in Table Olives. *J. Food Qual.* **2018**, *2018*, 3193185. [CrossRef]
12. Charoenprasert, S.; Mitchell, A. Factors influencing phenolic compounds in table olives (*Olea europaea*). *J. Agric. Food Chem.* **2012**, *60*, 7081–7095. [CrossRef] [PubMed]
13. Ghanbari, R.; Anwar, F.; Alkharfy, K.M.; Gilani, A.H.; Saari, N. Valuable nutrients and functional bioactives in different parts of olive (*Olea europaea* L.)-A review. *Int. J. Mol. Sci.* **2012**, *13*, 3291–3340. [CrossRef]
14. International Olive Oil Council (IOC). Production Data for Table Olives. 2016. Available online: https://www.internationaloliveoil.org/what-we-do/economic-affairs-promotion-unit/ (accessed on 17 January 2020).
15. Bach-Faig, A.; Berry, E.M.; Lairon, D.; Reguant, J.; Trichopoulou, A.; Dernini, S.; Medina, F.X.; Battino, M.; Belahsen, R.; Miranda, G.; et al. Mediterranean diet pyramid today. Science and cultural updates. *Public Health Nutr.* **2011**, *14*, 2274–2284. [CrossRef]
16. Uylaşer, V.; Yildiz, G. The Historical Development and Nutritional Importance of Olive and Olive Oil Constituted an Important Part of the Mediterranean Diet. *Crit. Rev. Food Sci. Nutr.* **2014**, *54*, 1092–1101. [CrossRef]
17. International Olive Oil Council (IOC). *Trade Standard Applying to Table Olives*; International Olive Oil Council: Madrid, Spain, 2004.
18. Marsilio, V.; Campestre, C.; Lanza, B. Phenolic compounds change during California-style ripe olive processing. *Food Chem.* **2001**, *74*, 55–60. [CrossRef]
19. Sakouhi, F.; Harrabi, S.; Absalon, C.; Sbei, K.; Boukhchina, S.; Kallel, H. α-Tocopherol and fatty acids contents of some Tunisian table olives (*Olea europea* L.): Changes in their composition during ripening and processing. *Food Chem.* **2008**, *108*, 833–839. [CrossRef] [PubMed]
20. Ribarova, F.; Zanev, R.; Shishkov, S.; Rizov, N. α-Tocopherol, fatty acids and their correlations in Bulgarian foodstuffs. *J. Food Compos. Anal.* **2003**, *16*, 659–667. [CrossRef]
21. Lanza, B.; Di Serio, M.G.; Iannucci, E.; Russi, F.; Marfisi, P. Nutritional, textural and sensorial characterisation of Italian table olives (*Olea europaea* L. cv. 'Intosso d'Abruzzo'). *Int. J. Food Sci. Technol.* **2010**, *45*, 67–74. [CrossRef]
22. López-López, A.; Cortés-Delgado, A.; Garrido-Fernández, A. Effect of green Spanish-style processing (Manzanilla and Hojiblanca) on the quality parameters and fatty acid and triacylglycerol compositions of olive fat. *Food Chem.* **2015**, *188*, 37–45. [CrossRef] [PubMed]
23. Cano-Lamadrid, M.; Girón, I.F.; Pleite, R.; Burló, F.; Corell, M.; Moriana, A.; Carbonell-Barrachina, A.A. Quality attributes of table olives as affected by regulated deficit irrigation. *LWT-Food Sci. Technol.* **2015**, *62*, 19–26. [CrossRef]
24. Sánchez-Rodríguez, L.; Cano-Lamadrid, M.; Carbonell-Barrachina, A.A.; Wojdyło, A.; Sendra, E.; Hernández, F. Polyphenol Profile in Manzanilla Table Olives As Affected by Water Deficit during Specific Phenological Stages and Spanish-Style Processing. *J. Agric. Food Chem.* **2019**, *67*, 661–670. [CrossRef]
25. Mastralexi, A.; Mantzouridou, F.T.; Tsimidou, M.Z. Evolution of Safety and Other Quality Parameters of the Greek PDO Table Olives "Prasines Elies Chalkidikis" During Industrial Scale Processing and Storage. *Eur. J. Lipid Sci. Technol.* **2019**, *121*, e1800171. [CrossRef]
26. Commission European. Regulation (EU) No 432/2012 of the 16 May 2012 Establishing a List of Permitted Health Claims Made on Foods, Other Than Those Referring to the Reduction of Disease Risk and to Children's Development and Health. *Official Journal European Union*, 16 May 2012.
27. Sánchez Gómez, A.H.; García García, P.; Rejano Navarro, L. Elaboration of table olives. *Grasas y Aceites* **2006**, *57*, 76–84. [CrossRef]
28. Boskou, G.; Salta, F.N.; Chrysostomou, S.; Mylona, A.; Chiou, A.; Andrikopoulos, N.K. Antioxidant capacity and phenolic profile of table olives from the Greek market. *Food Chem.* **2006**, *94*, 558–564. [CrossRef]
29. Pires-Cabral, P.; Barros, T.; Nunes, P.; Quintas, C. Physicochemical, nutritional and microbiological characteristics of traditional table olives from Southern Portugal. *Emirates J. Food Agric.* **2018**, *30*, 611–620.

30. Pires-Cabral, P.; Barros, T.; Mateus, T.; Prata, J.; Quintas, C. The effect of seasoning with herbs on the nutritional, safety and sensory properties of reduced-sodium fermented Cobrançosa cv. table olives. *AIMS Agric. Food* **2018**, *3*, 521–534. [CrossRef]
31. D'Antuono, E.; Garbetta, A.; Ciasca, B.; Linsalata, V.; Minervini, F.; Lattanzio, V.M.T.; Logrieco, A.F.; Cardinali, A. Biophenols from Table Olive cv Bella di Cerignola: Chemical Characterization, Bioaccessibility, and Intestinal Absorption. *J. Agric. Food Chem.* **2016**, *64*, 5671–5678. [CrossRef] [PubMed]
32. Romero, C.; Brenes, M.; Yousfi, K.; Garciä, P.; Garciä, A.; Garrido, A. Effect of Cultivar and Processing Method on the Contents of Polyphenols in Table Olives. *J. Agric. Food Chem.* **2004**, *52*, 479–484. [CrossRef] [PubMed]
33. Mantzouridou, F.; Tsimidou, M.Z. Microbiological quality and biophenol content of hot air-dried Thassos cv. table olives upon storage. *Eur. J. Lipid Sci. Technol.* **2011**, *113*, 786–795. [CrossRef]
34. Fernández-Poyatos, M.P.; Ruiz-Medina, A.; Llorent-Martínez, E.J. Phytochemical profile, mineral content, and antioxidant activity of *Olea europaea* L. cv. Cornezuelo table olives. Influence of in vitro simulated gastrointestinal digestion. *Food Chem.* **2019**, *297*, e124933. [CrossRef] [PubMed]
35. Rodríguez, G.; Lama, A.; Jaramillo, S.; Fuentes-Alventosa, J.M.; Guillén, R.; Jiménez-Araujo, A.; Rodríguez-Arcos, R.; Fernández-Bolanos, J. 3,4-Dihydroxyphenylglycol (DHPG): An Important Phenolic Compound Present in Natural Table Olives. *J. Agric. Food Chem.* **2009**, *57*, 6298–6304. [CrossRef]
36. Rodríguez, G.; Rodríguez, R.; Fernández-Bolaños, J.; Guillén, R.; Jiménez, A. Antioxidant activity of effluents during the purification of hydroxytyrosol and 3,4-dihydroxyphenyl glycol from olive oil waste. *Eur. Food Res. Technol.* **2007**, *224*, 733–741. [CrossRef]
37. Lanza, B.; Di Serio, M.G.; Russi, F.; Iannucci, E.; Giansante, L.; Di Loreto, G.; Di Giacinto, L. Evaluation of the nutritional value of oven-dried table olives (cv. Majatica) processed by the Ferrandina style. *Riv. Ital. Delle Sostanze Grasse* **2014**, *91*, 117–127.
38. Mousori, E.; Melliou, E.; Magiatis, P. Isolation of Megaritolactones and Other Bioactive Metabolites from 'Megaritiki' Table Olives and Debittering Water. *J. Agric. Food Chem.* **2014**, *62*, 660–667. [CrossRef]
39. Tokuşoğlu, Ö.; Alpas, H.; Bozoğlu, F. High hydrostatic pressure effects on mold flora, citrinin mycotoxin, hydroxytyrosol, oleuropein phenolics and antioxidant activity of black table olives. *Innov. Food Sci. Emerg. Technol.* **2010**, *11*, 250–258. [CrossRef]
40. Ben Othman, N.; Roblain, D.; Thonart, P.; Hamdi, M. Tunisian table olive phenolic compounds and their antioxidant capacity. *J. Food Sci.* **2008**, *73*, C235–C240. [CrossRef] [PubMed]
41. Valenčič, V.; Mavsar, D.B.; Bučar-Miklavčič, M.; Butinar, B.; Čadež, N.; Golob, T.; Raspor, P.; Možina, S.S. The impact of production technology on the growth of indigenous microflora and quality of table olives from Slovenian Istria. *Food Technol. Biotechnol.* **2010**, *48*, 404–410.
42. Zoidou, E.; Melliou, E.; Gikas, E.; Tsarbopoulos, A.; Magiatis, P.; Skaltsounis, A.L. Identification of Throuba Thassos, a Traditional Greek Table Olive Variety, as a Nutritional Rich Source of Oleuropein. *J. Agric. Food Chem.* **2010**, *58*, 46–50. [CrossRef] [PubMed]
43. Melliou, E.; Zweigenbaum, J.A.; Mitchell, A.E. Ultrahigh-Pressure Liquid Chromatography Triple-Quadrupole Tandem Mass Spectrometry Quantitation of Polyphenols and Secoiridoids in California-Style Black Ripe Olives and Dry Salt-Cured Olives. *J. Agric. Food Chem.* **2015**, *63*, 2400–2405. [CrossRef]
44. Servili, M.; Minnocci, F.; Veneziani, G.; Taticchi, A.; Urbani, S.; Esposto, S.; Sebastiani, L.; Valmorri, S.; Corestti, A. Compositional and Tissue Modifications Induced by the Natural Fermentation Process in Table Olives. *J. Agric. Food Chem.* **2008**, *56*, 6389–6396. [CrossRef]
45. Tataridou, M.; Kotzekidou, P. Fermentation of table olives by oleuropeinolytic starter culture in reduced salt brines and inactivation of Escherichia coli O157:H7 and Listeria monocytogenes. *Int. J. Food Microbiol.* **2015**, *208*, 122–130. [CrossRef]
46. Pistarino, E.; Aliakbarian, B.; Casazza, A.A.; Paini, M.; Cosulich, M.E.; Perego, P. Combined effect of starter culture and temperature on phenolic compounds during fermentation of Taggiasca black olives. *Food Chem.* **2013**, *138*, 2043–2049. [CrossRef]
47. Durante, M.; Tufariello, M.; Tommasi, L.; Lenucci, M.S.; Bleve, G.; Mita, G. Evaluation of bioactive compounds in black table olives fermented with selected microbial starters. *J. Sci. Food Agric.* **2018**, *98*, 96–103. [CrossRef]
48. D'Antuono, I.; Bruno, A.; Linsalata, V.; Minervini, F.; Garbetta, A.; Tufariello, M.; Mita, G.; Logrieco, A.F.; Bleve, G.; Cardinali, A. Fermented Apulian table olives: Effect of selected microbial starters on polyphenols composition, antioxidant activities and bioaccessibility. *Food Chem.* **2018**, *248*, 137–145. [CrossRef]

49. Tufariello, M.; Anglana, C.; Crupi, P.; Virtuosi, I.; Fiume, P.; Di Terlizzi, B.; Moselhy, N.; Attay, H.A.G.; Pati, S.; Logrieco, A.F.; et al. Efficacy of yeast starters to drive and improve Picual, Manzanilla and Kalamàta table olive fermentation. *J. Sci. Food Agric.* **2019**, *99*, 2504–2512. [CrossRef]
50. Sousa, A.; Casal, S.; Bento, A.; Malheiro, R.; Oliveira, M.B.P.P.; Pereira, J.A. Chemical characterization of "alcaparras" stoned table olives from northeast Portugal. *Molecules* **2011**, *16*, 9025–9040. [CrossRef] [PubMed]
51. Malheiro, R.; Casal, S.; Sousa, A.; de Pinho, P.G.; Peres, A.M.; Dias, L.G.; Bento, A.; Pereira, J.A. Effect of Cultivar on Sensory Characteristics, Chemical Composition, and Nutritional Value of Stoned Green Table Olives. *Food Bioprocess Technol.* **2012**, *5*, 1733–1742. [CrossRef]
52. López, A.; Montaño, A.; García, P.; Garrido, A. Fatty acid profile of table olives and its multivariate characterization using unsupervised (PCA) and supervised (DA) chemometrics. *J. Agric. Food Chem.* **2006**, *54*, 6747–6753. [CrossRef] [PubMed]
53. Moreno-Baquero, J.M.; Bautista-Gallego, J.; Garrido-Fernández, A.; López-López, A. Mineral and sensory profile of seasoned cracked olives packed in diverse salt mixtures. *Food Chem.* **2013**, *138*, 1–8. [CrossRef] [PubMed]
54. Saúde, C.; Barros, T.; Mateus, T.; Quintas, C.; Pires-Cabral, P. Effect of chloride salts on the sensory and nutritional properties of cracked table olives of the Maçanilha Algarvia cultivar. *Food Biosci.* **2017**, *19*, 73–79. [CrossRef]
55. Savaş, E.; Uylaşer, V. Quality improvement of green table olive cv. "domat" (*Olea europaea* L.) grown in turkey using different de-bittering methods. *Not. Bot. Horti Agrobot. Cluj-Napoca* **2013**, *41*, 269–275. [CrossRef]
56. Romero, C.; García, A.; Medina, E.; Ruíz-Méndez, M.V.; de Castro, A.; Brenes, M. Triterpenic acids in table olives. *Food Chem.* **2010**, *118*, 670–674. [CrossRef]
57. Jiménez, A.; Rodríguez, R.; Fernández-Caro, I.; Guillén, R.; Fernández-Bolaños, J.; Heredia, A. Dietary fibre content of table olives processed under different european styles: Study of physico-chemical characteristics. *J. Sci. Food Agric.* **2000**, *80*, 1903–1908. [CrossRef]
58. Arrojo-López, F.N.; Duran-Quintana, M.C.; Romero, C.; Rodríguez-Gómez, F.; Garrido-Fernández, A. Effect of Storage Process on the Sugars, Polyphenols, Color and Microbiological Changes in Cracked Manzanilla-Aloreña Table Olives. *J. Agric. Food Chem.* **2007**, *55*, 7434–7444. [CrossRef]
59. Habibi, M.; Golmakani, M.T.; Farahnaky, A.; Mesbahi, G.; Majzoobi, M. NaOH-free debittering of table olives using power ultrasound. *Food Chem.* **2016**, *192*, 775–781. [CrossRef]
60. Rastogi, N.K.; Raghavarao, K.S.M.S.; Balasubramaniam, V.M.; Niranjan, K.; Knorr, D. Opportunities and challenges in high pressure processing of foods. *Crit. Rev. Food Sci. Nutr.* **2007**, *47*, 69–112. [CrossRef] [PubMed]
61. López, A.; Montaño, A.; Garrido, A. Provitamin A carotenoids in table olives according to processing styles, cultivars, and commercial presentations. *Eur. Food Res. Technol.* **2005**, *221*, 406–411. [CrossRef]
62. López-López, A.; Montaño, A.; Cortés-Delgado, A.; Garrido-Fernández, A. Survey of vitamin B6 content in commercial presentations of table olives. *Plant Foods Hum. Nutr.* **2008**, *63*, 87–91. [CrossRef] [PubMed]
63. Strüh, C.; Jäger, S.; Schempp, C.; Scheffler, A.; Martin, S. Solubilized triterpenes from mistletoe show anti-tumor effects on skin-derived cell lines. *Planta Med.* **2008**, *74*. [CrossRef]
64. Lanza, B.; Di Serio, M.G.; Iannucci, E. Effects of maturation and processing technologies on nutritional and sensory qualities of Itrana table olives. *Grasas y Aceites* **2013**, *64*, 272–284. [CrossRef]
65. Wood, J.D.; Enser, M.; Fisher, A.V.; Nute, G.R.; Sheard, P.R.; Richardson, R.I.; Hughes, S.I.; Whittington, F.M. Fat deposition, fatty acid composition and meat quality: A review. *Meat Sci.* **2008**, *78*, 343–358. [CrossRef]
66. Kotzekidou, P.; Tsakalidou, E. Fermentation Biotechnology of Animal Based Traditional Foods of the Middle East and Mediterranean Region. In *Food Biotechnology*, 2nd ed.; Shetty, K., Paliyath, G., Pometto, A., Levin, R.E., Eds.; CRC Press: Boca Raton, FL, USA, 2005; pp. 1795–1828.
67. Ruiz-Barba, J.L.; Jiménez-Díaz, R. A novel Lactobacillus pentosus-paired starter culture for Spanish-style green olivefermentation. *Food Microbiol.* **2012**, *30*, 253–259. [CrossRef]
68. González, M.M.; Navarro, T.; Gómez, G.; Pérez, R.A.; De Lorenzo, C. Análisis sensorial de aceituna de mesa: II. Aplicabilidad práctica y correlación con el análisis instrumental. *Grasas y Aceites* **2007**, *58*, 231–236. [CrossRef]
69. Marsilio, V.; Russi, F.; Iannucci, E.; Sabatini, N. Effects of alkali neutralization with CO_2 on fermentation, chemical parameters and sensory characteristics in Spanish-style green olives (*Olea europaea* L.). *LWT-Food Sci. Technol.* **2008**, *41*, 796–802. [CrossRef]

70. Yilmaz, E.; Aydeniz, B. Sensory evaluation and consumer perception of some commercial green table olives. *Br. Food J.* **2012**, *114*, 1085–1094. [CrossRef]
71. Moreno-Baquero, J.M.; Bautista-Gallego, J.; Garrido-Fernández, A.; López-López, A. Mineral Content and Sensory Characteristics of Gordal Green Table Olives Fermented in Chloride Salt Mixtures. *J. Food Sci.* **2012**, *77*, S107–S114. [CrossRef] [PubMed]
72. López-López, A.; Bautista-Gallego, J.; Moreno-Baquero, J.M.; Garrido-Fernández, A. Fermentation in nutrient salt mixtures affects green Spanish-style Manzanilla table olive characteristics. *Food Chem.* **2016**, *211*, 415–422. [CrossRef] [PubMed]
73. Cano-Lamadrid, M.; Hernández, F.; Corell, M.; Burló, F.; Legua, P.; Moriana, A.; Carbonell-Barrachina, Á.A. Antioxidant capacity, fatty acids profile, and descriptive sensory analysis of table olives as affected by deficit irrigation. *J. Sci. Food Agric.* **2017**, *97*, 444–451. [CrossRef] [PubMed]
74. López-López, A.; Sánchez-Gómez, A.H.; Montaño, A.; Cortés-Delgado, A.; Garrido-Fernández, A. Sensory profile of green Spanish-style table olives according to cultivar and origin. *Food Res. Int.* **2018**, *108*, 347–356. [CrossRef] [PubMed]
75. López-López, A.; Sánchez-Gómez, A.H.; Montaño, A.; Cortés-Delgado, A.; Garrido-Fernández, A. Sensory characterisation of black ripe table olives from Spanish Manzanilla and Hojiblanca cultivars. *Food Res. Int.* **2019**, *116*, 114–125. [CrossRef]
76. Sánchez-Rodríguez, L.; Cano-Lamadrid, M.; Carbonell-Barrachina, Á.A.; Sendra, E.; Hernández, F. Volatile composition, sensory profile and consumer acceptability of hydrosostainable table olives. *Foods* **2019**, *8*, 470. [CrossRef]
77. Panagou, E.Z.; Sahgal, N.; Magan, N.; Nychas, G.J.E. Table olives volatile fingerprints: Potential of an electronic nose for quality discrimination. *Sens. Actuators B Chem.* **2008**, *134*, 902–907. [CrossRef]
78. Aponte, M.; Ventorino, V.; Blaiotta, G.; Volpe, G.; Farina, V.; Avellone, G.; Lanza, C.M.; Moschetti, G. Study of green Sicilian table olive fermentations through microbiological, chemical and sensory analyses. *Food Microbiol.* **2010**, *27*, 162–170. [CrossRef]
79. Fadda, C.; Del Caro, A.; Sanguinetti, A.M.; Piga, A. Texture and antioxidant evolution of naturally green table olives as affected by different sodium chloride brine concentrations. *Grasas y Aceites* **2014**, *65*, e002.
80. Lanza, B.; Amoruso, F. Sensory analysis of natural table olives: Relationship between appearance of defect and gustatory-kinaesthetic sensation changes. *LWT-Food Sci. Technol.* **2016**, *68*, 365–372. [CrossRef]
81. Lee, S.M.; Kitsawad, K.; Sigal, A.; Flynn, D.; Guinard, J.X. Sensory Properties and Consumer Acceptance of Imported and Domestic Sliced Black Ripe Olives. *J. Food Sci.* **2012**, *77*, S438–S448. [CrossRef] [PubMed]
82. García-García, P.; Sánchez-Gómez, A.H.; Garrido-Fernández, A. Changes of physicochemical and sensory characteristics of packed ripe table olives from Spanish cultivars during shelf-life. *Int. J. Food Sci. Technol.* **2014**, *49*, 895–903. [CrossRef]
83. Sánchez, A.H.; López-López, A.; Cortés-Delgado, A.; de Castro, A.; Montaño, A. Aroma profile and volatile composition of black ripe olives (Manzanilla and Hojiblanca cultivars). *Food Res. Int.* **2020**, *127*, e108733. [CrossRef] [PubMed]
84. Gambella, F.; Piga, A.; Agabbio, M.; Vacca, V.; D'hallewin, G. Effect of different pre-treatments on drying of green table olives (*Ascolana tenera var.*). *Grasas y Aceites* **2000**, *51*, 173–176.
85. Piga, A.; Mincione, B.; Runcio, A.; Pinna, I.; Agabbio, M.; Poiana, M. Response to hot air drying of some olive cultivars of the south of Italy. *Acta Aliment.* **2005**, *34*, 427–440. [CrossRef]
86. Değirmencioğlu, N.; Gürbüz, O.; Değirmencioğlu, A.; Şahan, Y.; Özbey, H. Effect of MAP and vacuum sealing on sensory qualities of dry-salted olive. *Food Sci. Biotechnol.* **2011**, *20*, 1307–1313. [CrossRef]
87. Galán-Soldevilla, H.; Ruiz Pérez-Cacho, P. Panel training programme for the Protected Designation of Origin "Aceituna Aloreña de Malaga. " *Grasas y Aceites* **2012**, *63*, 109–117. [CrossRef]
88. Rodríguez-Gómez, F.; Ruiz-Bellido, M.A.; Romero-Gil, V.; Benítez-Cabello, A.; Garrido-Fernández, A.; Arroyo-López, F.N. Microbiological and physicochemical changes in natural green heat-shocked Aloreña de Málaga table olives. *Front. Microbiol.* **2017**, *8*, e2209. [CrossRef]
89. Galán-Soldevilla, H.; Ruiz Pérez-Cacho, P.; Hernández Campuzano, J.A. Determination of the characteristic sensory profiles of Aloreña table-olive. *Grasas y Aceites* **2013**, *64*, 442–452. [CrossRef]
90. Bautista-Gallego, J.; Arroyo-López, F.N.; Romero-Gil, V.; Rodríguez-Gómez, F.; Garrido-Fernández, A. Evaluating the effects of zinc chloride as a preservative in cracked table olive packing. *J. Food Prot.* **2011**, *74*, 2169–2176. [CrossRef]

91. Bautista-Gallego, J.; Moreno-Baquero, J.M.; Garrido-Fernández, A.; López-López, A. Development of a novel Zn fortified table olive product. *LWT-Food Sci. Technol.* **2013**, *50*, 264–271. [CrossRef]
92. Romero-Gil, V.; Rodríguez-Gómez, F.; Ruiz-Bellido, M.; Cabello, A.B.; Garrido-Fernández, A.; Arroyo-López, F.N. Shelf-life of traditionally-seasoned aloreña de málaga table olives based on package appearance and fruit characteristics. *Grasas y Aceites* **2019**, *70*, e306. [CrossRef]
93. Lanza, B.; Di Serio, M.G.; Giansante, L.; Di Loreto, G.; Russi, F.; Di Giacinto, L. Effects of pasteurisation and storage on quality characteristics of table olives preserved in olive oil. *Int. J. Food Sci. Technol.* **2013**, *48*, 2630–2637. [CrossRef]
94. Alves, M.; Esteves, E.; Quintas, C. Effect of preservatives and acidifying agents on the shelf life of packed cracked green table olives from Maçanilha cultivar. *Food Packag. Shelf Life* **2015**, *5*, 32–40. [CrossRef]
95. Rodrigues, N.; Marx, Í.M.G.; Dias, L.G.; Veloso, A.C.A.; Pereira, J.A.; Peres, A.M. Monitoring the debittering of traditional stoned green table olives during the aqueous washing process using an electronic tongue. *LWT-Food Sci. Technol.* **2019**, *109*, 327–335. [CrossRef]
96. Marsilio, V.; D'Andria, R.; Lanza, B.; Russi, F.; Iannucci, E.; Lavini, A.; Morelli, G. Effect of irrigation and lactic acid bacteria inoculants on the phenolic fraction, fermentation and sensory characteristics of olive (*Olea europaea* L. cv. Ascolana tenera) fruits. *J. Sci. Food Agric.* **2006**, *86*, 1005–1013. [CrossRef]
97. Randazzo, C.L.; Fava, G.; Tomaselli, F.; Romeo, F.V.; Pennino, G.; Vitello, E.; Caggia, C. Effect of kaolin and copper-based products and of starter cultures on green table olive fermentation. *Food Microbiol.* **2011**, *28*, 910–919. [CrossRef]
98. Randazzo, C.L.; Todaro, A.; Pino, A.; Pitino, I.; Corona, O.; Mazzaglia, A.; Caggia, C. Giarraffa and Grossa di Spagna naturally fermented table olives: Effect of starter and probiotic cultures on chemical, microbiological and sensory traits. *Food Res. Int.* **2014**, *62*, 1154–1164. [CrossRef]
99. Campus, M.; Sedda, P.; Cauli, E.; Piras, F.; Comunian, R.; Paba, A.; Daga, E.; Schirru, S.; Angioni, A.; Zurru, R.; et al. Evaluation of a single strain starter culture, a selected inoculum enrichment, and natural microflora in the processing of Tonda di Cagliari natural table olives: Impact on chemical, microbiological, sensory and texture quality. *LWT-Food Sci. Technol.* **2015**, *64*, 671–677. [CrossRef]
100. Comunian, R.; Ferrocino, I.; Paba, A.; Daga, E.; Campus, M.; Di Salvo, R.; Cauli, E.; Piras, F.; Zurru, R.; Cocolin, L. Evolution of microbiota during spontaneous and inoculated Tonda di Cagliari table olives fermentation and impact on sensory characteristics. *LWT-Food Sci. Technol.* **2017**, *84*, 64–72. [CrossRef]
101. Martorana, A.; Alfonzo, A.; Settanni, L.; Corona, O.; la Croce, F.; Caruso, T.; Moschetti, G.; Francesca, N. An innovative method to produce green table olives based on "pied de cuve" technology. *Food Microbiol.* **2015**, *50*, 126–140. [CrossRef] [PubMed]
102. Martorana, A.; Alfonzo, A.; Settanni, L.; Corona, O.; La Croce, F.; Caruso, T.; Moschetti, G.; Francesca, N. Effect of the mechanical harvest of drupes on the quality characteristics of green fermented table olives. *J. Sci. Food Agric.* **2016**, *96*, 2004–2017. [CrossRef] [PubMed]
103. Campus, M.; Cauli, E.; Scano, E.; Piras, F.; Comunian, R.; Paba, A.; Daga, E.; Di Salvo, R.; Sedda, P.; Angioni, A.; et al. Towards Controlled Fermentation of Table Olives: LAB Starter Driven Process in an Automatic Pilot Processing Plant. *Food Bioprocess Technol.* **2017**, *10*, 1063–1073. [CrossRef]
104. Pino, A.; De Angelis, M.D.; Todaro, A.; Van Hoorde, K.V.; Randazzo, C.L.; Caggia, C. Fermentation of Nocellara Etnea table olives by functional starter cultures at different low salt concentrations. *Front. Microbiol.* **2018**, *9*, e1125. [CrossRef]
105. Randazzo, C.L.; Russo, N.; Pino, A.; Mazzaglia, A.; Ferrante, M.; Conti, G.O.; Caggia, C. Effects of selected bacterial cultures on safety and sensory traits of Nocellara Etnea olives produced at large factory scale. *Food Chem. Toxicol.* **2018**, *115*, 491–498. [CrossRef]
106. Romeo, F.V.; Timpanaro, N.; Intelisano, S.; Rapisarda, P. Quality evaluation of Aitana, Caiazzana and Nocellara del Belice table olives fermented with a commercial starter culture. *Emirates J. Food Agric.* **2018**, *30*, 604–610.
107. Pino, A.; Vaccalluzzo, A.; Solieri, L.; Romeo, F.V.; Todaro, A.; Caggia, C.; Arroyo-López, F.N.; Bautista-Gallego, J.; Randazzo, C.L. Effect of sequential inoculum of beta-glucosidase positive and probiotic strains on brine fermentation to obtain low salt sicilian table olives. *Front. Microbiol.* **2019**, *10*, e174. [CrossRef]
108. De Angelis, M.; Campanella, D.; Cosmai, L.; Summo, C.; Rizzello, C.G.; Caponio, F. Microbiota and metabolome of un-started and started Greek-type fermentation of Bella di Cerignola table olives. *Food Microbiol.* **2015**, *52*, 18–30. [CrossRef]

109. Chytiri, A.; Tasioula-Margari, M.; Bleve, G.; Kontogianni, V.G.; Kallimanis, A.; Kontominas, M.G. Effect of different inoculation strategies of selected yeast and LAB cultures on Conservolea and Kalamàta table olives considering phenol content, texture, and sensory attributes. *J. Sci. Food Agric.* **2020**, *100*, 926–935. [CrossRef]
110. Ciafardini, G.; Zullo, B.A. Use of selected yeast starter cultures in industrial-scale processing of brined Taggiasca black table olives. *Food Microbiol.* **2019**, *84*, 103250. [CrossRef]
111. Aponte, M.; Blaiotta, G.; La Croce, F.; Mazzaglia, A.; Farina, V.; Settanni, L.; Moschetti, G. Use of selected autochthonous lactic acid bacteria for Spanish-style table olive fermentation. *Food Microbiol.* **2012**, *30*, 8–16. [CrossRef]
112. Marsilio, V. Sensory analysis of table olives. *Olivae* **2002**, *90*, 32–41.
113. International Olive Oil Council (IOC). *Method for the Sensory Analysis of Table Olives*; International Olive Oil Council: Madrid, Spain, 2008.
114. International Olive Oil Council (IOC). *Sensory Analysis of Table Olives*; International Olive Oil Council: Madrid, Spain, 2010.
115. Vergara, J.V.; Blana, V.; Mallouchos, A.; Stamatiou, A.; Panagou, E.Z. Evaluating the efficacy of brine acidification as implemented by the Greek table olive industry on the fermentation profile of Conservolea green olives. *LWT-Food Sci. Technol.* **2013**, *53*, 113–119. [CrossRef]
116. International Olive Oil Council (IOC). *Method for the Sensory Analysis of Table Olives*; International Olive Oil Council: Madrid, Spain, 2011.
117. Bautista-Gallego, J.; Arroyo-López, F.N.; Romero-Gil, V.; Rodríguez-Gómez, F.; Garrido-Fernández, A. The effect of $ZnCl_2$ on green Spanish-style table olive packaging, a presentation style dependent behaviour. *J. Sci. Food Agric.* **2015**, *95*, 1670–1677. [CrossRef] [PubMed]
118. Piga, A.; Gambella, F.; Vacca, V.; Agabbio, M. Response of three Sardinian olive cultivars to Greek-style processing. *Ital. J. Food Sci.* **2001**, *13*, 29–40.
119. Piga, A.; Agabbio, M. Quality improvement of naturally green table olives by controlling some processing parameters. *Ital. J. Food Sci.* **2003**, *15*, 259–268.
120. Kanavouras, A.; Gazouli, M.; Tzouvelekis Leonidas, L.; Petrakis, C. Evaluation of black table olives in different brines. *Grasas y Aceites* **2005**, *56*, 106–115. [CrossRef]
121. Hurtado, A.; Reguant, C.; Bordons, A.; Rozès, N. Influence of fruit ripeness and salt concentration on the microbial processing of Arbequina table olives. *Food Microbiol.* **2009**, *26*, 827–833. [CrossRef]
122. UNE. *UNE EN ISO 4120:2008. Análisis Sensorial. Metodología. Prueba Triangular*; Asociación Española de Normalización: Madrid, Spain, 2008.
123. UNI. UNI 10957. In *Sensory Analysis—Method for Establishing a Sensory Profile in Foodstuffs and Beverages*; Ente Nazionale Italiano di Normazione: Roma, Italy, 2003.
124. Stone, E.; Sidel, J.L. Affective testing. In *Sensory Evaluation Practices*, 3rd ed.; Stone, E., Sidel, J.L., Eds.; Elsevier Academic Press: London, UK, 2004; pp. 262–264.
125. Lawless, H.T.; Heymann, H. *Sensory Evaluation of Food: Principles and Practices*, 2nd ed.; Springer Science + Business Media, LLC: New York, NY, USA, 2010.
126. Meilgaard, M.; Civille, G.V.; Carr, B.T. *Sensory Evaluation Techniques*, 2nd ed.; CRC Press: Boca Raton, FL, USA, 1991; pp. 201–235.
127. Panagou, E.Z.; Tassou, C.C.; Katsaboxakis, K.Z. Microbiological, physicochemical and organoleptic changes in dry-salted olives of Thassos variety stored under different modified atmospheres at 4 and 20 °C. *Int. J. Food Sci. Technol.* **2002**, *37*, 635–641. [CrossRef]
128. Panagou, E.Z. Effect of different packing treatments on the microbiological and physicochemical characteristics of untreated green olives of the Conservolea cultivar. *J. Sci. Food Agric.* **2004**, *84*, 757–764. [CrossRef]
129. Pradas, I.; Del Pino, B.; Peña, F.; Ortiz, V.; Moreno-Rojas, J.M.; Fernández-Hernández, A.; García-Mesa, J.A. The use of high hydrostatic pressure (HHP) treatments for table olives preservation. *Innov. Food Sci. Emerg. Technol.* **2012**, *13*, 64–68. [CrossRef]
130. López-López, A.; Garrido-Fernández, A. *Producción, Elaboración, Composición y Valor Nutricional de la Aceituna Aloreña de Málaga*; Edita Redagua: Málaga, Spain, 2010.
131. Hibbert, D.B. Chemometric Analysis of Sensory Data. In *Comprehensive Chemometrics. Chemical and Biochemical Data Analysis*; Brown, S., Tauler, R., Walczak, B., Eds.; Elsevier: Oxford, UK, 2009; Volume 4, pp. 377–424.
132. Panagou, E.Z.; Tassou, C.C.; Skandamis, P.N. Physicochemical, microbiological, and organoleptic profiles of Greek table olives from retail outlets. *J. Food Prot.* **2006**, *69*, 1732–1738. [CrossRef] [PubMed]

133. Lanza, B.; Amoruso, F. Panel performance, discrimination power of descriptors, and sensory characterization of table olive samples. *J. Sens. Stud.* **2020**, *35*, e12542. [CrossRef]
134. ISO. ISO 8586:2012. In *Sensory Analysis—General Guidelines for the Selection, Training and Monitoring of Selected Assessors and Expert Sensory Assessors*, 1st ed.; International Organization for Standardization: Geneva, Switzerland, 2012.
135. Martorana, A.; Alfonzo, A.; Gaglio, R.; Settanni, L.; Corona, O.; La Croce, F.; Vagnoli, P.; Caruso, T.; Moschetti, G.; Francesca, N. Evaluation of different conditions to enhance the performances of Lactobacillus pentosus OM13 during industrial production of Spanish-style table olives. *Food Microbiol.* **2017**, *61*, 150–158. [CrossRef]
136. de Castro, A.; Sánchez, A.H.; Cortés-Delgado, A.; López-López, A.; Montaño, A. Effect of Spanish-style processing steps and inoculation with Lactobacillus pentosus starter culture on the volatile composition of cv. Manzanilla green olives. *Food Chem.* **2019**, *271*, 543–549. [CrossRef]
137. Marsilio, V.; Seghetti, L.; Iannucci, E.; Russi, F.; Lanza, B.; Felicioni, M. Use of a lactic acid bacteria starter culture during green olive (*Olea europaea* L cv Ascolana tenera) processing. *J. Sci. Food Agric.* **2005**, *85*, 1084–1090. [CrossRef]

foods

MDPI

Article

Benefits of the Use of Lactic Acid Bacteria Starter in Green Cracked Cypriot Table Olives Fermentation

Dimitrios A. Anagnostopoulos, Vlasios Goulas, Eleni Xenofontos, Christos Vouras, Nikolaos Nikoloudakis and Dimitrios Tsaltas *

Department of Agricultural Sciences, Biotechnology and Food Science, Cyprus University of Technology, Limassol 3036, Cyprus; da.anagnostopoulos@edu.cut.ac.cy (D.A.A.); vlasios.goulas@cut.ac.cy (V.G.); eleni_xenofontos@hotmail.com (E.X.); ChrisVour@hotmail.com (C.V.); n.nikoloudakis@cut.ac.cy (N.N.)
* Correspondence: dimitris.tsaltas@cut.ac.cy

Received: 11 November 2019; Accepted: 20 December 2019; Published: 23 December 2019

Abstract: Table olives are one of the most established Mediterranean vegetables, having an exponential increase consumption year by year. In the natural-style processing, olives are produced by spontaneous fermentation, without any chemical debittering. This natural fermentation process remains empirical and variable since it is strongly influenced by physicochemical parameters and microorganism presence in olive drupes. In the present work, Cypriot green cracked table olives were processed directly in brine (natural olives), using three distinct methods: spontaneous fermentation, inoculation with lactic acid bacteria at a 7% or a 10% NaCl concentration. Sensory, physicochemical, and microbiological alterations were monitored at intervals, and major differences were detected across treatments. Results indicated that the predominant microorganisms in the inoculated treatments were lactic acid bacteria, while yeasts predominated in control. As a consequence, starter culture contributed to a crucial effect on olives fermentation, leading to faster acidification and lower pH. This was attributed to a successful lactic acid fermentation, contrasting the acetic and alcoholic fermentation observed in control. Furthermore, it was established that inhibition of enterobacteria growth was achieved in a shorter period and at a significantly lower salt concentration, compared to the spontaneous fermentation. Even though no significant variances were detected in terms of the total phenolic content and antioxidant capacity, the degradation of oleuropein was achieved faster in inoculated treatments, thus, producing higher levels of hydroxytyrosol. Notably, the reduction of salt concentration, in combination with the use of starter, accented novel organoleptic characteristics in the final product, as confirmed from a sensory panel; hence, it becomes obvious that the production of Cypriot table olives at reduced NaCl levels is feasible.

Keywords: fermentation; table olives; microbiological changes; organoleptic; physicochemical

1. Introduction

Table olives are an essential element, which is closely related to Mediterranean history. Nowadays, they are considered as the most important vegetables worldwide, with a gross production exceeding 2.7 million tonnes/year [1]. The main purpose of table olive fermentation is to achieve a preservation effect and, in parallel, enhancing the organoleptic attributes of the processed product, hence, meeting consumer's needs. However, in order to standardize this process and consequently secure the quality of the final product, the study of microbiological and physicochemical descriptors for monitoring the fermentation is a pre-request [2]. Three styles (Spanish, Greek, and Californian) are the most well-known and established commercial types globally [3].

Natural fermentation is mainly driven by yeasts and lactic acid bacteria (LAB), present on olive drupes [4,5]. It has been noted that the LAB is responsible for the fermentation of treated olives (Spanish style). However, in a natural process, LAB and yeasts compete, and in some cases, yeasts can exclusively direct fermentation [6]. Except from these two dominant microorganisms, diverse microbial populations are also participating during olive fermentation, such as several species of *Enterobacteriaceae, Clostridium, Pseudomonas, Staphylococcus,* and molds [7]. These microorganisms via their metabolic activities contribute to crucial aspects, such as organoleptic characteristics (color, texture, flavor, etc.) and safe consumption [8]. In general, LAB activity results in brine acidification, via the production of lactic and other acids, using the fermentable substrates, resulting in pH decrease, providing microbiological control to the final product, hence, extending its shelf life [9,10]. Oppositely, yeasts conduce to the flavor and aroma formation via the production of volatile and other desirable compounds, while, at the same time, they enhance LAB growth and the degradation of phenolic and secosteroid compounds, such as oleuropein [11]. However, the microbiota formation also heavily depends on olive cultivar type since different fruit dimension and composition can affect the microbial dynamics responsible for olive fermentation and sway the sensorial attitudes of the product [12].

During fermentation, major physicochemical changes are taking place. Water-soluble compounds are diffused from olives to the brine, while salt follows the opposite direction, until equilibrium at the end of the brining process [13]. Fermentable sugars are the main source of carbon for microorganisms, providing organic acids, which are essential for the stability and succession of the fermentation process.

Although the physicochemical maturation of olives and brines, during processing, has been thoroughly investigated [2,14–20], there is no information about Cypriot green naturally fermented olives. Furthermore, significant organoleptic parameters, such as texture and color, are understudied. Both are the main attributes that most affect the consumer's acceptance and may be strongly affected during processing [19,20].

During olive fermentation, a significant amount of salt is added as a preservative in order to prevent undesirable growth of pathogens and improve the organoleptic characteristics of the final product [18]. However, according to the World Health Organization [21], the daily proposed sodium intake has been set at 5 g. Therefore, one of the main goals of the food industry is to harmonize the global nutritional policies according to this guideline. However, the potential NaCl replacement depends on a plethora of factors, linked to cultivar type, drupe composition, as well as the processing and technological parameters [19,22]. All these parameters should be well inquired before implementation at the industrial scale. Furthermore, the final product must be safe from the microbiological point of view.

Several studies reported microbiological and chemical changes in table olives during spontaneous or controlled fermentation employing different cultivars [2,9,14,23,24]; however, the 'fermentation map' of Cypriot green cracked olives have not been charted yet. For the above-cited reasons, the aims of this work were (a) to study the microbial and physicochemical changes of Cypriot green cracked table olives during fermentation process at industrial scale, (b) to identify potential markers associated with the fermentation progress, (c) to accelerate the fermentative process by adding a starter culture, and (d) to study the effect of reducing NaCl concentration in combination with starter culture, in order to produce a secure, nutritious, and healthy final product.

2. Materials and Methods

2.1. Olives Samples and Fermentation Procedure

Olive fruits (*Olea europaea*) were harvested from a commercial orchard (Novel Agro, Nicosia). All fruits were harvested at the green stage of ripening, based on size uniformity criteria and even external color. After the elimination of the defective fruits, drupes were thoroughly rinsed with tap water to eliminate contaminants.

Subsequently, the olives were cracked and subjected to three different types of fermentation, in duplicate (Biological replicate). A particular amount of olive fruits (20 kg) were placed in plastic tanks of 25 kg capacity filled with brines supplemented with 0.33% *w*/*v* citric acid. The citric acid was added in accordance with the Cypriot industrial standard production procedure of table olives. The process was monitored for 365 days (23 ± 2 °C). The three types of treatments were: (i) spontaneous fermentation in 10% *w/v* NaCl, (Control, Olive 7 [OL7], and Brine 7 [AL7]), (ii) and (iii) fermentation inoculated with a starter culture of *Lactobacillus plantarum* (Vege-Start 60′, Chr. Hansen A/S, Copenhagen, Denmark) in (ii) 10% *w/v* NaCl (Olive 8 [OL8] and Brine 8 [AL8]) and (iii) 7% NaCl *w/v* (Olive 9 [OL9] and Brine 9 [AL9]). The amount of NaCl content to 7% was selected because the aim of the Cypriot olives industry is to reduce the sodium content close to 7%.

2.2. Microbiological Analysis

Samples were analyzed at regular time intervals (Days 0, 8, 15, 22, 29, 45, 60, 90, 120, 150, 210, 281, 365) throughout fermentation. They were determined for the total viable count (TVC), *Enterobacteriaceae*, LAB, yeasts, *Coliforms*, *Staphylococci*, using the standard pour and spread plate methods after serial dilutions in 0.85% *w/v* saline water (Table 1). In the case of olives, before serial dilutions, 10 g were aseptically transferred to stomacher bags filled with 90 mL of saline solution (0.85% *w/v* NaCl) and homogenized for 2 min using a Stomacher at 220 rpm speed (Bug Mixer, Interscience, Saint Nom, France). Volumes of 0.1 mL or 1 mL (spread and pour plate, respectively) of serial dilutions in saline solution were placed in Petri dishes for enumeration of the microorganisms. All samples were analyzed in triplicates.

Table 1. Microbiological media used for microflora enumeration.

Growth Media	Microorganisms	Method	Incubation Conditions
Plate count agar (PCA) (Merck, Darmstadt, Germany)	Total viable count	Spread plate	30 °C/72 h
De Man-Rogosa-Sharpe agar (MRS) (Oxoid, Basingstoke, UK) + natamycin 0.1%	Lactic acid bacteria	Pour plate/Overlay	30 °C/72 h
Sabouraud agar (Oxoid, Basingstoke, UK)	Yeast and Molds	Spread plate	25 °C/5 d
Violet red bile glycose agar (VRBGA) (BD, Sparks, MD)	*Enterobacteriacae*	Pour plate/Overlay	37 °C/24 h
Violet red bile lactose agar (VRBLA) (Oxoid, Basingstoke, UK)	*Coliforms*	Pour plate/Overlay	30 °C/24 h
Mannitol salt agar (MSA) (Oxoid, Basingstoke, UK)	*Staphylococci*	Spread plate	30 °C/48 h

2.3. Physicochemical Analysis

Titratable acidity (TA) was determined by potentiometric titration with 0.1 mol L^{-1} NaOH up to pH 8.3, and results were expressed as a percentage of lactic acid (*w/v*). pH was calculated using a pH meter (Hanna Instruments, Luton, UK). The salinity of brines was determined using a salinometer. Electrical conductivity was calculated using a conductivity meter (Mettler Toledo, Zürich, Switzerland). Finally, water potential was determined using a WP4C dewpoint potentiometer, following the manufacturer's instructions. All measurements were performed in triplicates.

Sugars (glucose and fructose), organic acids (lactic, succinic, tartaric, acetic, citric, and malic), and alcohol (ethanol, glycerol) levels were determined during the fermentation, as described in previous studies [9,16], with some modifications. In 1 mL of brine, 100 μL of $HClO_4$ was added, and the samples remained at 4 °C for 24 h, following by centrifugation at 12,000 rpm for 60 min at 4 °C. Then, the supernatants were stored at −20 °C for further analysis. Just prior to the analysis, samples were filtered (using 0.22 μm pore diameter filters). Chromatographic analysis conditions were applied as follows: Column: Aminex HPX-87H, 4.6 mm × 250 mm × 3.5 μm (Bio-Rad, Hercules, CA, USA), solvent: 4.5 mM H_2SO_4 in H_2O, isocratic flow rate: 0.5 mL min^{-1}, assay temperature: 65 °C; detectors: refractive index detector for sugars and alcohols, and fluorescence at 210 nm for organic acids, injection sample volume in HPLC: 20 μL. Quantitation (mM) was performed by standard curves generated by chromatographic analysis of the standard solutions of the respective substances at various concentrations.

Total polyphenols and antioxidant capacity of brines and fruits were quantified, followed by the identification and quantification of the main polyphenols (oleuropein, hydroxytyrosol) by high-pressure liquid chromatography (HPLC) (Waters 1525) analysis at regular time intervals throughout fermentation. The extraction of the phenolic compounds was carried out, as reported by Tataridou and Kotzekidou [25]. The determination of total phenolic components using the Folin–Ciocalteu (F.C) reagent was based on the method described previously [26]. The reaction products were measured spectrophotometrically at 765 nm. The results were expressed as mg/g or mg/mL of gallic acid equivalent (GAE).

The antioxidant activity was determined using the stable radical 2,2-diphenyl-1-picrylhydrazyl DPPH (Sigma-Aldrich, Taufkirchen, Germany), according to a procedure described previously [27]. Trolox equivalent antioxidant capacity (TEAC) was used as standard. The results were expressed as mg/g TEAC fresh weight, using the standard curve of Trolox. The measurements were performed three times.

Chromatographic analysis in the extracts was performed using HPLC (Waters 1525). The solvents (mobile phase) used were: Solvent A: 1% acetic acid HPLC grade, Solvent B: 100% acetonitrile HPLC grade, Solvent C: 100% methanol grade HPLC. Chromatographic analysis conditions were as follows: Column C18, 4.6 mm × 250 mm × 5 µm (Sigma-Aldrich, Taufkirchen, Germany), 0–20 min: 95% solvent A + 5% solvent B, 20–40 min: 75% solvent A + 25% solvent B, 40–50 min: 50% solvent A + 50% solvent B, 50–60 min: 5% solvent A + 95% solvent B, 60 min: 95% solvent A + 5% solvent B. Phenolic compounds (oleuropein and hydroxytyrosol) were estimated at the ultraviolet spectrum (254 and 280 nm) using the respectively standards. Results were expressed as means (mg/g or mg/mL) and standard deviations of three replicates.

2.4. Color and Texture Analysis

Color determination of table olives was performed during the whole process using a CR200 Chroma Meter (Konica Minolta, Nicosia, Cyprus). The instrument was set to the standard white color (Y = 93.9, X = 0.313 and y = 0.3209 or L * = 94.11, a * = −0.99 and b * = 0.89). The olive color was assessed by taking at least 10 random measurements from the surface of different olives [17]. The color was expressed as L * (bright, dark-low dark color values), a * (negative values indicate green, while positive values indicate redness), and b * (negative values indicate blue, and positive values indicate yellow). Furthermore, reduction in parameter hue angle (h *) corresponded to change in color from green to yellow. Finally, an increase in C * corresponded to a stronger color.

Texture analysis was monitored in whole fermentation by taking at least 10 random measurements of different olives, using a dynamometer (John Chatillon and Sons, New Gardens, NY, USA) carrying a 9.5 mm (length) and 3.2 mm (diameter) piston with a 2 mm cylindrical probe [20]. The test speed was 1.5 mm/s, and the penetration force was expressed in N.

2.5. Sensory Evaluation

Olive samples were evaluated organoleptically after 4 months of fermentation by a thirteen-member taste-certified panel (5 males and 8 females, aged from 20 to 45 years old) according to International Olive Oil Council (Regulation COI/OT/MO No 1/Rev.1). Texture, flavor, saltiness, bitterness, acidity, off flavors, and overall acceptance were assessed. Each of these features was rated as follows:

- Texture: 0 = soft, 5 = intermediate, 10 = coherent
- Flavor: 0 = absence, 5 = moderate, 10 = strong
- Salty: 0 = no, 5 = moderate, 10 = very much
- Bitterness: 0 = No, 5 = moderate, 10 = high
- Acidity: 0 = no, 5 = moderate, 10 = high
- Off flavors: 0 = absence, 5 = moderate, 10 = strong
- Overall acceptance: 0 = reject, 5 = moderate, 10 = strongly accept

2.6. Isolation of the Predominant Microflora

Representative colonies growing on De Man-Rogosa-Sharpe agar (MRS) (LAB) and Sabouraud (yeasts) agar plates were isolated at different stages of fermentation. The isolates were purified by streaking twice on the same medium after phenotypic observation using a light microscope. Pure bacterial and yeast cells were stored at −80 °C using glycerol (20%) for future use.

Finally, in order to detect the presence of the starter culture, rep-PCR genomic fingerprinting was performed on 17 random strains isolated from MRS agar, from brines AL7, AL8, and AL9 at 120 days of fermentation, using the $(GTG)_5$-primer (5′-GTG GTG GTG GTG GTG-3′). DNA from each strain was obtained according to Bautista-Gallego [28] and stored at −80 °C. PCR reaction and amplification conditions were applied following a method previously described [29].

2.7. Statistical Analysis

Data were subjected to an analysis of variance (one-way ANOVA), using the SPSS 20 software (StatSoft Inc., Tulsa, OK, USA), in order to identify statistically significant differences of microbiological, physicochemical, and sensory characteristics across fermentation treatments. Differences between means were determined by the statistical LSD test at $p \leq 0.05$. In order to study the correlations between variables and treatments, principal components analysis (PCA) was performed (SPSS 20). Furthermore, two of Pearson's correlation matrices (among components and between components-treatments) were calculated, and an optimal Kaiser–Meyer–Olkin (KMO) measure of sampling adequacy was established. A hierarchical clustering analysis (HCA) of the correlation coefficients was depicted using the gplots version 3.0.1 (heatmap.2 command; R Foundation for Statistical Computing, Vienna, Austria). Finally, matrices of the original component data were standardized in order to depict (via a hierarchical clustering analysis heatmap) differences in the content of the relative variables.

3. Results and Discussion

3.1. Microbiological Analyses

Microbial enumeration was determined in all treatments during fermentation (Figure 1); In general, LAB and yeast numbers were steadily increased and predominated across treatments. On the contrary, *Enterobacteriaceae* and *Coliforms* species were decreased, while *Staphylococci* were not detected during the whole process.

The population size of *Enterobacteriaceae* and *Coliforms* was very similar (no statistical differences were detected) between brines AL8 and AL9 and differed compared to the control during the first days of the fermentation. Specifically, they were detected at an average of 3.5 log cfu/mL at the beginning of the process, but they decreased rapidly and could not be detected after 15 days of fermentation in AL8 and AL9, and after 22 days in AL7, indicating the usefulness of the starter. Indeed, according to Rodriguez-Gomez et al. [30], the use of selected *Lactobacillus pentosus* strains as starters decreased the *Enterobacteriaceae* population faster than in the control treatment. The inoculation contribution to the inactivation of *Enterobacteriaceae* has also been previously noticed [9,17,31]. However, it is obvious that in the present study, *Enterobacteriaceae* decreased more swiftly (about half the time). This could be justified by the use of citric acid at the beginning of the process that led to an early pH decrement at the initial stage of the process, resulting in *Enterobacteriaceae* suppression.

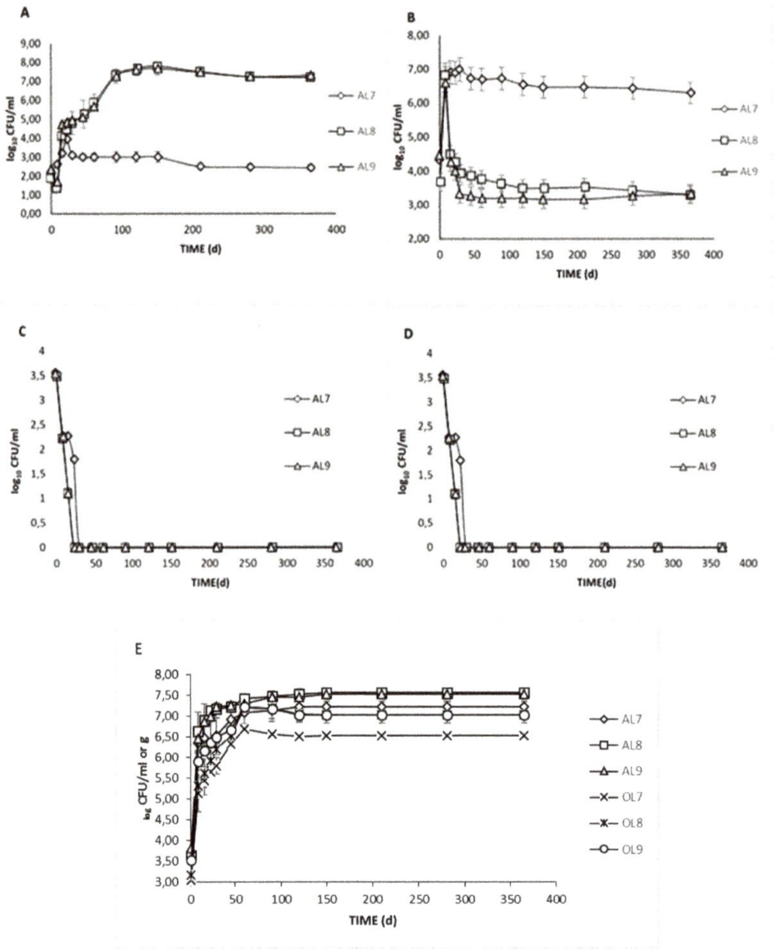

Figure 1. Evolution of microbial changes of spontaneous (◊), inoculated (10% NaCl) (□), and inoculated (7% NaCl) (Δ) fermentation of Cypriot green cracked table olives. LAB (**A**), Yeasts (**B**), *Enterobacteriaceae* (**C**), *Coliforms*, and (**D**) TVC (total viable count) (**E**). Data points expressed as log10 CFU/mL of 3 replicates ± standard deviation.

The population of LAB in brine samples changed significantly across the different treatments during fermentation. More specifically, there was an initial increase in LAB counts of AL7 (control) until the 22nd day of fermentation, reaching an average value of 3.95 log cfu/mL. After that peak, a slight decrease was observed, and numbers were retained until the end of the process. The low values of the LAB population in control was in accordance with the literature. LAB populations were limited in spontaneous fermentation, and this was linked to several factors that could have limited the adaptation of LAB in naturally fermented table olives. Some of these factors were the ambient temperature, high salt content, the availability of a source of energy, and natural inhibitory compounds presented in drupes since the fruits were not subjected to lye treatment [23]. On the other hand, in AL8 and AL9, a slight decrease was observed during the first 8 days, followed by a major increase of LAB population, reaching a maximum rate at 120 days (7.67 log cfu/mL and 7.6 log cfu/mL, respectively). No significant differences between these two treatments were observed, except that during the first days of fermentation, LAB populations in AL9 were initially higher. This could be related to the higher diffusion of sugars from olives to the brines due to the reduced NaCl concentration. Moreover, it must be mentioned that the reduction of the LAB population for AL8 and AL9 in the first days of the process indicated an intense competition between the starter with indigenous microflora nutrient assimilation. Indeed, a similar trend was noted in previous studies [17,32], attributed to the lack of nutritional substrates, as well as the presence of inhibitor compounds. According to our data, it was demonstrated that the starter culture withstood the competition with the natural microflora and was not affected by a high salt concentration while predominated in a short period, in contrast to the control treatment. Still, the prevalence of a population during fermentation is a multifactorial process and cannot be always accurately projected. Contrasting to the data of the current study, Rodríguez-Gómez et al. [30] reported that LAB population numbers between inoculated and control treatments had no significant differences. Finally, during the second half of fermentation, a decreasing tendency was noted but always over 7 log cfu/mL, while the population in control was close to the detection limit (2 log cfu/mL).

Yeast growth had an initial lag phase in all cases, occurring across treatments and reaching the maximum level approximately at circa 8 days (7 log cfu/mL), which was in agreement with the literature [2,29,30]. Following, a major decrease in AL8 and AL9 was observed, reaching a value of 3.7 log cfu/mL and 3.5 log cfu/mL at day 60, respectively (Figure 1B). From that point, population levels were maintained steadily until the completion of the process. On the other hand, in AL7, a major increase was observed until the 8th day, and after that, the population reached and retained 7 log cfu/mL level up to the end. As a result, yeasts were the predominant microorganisms in the control treatment. According to the literature, the dominance of LAB in Spanish-style olives has been extremely reported. On the other hand, yeasts are the main organisms driving the fermentation of naturally processing olives [33], although there are studies that have reported the presence of LAB [34]. In the current study, LAB growth in AL7 might have been hampered by salt-tolerant yeast species, resulting in a less acidic product, which was in accordance with the literature [29,35]. Nevertheless, yeast growth is not considered to present any consumption risk during the fermentation of green olives; On the contrary, yeast can metabolize ingredients that enrich the sensory palette and determine the quality of the final product [36]. It is worth noticing that the presence of yeasts, especially in the first days of the process, might have led to the enhancement of the starter culture in inoculated treatments due to their potential production of vitamins and other nutrients, which are mandatory for LAB growth [37].

The microbial composition of TVC in fruits and brines was also depicted (Figure 1E). Overall, the number of microorganisms detected in olive fruits was 1 or 1.5 log lower compared to brines, throughout fermentation. At the early stages, total aerobic counts ranged from 5.2 (OL7) to 5.7 (OL8, OL9) log cfu/g in pulps and from about 6.3 to 6.6 log cfu/mL in brines. The population was increased in all treatments until the 60th day, reaching a maximum value of 7.1, 7.4, 7.3 log cfu/mL for AL7, AL8, and AL9, respectively, while the populations of brined fruits were about 1 log lower than their brines. This magnitude of deviation, among fruits and brine, was also reported in a study carried out on commercialized table olives in Portugal [38]. This finding could be related to the high presence of phenolic compounds in olives, thus, high antimicrobial activity, leading to microbial inhibition, especially in the first 45 days of fermentation.

3.2. Physicochemical Analyses

The changes in pH in the brines during fermentation of all varieties are presented in Figure 2A. The initial values (Day 0) in all treatments were very low (ca. 3.3) due to the use of citric acid at the beginning of fermentation. After that, there was an increase of about 1–1.5 units until day 22, followed by a major decrease in inoculated treatments (3.5 and 3.3 for AL8 and AL9, respectively). In control, pH remained stable, reaching finally a value of 4 on the 90th day. In all treatments, a slight increase (4, 3.8, and 3.7 for AL7, AL8, and AL9, respectively) was observed, which was stabilized thereafter at about pH 4 at day 365. No differences between treatments were observed at this time point. It was crucial to mention that the fast acidification in the brine matrix was a crucial preliminary step for the succession of fermentation process; pH in brines below 4.5 preserved table olives from spoilage and pathogen microbial growth during fermentation. Furthermore, it had to be noted that the positive effect of the starter in pH drop was profound, especially in the first days, which was in agreement with previous studies [9,18,39].

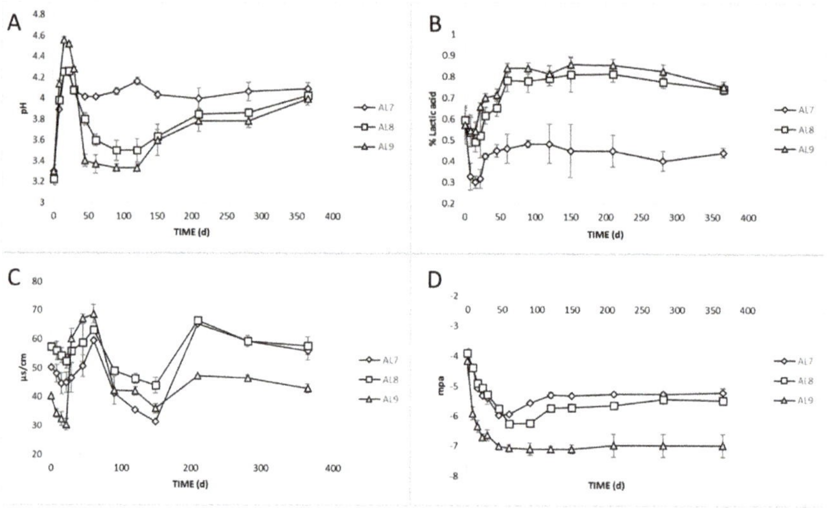

Figure 2. Changes in pH (**A**), titratable acidity (**B**), electrical conductivity (**C**), and water potential (**D**) throughout the fermentation of spontaneous (◊), inoculated (10% NaCl) (□), and inoculated (7% NaCl) (Δ) of Cypriot green cracked table olives. Results are expressed as means and standard deviations of three replicates.

The reverse change was followed on titratable acidity, as expected (Figure 2B). The highest values were recorded in AL8 and AL9 due to the dominance of LAB (0.81 and 0.86% lactic acid, respectively). It was notable that the effect of initial acidification with citric acid was evident for high values of titratable acidity in the brines during the first days. The titratable acidity was higher in AL9 until the 29th day, and, thereafter, no significant differences were observed between inoculated samples. The higher values in AL9 at the first days of fermentation could be explained due to the low salt concentration, which allowed the faster diffusion of sugars from olives to brines, and thus the faster start of fermentation from LAB. The titratable acidity levels found were in accordance with the LAB enumeration and pH values, described above. Moreover, it is noteworthy that a value of more than 0.48% lactic acid in AL7 was not reached at any time, probably due to the dominance of yeasts, in combination with the weakness of LAB to produce lactic acid due to their low population. In another study [16], fermentation of table olives driven by yeasts attained a final pH close to 4.2–4.3, which was in good agreement with the final pH values of AL7 reported in the present study. However, even though yeasts were the dominant microbial group, the final values for pH and acidity were within the limits of the trade standard applying to table olives of the IOC (2004), where for natural fermentation, the maximum limit for pH and minimum acidity should be 4.3 and 0.3%, respectively. Notably, the higher acidic environment in AL8 and AL9 samples are enough to prevent the growth of spoilage and/or pathogen microorganisms, and thus they may provide an added value to the product. The latter could be confirmed by the faster elimination of such microorganisms, as mentioned above. Thus, our findings suggested that the use of LAB starter culture had a significant effect on the acidification of the brines, achieving a more controllable and successful fermentation.

During fermentation, the production of higher acidity in inoculated treatments caused an increase in electrical conductivity. The pH curve represents the kinetics of the production of H^- ions, while that of electrical conductivity represents the production of all ionic species [40]. Figure 2C shows the changes in electrical conductivity during the whole process. As was clearly observed, there was an initial decrease in all treatments until day 22, followed by a major increase until the 60th day. Significant differences were observed in all treatments, while AL9 had the highest values from day 29 to day 60, followed by AL8 and AL7. This was in accordance with pH and acidity scores at this time point, as described above, indicating a clear correlation between the three parameters, confirmed by HCA, as well. In a previous study [41], it was also demonstrated a curvilinear relationship between pH and conductivity during mixed coagulation of milk. However, according to our knowledge, this is the first study indicating the correlation between these parameters in table olives fermentation. Thus, electrical conductivity could be used as a potentially useful tool for table olives monitoring during the fermentation process.

Regarding the water potential of olive fruits, there was a clear difference between OL9 and the other two treatments during the whole process (Figure 2D). The initial values of the three treatments were about −3.9 Mpa. There was a decrease during the first 60 days, where OL9 was statistically higher compared to OL7 and OL8 olive fruits. After that period, the values of all treatments started to have a slightly increasing trend up to the 120th day and then remained stable until the end of the process. Water potential expressed the tendency of the water to move from the fruit to the brine and was related to the expression of osmosis. Thus, it was clear that osmosis pressure in OL9 was higher than in the other two treatments, allowing the faster diffusion of flesh tissue components (sugars, organic acids, polyphenols, etc.) to the brines. Indeed, Papadelli et al. [9] reported that the slow extraction of soluble components from the olives to the brine was related to high NaCl concentration. This is the first report proposing the use of water potential as a tool for soluble component kinetic estimations of table olives during the fermentation process.

Finally, salinity in the brines was monitored throughout fermentation, and adjusted to the initial values of 10% and 7% for each treatment, by periodic dry salt additions in the brines. Salt equilibrium was reached in ca. 3 months and until the end of the process, salt concentration was maintained to its initial values.

The total phenolic evolution of fruits and brine samples is presented in Figure 3A,B, respectively. As clearly observed, the profiles of total phenolic content were quite similar across treatments. Olives exhibited a major loss in total phenolic content during fermentation mainly due to their degradation by LAB and yeasts and secondary due to their diffusion to the brine, as well. A similar trend has been noted by other studies [13,42]. During the first 45 days of fermentation, the decrease rates of phenolic contents were estimated to 37%, 68%, and 75% for OL7, OL8, and OL9, respectively. After 120 days of brining, phenolic content attained a steady-state in traces, with no differences between treatments. The total reduction of phenolic contents was about 88% for all treatments. In fact, the diffusion of phenolic compounds from olive flesh to the brine depended on several parameters, such as cultivar characteristics, fruit skin permeability, type of polyphenols presented in olive flesh, brine concentration, and their ability to diffuse outside the fruit due to accidental or purposely made fruit damage (cracked or razor slitting). As expected, the total phenolic contents in brines increased gradually in all fermenters to rich maximum concentrations of 3.24, 3.85, and 4 g GAE/g after 29, 22, and 22 days of fermentation for AL7, AL8, and AL9, respectively. After the 29th day of brining, the phenolic content started to decrease. This decline might be due to the degradation of phenolic acids by *Lactobacillus plantarum*. It has been demonstrated that *Lactobacillus plantarum* contains phenolic acid decarboxylases, which decarboxylate different phenolic compounds to their corresponding vinyl derivatives [43]. However, it is obviously an analogous reduction of total phenols in control treatment, in which, as previously mentioned, yeasts were the leading microorganisms. This indicates a high enzymatic activity of indigenous yeasts in the degradation of phenolic compounds, which is in accordance with the literature, where it has been reported the important role of yeasts in the olive debittering process [2]. This finding could explain the similarities of total polyphenols loss between inoculated and control samples from the 60th day of fermentation and thereafter.

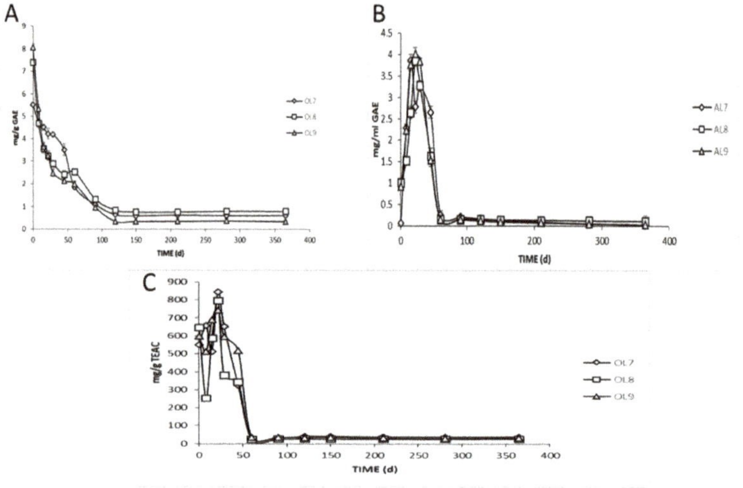

Figure 3. Total phenolic content of olive pulps (**A**) and brines (**B**) and antioxidant capacity of olives (**C**) during spontaneous (◊), inoculated (10% NaCl) (□), and inoculated (7% NaCl) (Δ) fermentation of Cypriot green cracked table olives. Results are expressed as means and standard deviations of three replicates, equivalent of mg/g or mL.

Additionally, the loss in phenolic compounds resulted in a remarkable loss of antioxidant capacity in olive fruits, as well (Figure 3C). No significant differences between different fermentations were observed after 60 days. The loss of antioxidant capacity transcended to 90% for all treatments.

A major decrease of oleuropein was observed in olive fruits, mainly due to its diffusion to brines and its degradation caused by enzymatic activity (Figure 4A). No significant differences were observed between treatments at the end of the process, which agreed with the trends in total polyphenols values described above. However, the faster-decreasing values of oleuropein in inoculated treatments until the 45th day were notable. At this time point, the reduction reached 62% and 69% for OL8 and OL9, respectively, while, in control, it was no more than 24%. This finding indicated that the inoculated samples were ready to eat in a shorter period. Furthermore, the reduction of oleuropein was also accompanied by an increase in its hydrolysis product in brines (hydroxytyrosol), where the inoculated treatments had significantly higher values after 90 days of fermentation (Figure 4B). This finding confirmed that the enzymatic activity of the starter culture was higher, affecting the secosteroid glucosides and their aglycon derivatives. In line with our findings, in previous studies [43,44], hydroxytyrosol was referred to as the main phenolic compound found in the brines inoculated with the commercial starter. This compound has been mainly linked to the hydrolysis of oleuropein [13] being an important biophenol belonging to the odiphenol group with special antioxidant activity [45], and it has been considered as a marker for the determination of olive debittering [46].

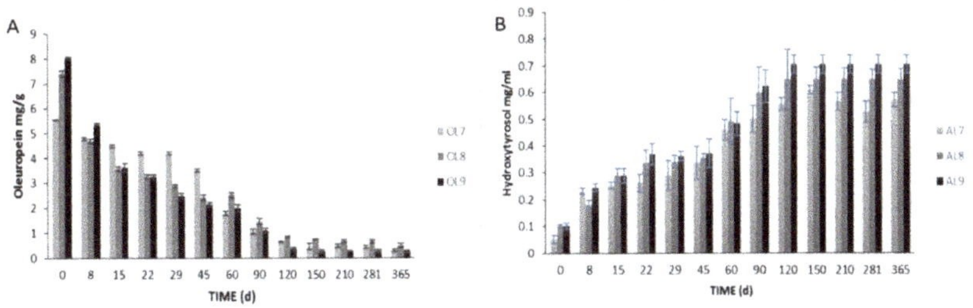

Figure 4. Evolution of oleuropein (**A**) and hydroxytyrosol (**B**) during spontaneous (7), inoculated (10% NaCl) (8), and inoculated (7% NaCl) (9) fermentation of Cypriot green cracked table olives. Results are expressed as means (mg/g or mg/mL) and standard deviations at different times of fermentations.

The changes in the concentration of organic acids in the brines are shown in Figure 5. Significant differences between the three treatments were observed during the whole process, as detected by HPLC analysis. More specific, citric acid was the main acid at the initial stage in all treatments due to its use at the beginning of the fermentation. After 22 days, in AL7, acetic acid became the main acid with considerable presence, while a slight reduction of citric from day 45 to day 120 was observed. However, the concentration of acetic was increasing until day 120 and then remained stable. Its presence in control could be related to yeast metabolism, as well as to the potential of heterofermentative LAB able to produce acetic acid under particular conditions of environmental stress as well as from the metabolism of citric acid [47]. Furthermore, malic and tartaric acids were also found in brines and were increased until day 22, indicating their presence in olive fruits and diffusion to the brines the first days of the process. The latter was in line with the literature, as well [9]. Afterward, these two acids remained unchanged until the end of the process. Thus, there was not any metabolic activity for those acids in AL7. Finally, lactic acid was also detected in the brine AL7 in concentrations not exceeding 32 mM throughout the process, which was related to the low populations of LAB found in the microbial enumeration. The low values found for lactic acid were in accordance with previous studies [16,37]. However, as expected, lactic acid was the main acid in inoculated treatments due to the predominance of LAB. Significant differences between these two treatments were observed from day 45 to day 120, where the concentration of lactic acid in AL9 was higher than in AL8. This was in a combination of pH and titratable acidity values described above. Lactic acid presented a gradual increase until day 120, reaching statistically significant higher values in AL9, followed by a steady decline thereafter. This indicated potential assimilation of lactic acid from yeasts after 120th day due to their high population, which was in accordance with the literature [6]. Citric acid was the main acid at the initial stage due to its use at the beginning of the fermentation. From day 22 to day 60, a major decrease was observed in both treatments (no significant differences), which was related to its microbial degradation to acetic acid [9]. Succinic acid was also determined in the inoculated treatments, the evolution of which was found to be similar to that of the acetic acid, for the same reason, as well (no differences between treatments). Furthermore, regarding malic and tartaric acids, there was an obvious initial increase during the first 22 days, followed by a major decrease and total disappearance of malic acid in about 120 days, while the tartaric acid remained steady until the end. This finding was in agreement with results reported previously [14,24], where malic acid detected in traces at the beginning of the process and decreased at the end of the fermentation period. Moreover, the gradual decrease of malic acid in brines observed during the fermentation of green olives is attributed to its microbial degradation to lactic acid and CO_2 [9,31]. Finally, for tartaric acid, it has been reported that yeast and other microorganisms are unable to metabolize it [48].

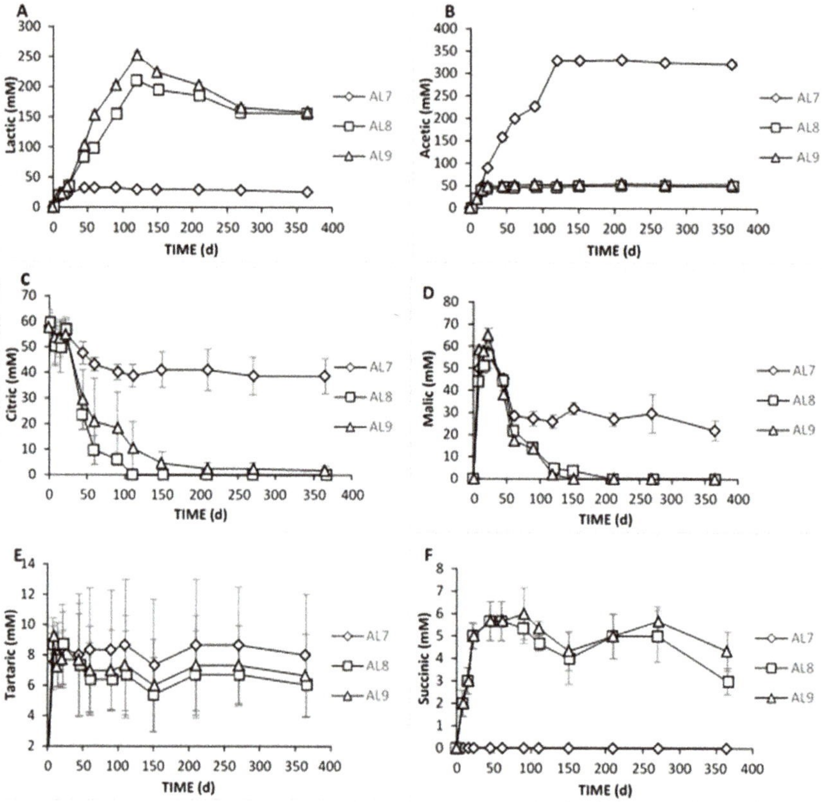

Figure 5. Changes in the concentration (mM) of organic acids (lactic, **A**; acetic, **B**; citric, **C**; malic, **D**; tartaric, **E**; and succinic, **F**) during spontaneous (◊), inoculated (10% NaCl) (□), and inoculated (7% NaCl) (Δ) fermentation of Cypriot green cracked table olives. Data points are expressed as means and standard deviations of three replicates.

Sugars diffused from fruits into the brine are the main nutritional elements for microbial growth and fermentation. According to our results, glucose and fructose were the main sugars found in the brines as it emerged by HPLC analysis (Figure 6). Glucose was steadily increasing the first days of fermentation, exhibiting the highest value at day 22 for AL9 and day 29 for AL8 and AL7. This could be confirmed by the faster diffusion of the sugar observed from olives to brine in AL9 because of the lower NaCl concentration. A major decrease was recorded thereafter since it was consumed for microbial growth. In fermentation AL9, this decrease was observed earlier (at day 45) compared to fermentation AL8 (day 60) and AL7 (day 90), which was in accordance with previous research, where it was reported that in the inoculated olives, the decrease of glucose was faster than in control [9]. It was notable that at the end of the process, glucose disappeared, but there was a remaining amount of ca. 0.5 mM in the AL7. A similar trend was found for fructose content in AL7. Its amount never exceeded 12 mM, and it was not found after 120 days of fermentation, while it was detectable in the same concentrations (ca. 7 mM) in AL8 and AL9, until the end of the fermentation. The total depletion of fructose in AL7 could be related to some fructophilic yeast species, a fact that agreed with the results of control treatment in the present study [9].

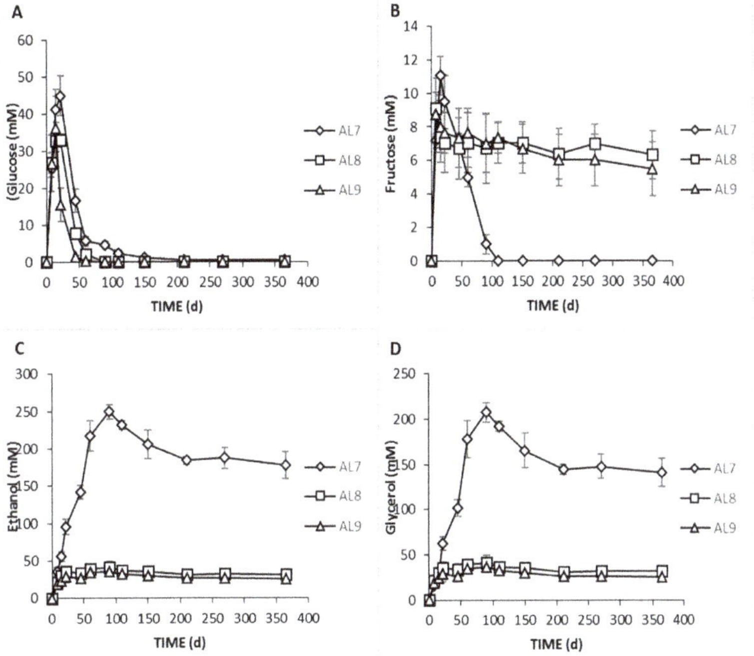

Figure 6. Changes in the concentration (mM) of soluble sugars (glucose, **A**; fructose, **B**) and alcohols (ethanol, **C**; glycerol, **D**) in the brines during processing of Cypriot green cracked table olives of Spontaneous (◊), inoculated (10% NaCl) (□), and inoculated (7% NaCl) (Δ) fermentations. Data points are expressed as means and standard deviations in triplicate.

Concerning ethanol, it is mainly related to yeast production activity, having a crucial impact on the sensory properties of naturally fermented olives [29]. Its concentration in AL7 increased gradually until day 90 of fermentation, reaching values 250 mM, followed by a minor decrease thereafter until the end of the process, where it was maintained at ca 178 mM. Similar trends were previously reported [2]. However, ethanol was detected in traces in inoculated treatments because of LAB dominance, confirming that yeast metabolic activity was affected by the inoculation of table olives. Another important product from yeast activity is glycerol [49]. Its presence has been linked with cell protection from osmotic stress [29]. According to our results, it was present in high concentration in control, as expected, while it was very limited in the other treatments (ca. 30 mM during the whole process). In AL7, its concentration increased gradually until the 90th day of fermentation in levels exceeding 207 mM, followed by a decrease afterward, reaching final values of ca. 140 mM. The presence of this compound in naturally fermented table olives has been reported previously [16,37], which was in good line with the present study. It has been noted that the presence of both compounds (ethanol and glycerol) can, in turn, affect crucial organoleptic characteristics, such as texture maintenance and aroma formation [36,50].

3.3. Firmness and Color Evolution of Olives

The texture has a great impact on consumer's acceptance of a product, while in main cases, it is considered to be the most important property [51]. The results of textural analysis during the whole process are presented in Figure 7. It could be observed that the values were being decreased as time passed until the 60th day, and after that remained steady until the end of fermentation, in all treatments. The values of OL9 were significant lower until day 29. This could be explained by the lower NaCl concentration. However, no significant differences were observed thereafter, indicating that neither lower NaCl nor brine inoculation affected the texture profile of the final product. Similarly, Fadda et al. [20] investigated the effect of brining time on the texture of naturally fermented green olives, reporting a texture decrease after 30 days of brining. Texture loss is strongly influenced by the enzymatic activity of dominant microflora and, in some cases, may cause softening due to the degradation of pectic substances of the cell wall and middle lamella [52]. The latter depends on crucial brine conditions, such as sodium content and pH [20].

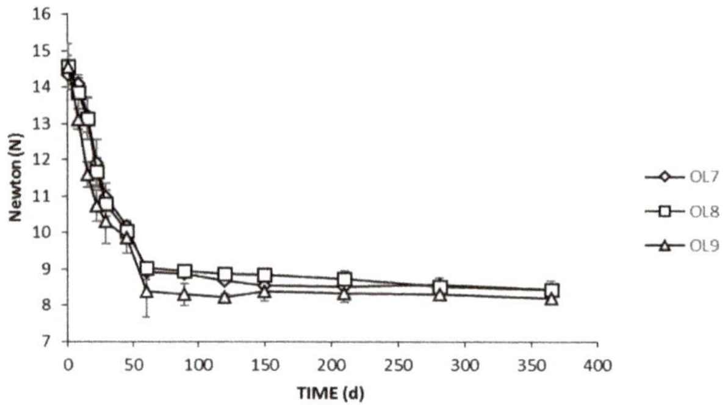

Figure 7. Evolution of texture of olive fruits during spontaneous (◊), inoculated (10% NaCl) (□), and inoculated (7% NaCl) (Δ) fermentation of Cypriot green cracked table olives. Data points are expressed as means (N) and standard deviations of 10 random measurements.

The color attribute of food products is another crucial factor in the acceptance of a food product. The color parameters of olives are listed in Supplementary Table S1. In general, no significant differences were observed between treatments in any of the parameters. Exceptions were the h* and C* parameters. The latter, in the inoculated treatments, were significantly lower than control. This parameter was related to the volume of color, which accounted for a shift to the dark-green zone. Furthermore, there were no significant differences in b* parameter for control and inoculated samples starting from a value of 33 ± 4.4, reaching a decreased value until 45th day (22.6 ± 6), and thereafter remained unchanged until the end. The decreasing values indicated lowering in yellow color. A similar tendency was observed for the parameter L* (no differences), which decreased until day 45 after the fermentation process. The value of the parameter L* was an indicator of the degree of lightness. However, according to the literature [53], light color is associated with a low pH value, which was not in agreement with the present study; otherwise, the inoculated samples should have higher lightness than control. This could be explained by the fact that the lightness parameter is mainly variety dependent. Finally, a major increase in a* parameter was observed in all treatments, demonstrating a distinct toning from green to red, starting for values of about −13 at the beginning of the process, while values of about −1.9 ± 1 were reached at 120 days and thereafter no changes were recorded. This effect could be attributed to the presence of chlorophyllase in the first days of fermentation, leading to hydrolysis of phytol or chemical oxidation reactions [54]. In general, natural green olives had high values for a* parameter, resulting in reddish tones. Finally, the loss of h* was faster in control and lowered significantly until the 150th day, indicating faster brownish coloration. The latter, among other organoleptic characteristics, makes this product less attractive [53], and thus this is another positive effect of inoculated samples.

3.4. Sensory Evaluation

The organoleptic profile of the samples is presented in Figure 8. In general, the samples were characterized by low remaining bitterness, good texture, and satisfactory acidic taste and odor. No off flavors were noticed in any samples. Overall, differences among treatments were detected on a bitterness descriptor, in which OL7 had a higher score from the other two. The higher contents of both ethanol and glycerol in the control sample was in line with the higher score to the bitterness and lower score of the acid descriptor, according to panelist evaluation. A similar trend was observed for the saltiness score, with a lower value scored in OL9 samples. However, no differences were recorded to a flavor descriptor. Regarding texture, OL8 and OL9 had lower scores. However, they received an equal value of acidity, which was higher than control, while they had the highest score for the overall acceptability descriptor. The most important attribute that influenced the judgment of the panelists was salt content, acidity, and bitterness to a lesser extent, as could be concluded by the scores of those parameters, in combination with the overall acceptability scores.

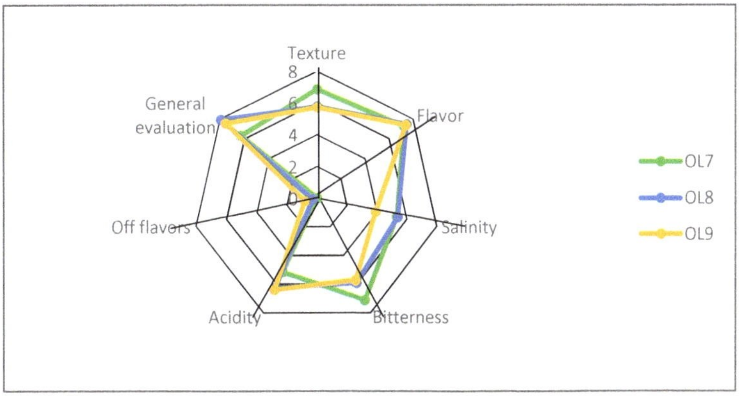

Figure 8. Sensory profiles of spontaneous (OL7), inoculated (10% NaCl) (OL8), and inoculated (7% NaCl) (OL9) fermentation of Cypriot green cracked table olives at 120 days of fermentation.

3.5. Multivariate Analysis

PCA between all studied variables resulted in four eigenvalues greater than 1, explaining an overall 88.84% of the total variance in the dataset, while the first two components explained 71.5% of the distribution (Figure 9B; Supplementary Table S2). PC1 was correlated with LAB, *Enterobacteriaceae*, texture, all color parameters, polyphenols, antioxidant capacity, lactic, citric, succinic acids, and glucose, while PC2 dealt with yeasts, acetic, tartaric acids, ethanol, and glycerol. PC3 was linked with the pH, fructose, malic, and tartaric acids. Finally, PC4 was related to conductivity and water potential. Furthermore, correlations between microbial and physicochemical data are shown in Figure 9A. Among organic acids, lactic and succinic acids were negatively correlated with yeasts and positively correlated with LAB. Oppositely, acetic, tartaric, and malic were positively correlated with yeasts and negatively with LAB. Zooming on the metabolomics, the acetic acid was positively correlated with ethanol and glycerol, confirming the results for control treatment, described above. Oleuropein, antioxidant capacity, total phenols, texture, and color parameters L*, b*, h* were closely related to each other, and all of them were negatively related to LAB and positively related with yeasts. Finally, hydroxytyrosol seemed to be highly correlated with LAB, confirming our results described above.

Regarding correlations between treatments, PCA grouped them into three clusters, clearly characterized based on inoculated treatments versus the control one during fermentation time, as control treatments being separated from inoculated treatments from the 45th day and thereafter (Figure 9B). Inoculation was apparently the most important treatment in sample distribution throughout fermentation. PC1 could be related to fermentation time since a gradual transition of time was noticeable from the right to the left part of the plot. It is crucial to mention that the reduction of NaCl concentration (AL9) did not affect the groups' distribution. This was a very promising finding, indicating that the NaCl reduction was an achievable goal for the table olives industry.

Furthermore, similarities in the observed microbial and physicochemical profiles between samples are presented in Figure 9C. In detail, in agreement with PCA, inoculated samples had similar profiles to each other after the 45th day, showing a negative correlation with total polyphenols, oleuropein, texture, color h*, color L*, color C*, and malic and citric acid, while they were positively correlated with LAB, titration, lactic and succinic acid, color a*, pH, glucose, and hydroxytyrosol. On the other hand, control treatment was closely related to yeasts, texture, acetic acid, ethanol, and glycerol, while it was negatively related to the positive parameters of inoculated treatments and fructose. Therefore, the multivariate analysis confirmed the different metabolic pathways between non-inoculated and inoculated treatments during fermentation.

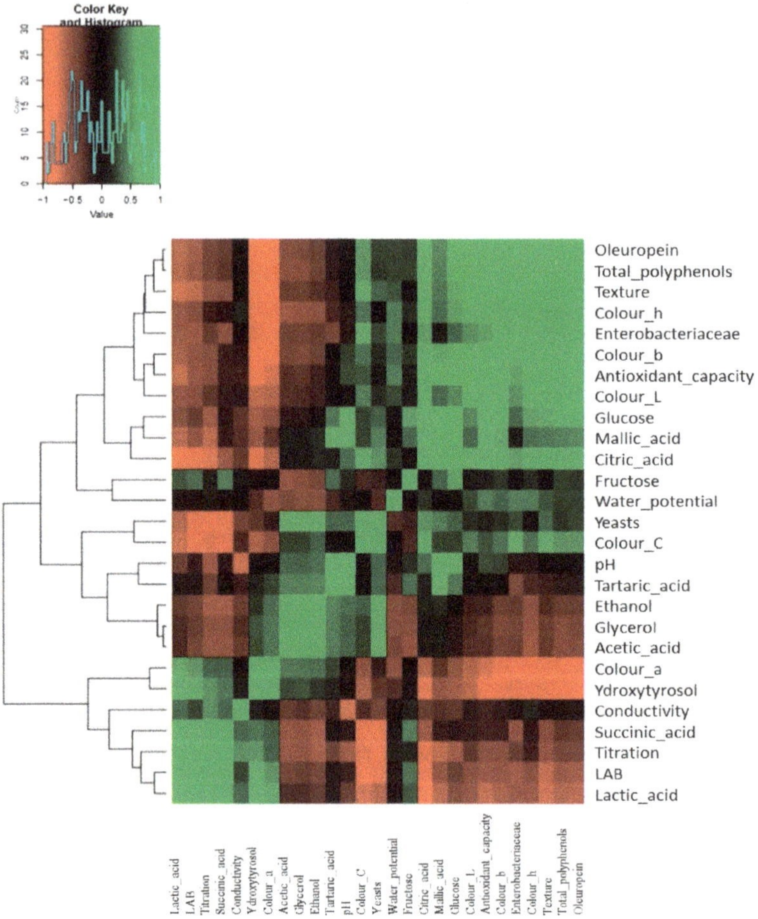

(A)

Figure 9. *Cont.*

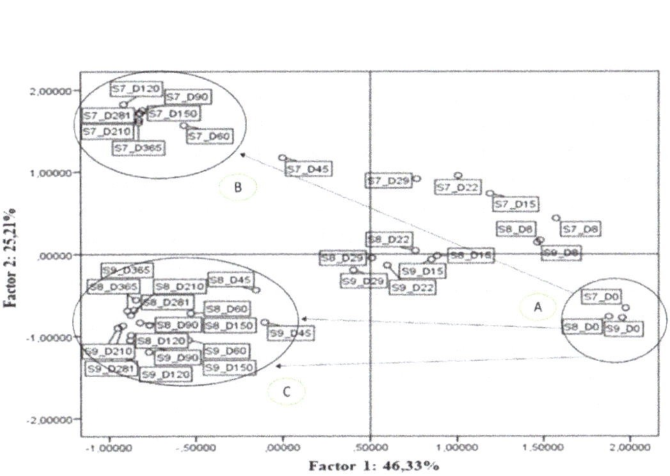

(**B**)

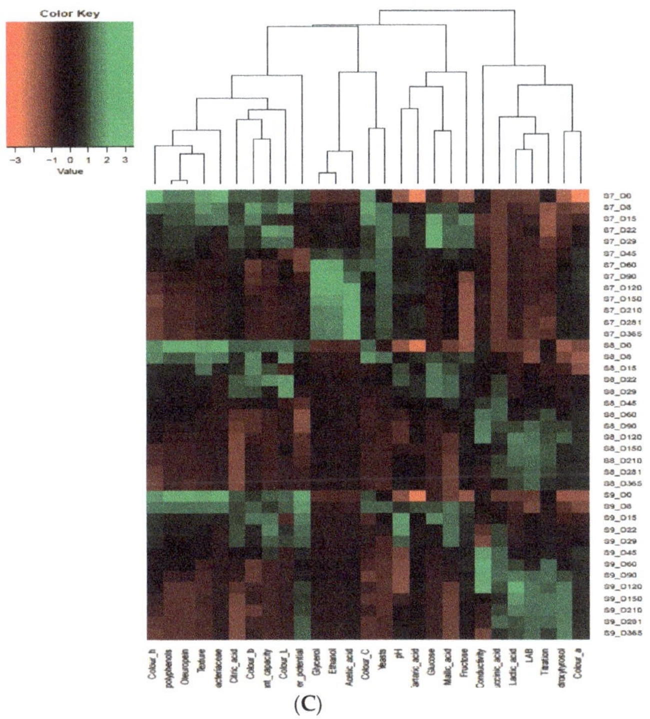

(**C**)

Figure 9. (**A**) PermutMatrixEN analysis between microbial and physicochemical profiles of spontaneous, inoculated (10% NaCl), and inoculated (7% NaCl) fermentation of Cypriot green cracked table olives. (**B**) The plot of scores and loadings between treatments formed by the first two principal components from the PCA (principal component analysis) analysis. Labeling of data points indicates the processing treatment of olives (S9: inoculated and 7% NaCl concentration, S8: inoculated and 10% NaCl concentration, S7: control) and fermentation time (D: Days). (**C**) Heatmap showing the similarities in the observed microbial and physicochemical profiles between the three experiments during fermentation days. Labeling of data points indicates the processing treatment of olives (S9: inoculated and 7% NaCl concentration, S8: inoculated and 10% NaCl concentration, S7: control) and fermentation time (D: Days).

3.6. Detection of the Presence of the Starter Culture

The presence of inoculated strain was monitored after 4 months of fermentation by rep-PCR on a pool of 17 strains from MRS agar (seven from AL8, eight from AL9, and two from AL7). Preliminarily, the repeatability of the method was confirmed using gDNA from the starter strain (Vegestart 60) as an internal control in four different gels from four different PCR reactions, obtaining a similarity of 91.8%. This value was retained as a threshold to establish the identity of isolates compared to the rep-PCR profile. The produced dendrogram clearly separated the studied strains into three clusters (Figure 10). More specifically, the first cluster related to the starter strain profile containing all isolates from AL9 (7/7, 100%) and many isolates from AL8 (5/8, 62.5%). The isolates of the second cluster belonged to AL8 (3/3), indicating that there were different strains, while isolates belonged to the third cluster came from AL7, in order to prove the distance between indigenous and starter LAB molecular profile.

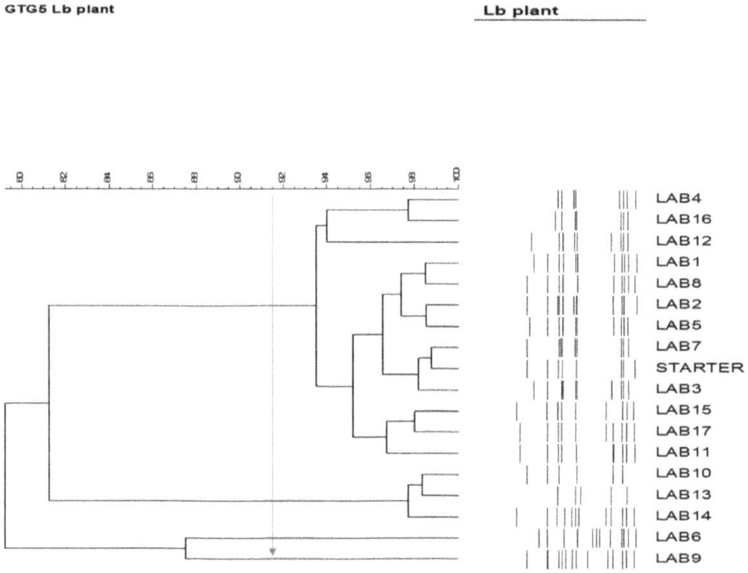

Figure 10. Dendrogram generated after cluster analysis of the digitized GTG5-PCR fingerprints of LAB (lactic acid bacteria) strains isolated from AL7 (LAB 6,9), AL8 (LAB 1,2,3,4,5,7,8), and AL9 (LAB 10,11,12,13,14,15,16,17) brine samples at 120 days of fermentation.

4. Conclusions

According to the results of the present study, microbial, biochemical, and sensorial attitudes were strongly affected by brines inoculation, although a minor influence of salt content was also noted. The use of starter culture changed the microbial dominance and led to faster acidification of brines and faster degradation of oleuropein, indicating the faster fermentation completion. Moreover, the reduction of sodium content resulted in a successful lactic fermentation of Cypriot green cracked table olives. The final products fulfilled microbiological criteria and exhibited more appreciated sensorial characteristics. In addition, the formulation of table olives with low salt content is healthier and more suitable for consumers at risk of hypertension, opening a new era for table olives industry.

It must be mentioned that according to our findings, Cypriot olives were ready to eat after 120 days of fermentation in all treatments. This could be supported by the elimination of oleuropein, as well as the depletion of sugars (glucose and fructose), at this time point. Furthermore, the minor changes occurred after 120 days, confirming the above conclusion. Thus, the subsequent period was considered a preservation stage until selling, which was also important to be studied, in order to avoid any risk regarding the final product.

New methodologies used in olive and brine analysis (water potential and electrical conductivity) provided us with strong indications that they could be of interest as potential tools for the monitoring of the fermentation progress. However, further studies are required to establish a validated protocol.

Concluding, the effect of the inoculation of table olives on the production of stable quality and the final product was not dependent on the alternating indigenous microflora. The use of starter culture could lead to the modernization of fermentation, with healthier products of high quality. The study of the table olive indigenous microflora might lead to further research concerning its beneficial effects during olive processing with additional biotechnological and probiotic potential. Thus, further work is underway in order to study the multifunctional features of the isolated LAB and yeasts since indigenous microorganisms may be more adjusted to the harsh southeast Mediterranean environmental conditions while adding further to locally appreciated organoleptic characteristics.

Supplementary Materials: The following are available online at http://www.mdpi.com/2304-8158/9/1/17/s1, Supplementary Table S1: Evolution of color parameters (a*, b*, L*, h*, and C*) of olive fruits during spontaneous (OL7), inoculated (10% NaCl) (OL8), and inoculated (7% NaCl) (OL9) fermentation of Cypriot green cracked table olives. Data points are expressed as means and standard deviations of 10 random measurements. Supplementary Table S2: Contribution of all studied variables to the factors in the PCA based on correlations.

Author Contributions: Conceptualization, D.A.A. and D.T.; Methodology, D.A.A. and V.G.; Software, D.A.A. and N.N.; Validation, D.T.; Formal analysis, D.A.A., C.V., and E.X.; Investigation, D.A.A. and D.T.; Resources D.T.; Data curation, D.A.A. and N.N.; Writing—Original draft preparation, D.A.A. and V.G.; Writing—Review and editing, N.N. and D.T.; Visualization, D.A.A. and D.T.; Supervision, D.T.; Project administration, D.T. All authors have read and agreed to the published version of the manuscript.

Funding: This research was funded by the project AGROID, an INTERREG Greece—Cyprus 2014–2020 Program, which is co-funded by the European Union (ERDF) and National Resources of Greece and Cyprus.

Acknowledgments: The authors would like to acknowledge the technical support and welcoming environment from Novel Agro Ltd. for the olive's fermentations.

Conflicts of Interest: None of the authors has a financial or personal relationship with other people or organizations that could inappropriately influence or bias this publication.

References

1. (IOC), I.O.O.C. Updates Series of World Statistics on Production, Imports, Exports and Consumption. 2018. Available online: https://www.internationaloliveoil.org/wp-content/uploads/2019/11/production3_ang.pdf (accessed on 11 November 2019).
2. Bleve, G.; Tufariello, M.; Durante, M.; Perbellini, E.; Ramires, F.A.; Grieco, F.; Cappello, M.S.; de Domenico, S.; Mita, G.; Tasioula-Margari, M.; et al. Physico-chemical and microbiological characterization of spontaneous fermentation of Cellina di Nardò and Leccino table olives. *Front. Microbiol.* **2014**, *5*, 1–18. [CrossRef] [PubMed]
3. Rodríguez-Gómez, F.; Ruiz-Bellido, M.A.; Romero-Gil, V.; Benítez-Cabello, A.; Garrido-Fernández, A.; Arroyo-López, F.N. Microbiological and physicochemical changes in natural green heat-shocked Aloreña de Málaga table olives. *Front. Microbiol.* **2017**, *8*. [CrossRef] [PubMed]
4. Corsetti, A.; Perpetuini, G.; Schirone, M.; Tofalo, R.; Suzzi, G. Application of starter cultures to table olive fermentation: An overview on the experimental studies. *Front. Microbiol.* **2012**, *3*, 1–6. [CrossRef] [PubMed]
5. Campus, M.; Degirmencioglu, N.; Comunian, R. Technologies and trends to improve table olive quality and safety. *Front. Microbiol.* **2018**, *9*. [CrossRef]
6. Hurtado, A.; Reguant, C.; Bordons, A.; Rozès, N. Lactic acid bacteria from fermented table olives. *Food Microbiol.* **2012**, *31*, 1–8. [CrossRef]
7. Bonatsou, S.; Tassou, C.; Panagou, E.; Nychas, G.-J. Table Olive Fermentation Using Starter Cultures with Multifunctional Potential. *Microorganisms* **2017**, *5*, 30. [CrossRef]
8. Heperkan, D. Microbiota of table olive fermentations and criteria of selection for their use as starters. *Front. Microbiol.* **2013**, *4*, 1–11. [CrossRef]

9. Papadelli, M.; Zoumpopoulou, G.; Georgalaki, M.; Anastasiou, R.; Manolopoulou, E.; Lytra, I.; Papadimitriou, K.; Tsakalidou, E. Evaluation of two lactic acid bacteria starter cultures for the fermentation of natural black table olives (Olea Europaea L cv Kalamon). *Polish J. Microbiol.* **2015**, *64*, 265–271. [CrossRef]
10. Sorrentino, G.; Muzzalupo, I.; Muccilli, S.; Timpanaro, N.; Russo, M.P.; Guardo, M.; Rapisarda, P.; Romeo, F.V. New accessions of Italian table olives (Olea europaea): Characterization of genotypes and quality of brined products. *Sci. Hortic. (Amsterdam)* **2016**, *213*, 34–41. [CrossRef]
11. Anagnostopoulos, D.; Bozoudi, D.; Tsaltas, D. Yeast ecology of fermented table olives: A tool for biotechnlogical applications. In *Yeast: Industrial Applications*; IntechOpen: Rijeka, Croatia, 2017; pp. 135–152.
12. Romeo, F.V. Microbiological Aspects of Table Olives. In *Olive Germplasm: The Olive Cultivation, Table Olive and Olive Oil Industry in Italy*; BoD–Books on Demand: Norderstedt, Germany, 2012.
13. Kiai, H.; Hafidi, A. Chemical composition changes in four green olive cultivars during spontaneous fermentation. *LWT Food Sci. Technol.* **2014**, *57*, 663–670. [CrossRef]
14. Panagou, E.Z.; Schillinger, U.; Franz, C.M.A.P.; Nychas, G.J.E. Microbiological and biochemical profile of cv. Conservolea naturally black olives during controlled fermentation with selected strains of lactic acid bacteria. *Food Microbiol.* **2008**, *25*, 348–358. [CrossRef] [PubMed]
15. Bautista-Gallego, J.; Arroyo-López, F.N.; Durán-Quintana, M.C.; Garrido-Fernández, A. Fermentation profiles of Manzanilla-Aloreña cracked green table olives in different chloride salt mixtures. *Food Microbiol.* **2010**, *27*, 403–412. [CrossRef] [PubMed]
16. Bleve, G.; Tufariello, M.; Durante, M.; Grieco, F.; Ramires, F.A.; Mita, G.; Tasioula-Margari, M.; Logrieco, A.F. Physico-chemical characterization of natural fermentation process of Conservolea and Kalamàta table olives and developement of a protocol for the pre-selection of fermentation starters. *Food Microbiol.* **2015**, *46*, 368–382. [CrossRef] [PubMed]
17. Chranioti, C.; Kotzekidou, P.; Gerasopoulos, D. Effect of starter cultures on fermentation of naturally and alkali-treated cv. Conservolea green olives. *LWT Food Sci. Technol.* **2018**, *89*, 403–408. [CrossRef]
18. Pino, A.; De Angelis, M.D.; Todaro, A.; Van Hoorde, K.V.; Randazzo, C.L.; Caggia, C. Fermentation of Nocellara Etnea table olives by functional starter cultures at different low salt concentrations. *Front. Microbiol.* **2018**, *9*. [CrossRef]
19. Bautista-Gallego, J.; Arroyo-López, F.N.; López-López, A.; Garrido-Fernández, A. Effect of chloride salt mixtures on selected attributes and mineral content of fermented cracked Aloreña olives. *LWT Food Sci. Technol.* **2011**, *44*, 120–129. [CrossRef]
20. Fadda, C.; Del Caro, A.; Sanguinetti, A.M.; Piga, A. Texture and antioxidant evolution of naturally green table olives as affected by different sodium chloride brine concentrations. *Grasas Aceites* **2014**, *65*, e002.
21. WHO. *2012 Guideline: Sodium Intake for Adults and Children*; WHO: Geneva, Switzerland, 2012.
22. Rantsiou, K. Salt Reduction in Vegetable Fermentation: Reality or Desire? *J. Food Sci.* **2013**, *78*.
23. Aponte, M.; Ventorino, V.; Blaiotta, G.; Volpe, G.; Farina, V.; Avellone, G.; Lanza, C.M.; Moschetti, G. Study of green Sicilian table olive fermentations through microbiological, chemical and sensory analyses. *Food Microbiol.* **2010**, *27*, 162–170. [CrossRef]
24. Tofalo, R.; Schirone, M.; Perpetuini, G.; Angelozzi, G.; Suzzi, G.; Corsetti, A. Microbiological and chemical profiles of naturally fermented table olives and brines from different Italian cultivars. *Antonie Leeuwenhoek Int. J. Gen. Mol. Microbiol.* **2012**, *102*, 121–131. [CrossRef]
25. Tataridou, M.; Kotzekidou, P. Fermentation of table olives by oleuropeinolytic starter culture in reduced salt brines and inactivation of Escherichia coli O157:H7 and Listeria monocytogenes. *Int. J. Food Microbiol.* **2015**, *208*, 122–130. [CrossRef] [PubMed]
26. Uylaşer, V. Changes in phenolic compounds during ripening in Gemlik variety olive fruits obtained from different locations. *CYTA J. Food* **2015**, *13*, 167–173. [CrossRef]
27. D'Antuono, I.; Bruno, A.; Linsalata, V.; Minervini, F.; Garbetta, A.; Tufariello, M.; Mita, G.; Logrieco, A.F.; Bleve, G.; Cardinali, A. Fermented Apulian table olives: Effect of selected microbial starters on polyphenols composition, antioxidant activities and bioaccessibility. *Food Chem.* **2018**, *248*, 137–145. [CrossRef] [PubMed]
28. Bautista-Gallego, J.; Arroyo-López, F.N.; Rantsiou, K.; Jiménez-Díaz, R.; Garrido-Fernández, A.; Cocolin, L. Screening of lactic acid bacteria isolated from fermented table olives with probiotic potential. *Food Res. Int.* **2013**, *50*, 135–142. [CrossRef]

29. Bonatsou, S.; Paramithiotis, S.; Panagou, E.Z. Evolution of yeast consortia during the fermentation of Kalamata natural black olives upon two initial acidification treatments. *Front. Microbiol.* **2018**, *8*, 1–13. [CrossRef] [PubMed]
30. Rodríguez-Gómez, F.; Bautista-Gallego, J.; Arroyo-López, F.N.; Romero-Gil, V.; Jiménez-Díaz, R.; Garrido-Fernández, A.; García-García, P. Tableolive fermentation with multifunctional Lactobacillus pentosus strains. *Food Control* **2013**, *34*, 96–105. [CrossRef]
31. Panagou, E.Z.; Tassou, C.C. Changes in volatile compounds and related biochemical profile during controlled fermentation of cv. Conservolea green olives. *Food Microbiol.* **2006**, *23*, 738–746. [CrossRef]
32. Segovia Bravo, K.A.; Arroyo López, F.N.; García García, P.; Durán Quintana, M.C.; Garrido Fernández, A. Treatment of green table olive solutions with ozone. Effect on their polyphenol content and on Lactobacillus pentosus and Saccharomyces cerevisiae growth. *Int. J. Food Microbiol.* **2007**, *114*, 60–68. [CrossRef]
33. Medina, E.; Brenes, M.; García-García, P.; Romero, C.; de Castro, A. Microbial ecology along the processing of Spanish olives darkened by oxidation. *Food Control* **2018**, *86*, 35–41. [CrossRef]
34. Abriouel, H.; Benomar, N.; Cobo, A.; Caballero, N.; Fernández Fuentes, M.Á.; Pérez-Pulido, R.; Gálvez, A. Characterization of lactic acid bacteria from naturally-fermented Manzanilla Aloreña green table olives. *Food Microbiol.* **2012**, *32*, 308–316. [CrossRef]
35. Marsilio, V.; Seghetti, L.; Iannucci, E.; Russi, F.; Lanza, B.; Felicioni, M. Use of a lactic acid bacteria starter culture during green olive (Olea europaea L cv Ascolana tenera) processing. *J. Sci. Food Agric.* **2005**, *85*, 1084–1090. [CrossRef]
36. Arroyo-López, F.N.; Romero-Gil, V.; Bautista-Gallego, J.; Rodríguez-Gómez, F.; Jiménez-Díaz, R.; García-García, P.; Querol, A.; Garrido-Fernández, A. Yeasts in table olive processing: Desirable or spoilage microorganisms? *Int. J. Food Microbiol.* **2012**, *160*, 42–49. [CrossRef] [PubMed]
37. Tufariello, M.; Durante, M.; Ramires, F.A.; Grieco, F.; Tommasi, L.; Perbellini, E.; Falco, V.; Tasioula-Margari, M.; Logrieco, A.F.; Mita, G.; et al. New process for production of fermented black table olives using selected autochthonous microbial resources. *Front. Microbiol.* **2015**, *6*, 1–15. [CrossRef] [PubMed]
38. Pereira, A.P.; Pereira, J.A.; Bento, A.; Estevinho, M.L. Microbiological characterization of table olives commercialized in Portugal in respect to safety aspects. *Food Chem. Toxicol.* **2008**, *46*, 2895–2902. [CrossRef] [PubMed]
39. Perpetuini, G.; Caruso, G.; Urbani, S.; Schirone, M.; Esposto, S.; Ciarrocchi, A.; Prete, R.; Garcia-Gonzalez, N.; Battistelli, N.; Gucci, R.; et al. Changes in polyphenolic concentrations of table olives (cv. Itrana) produced under different irrigation regimes during spontaneous or inoculated fermentation. *Front. Microbiol.* **2018**, *9*. [CrossRef]
40. Liebeherr, J. Chapter 1: Introduction To Chapter 1: Introduction To. *Semant. Web Educ.* **2006**, *1*, 1–15.
41. Cais-Sokolińska, D. Analysis of metabolic activity of lactic acid bacteria and yeast in model kefirs made from goat's milk and mixtures of goat's milk with mare's milk based on changes in electrical conductivity and impedance. *Mljekarstvo* **2017**, *67*, 277–282. [CrossRef]
42. Álvarez, D.M.E.; López, A.; Lamarque, A.L. Industrial improvement for naturally black olives production of Manzanilla and Arauco cultivars. *J. Food Process. Preserv.* **2014**, *38*, 106–115. [CrossRef]
43. Rodríguez, H.; Landete, J.M.; Curiel, J.A.; De Las Rivas, B.; Mancheño, J.M.; Muñoz, R. Characterization of the p-coumaric acid decarboxylase from Lactobacillus plantarum CECT 748T. *J. Agric. Food Chem.* **2008**, *56*, 3068–3072. [CrossRef]
44. Othman, N.B.; Roblain, D.; Chammen, N.; Thonart, P.; Hamdi, M. Antioxidant phenolic compounds loss during the fermentation of Chétoui olives. *Food Chem.* **2009**, *116*, 662–669. [CrossRef]
45. Pistarino, E.; Aliakbarian, B.; Casazza, A.A.; Paini, M.; Cosulich, M.E.; Perego, P. Combined effect of starter culture and temperature on phenolic compounds during fermentation of Taggiasca black olives. *Food Chem.* **2013**, *138*, 2043–2049. [CrossRef] [PubMed]
46. Randazzo, C.L.; Fava, G.; Tomaselli, F.; Romeo, F.V.; Pennino, G.; Vitello, E.; Caggia, C. Effect of kaolin and copper based products and of starter cultures on green table olive fermentation. *Food Microbiol.* **2011**, *28*, 910–919. [CrossRef] [PubMed]
47. Laëtitia, G.; Pascal, D.; Yann, D. The Citrate Metabolism in Homo- and Heterofermentative LAB: A Selective Means of Becoming Dominant over Other Microorganisms in Complex Ecosystems. *Food Nutr. Sci.* **2014**, *5*, 953–969. [CrossRef]

48. Chidi, B.S.; Bauer, F.F.; Rossouw, D. Organic acid metabolism and the impact of fermentation practices on wine acidity—A review. *S. Afr. J. Enol. Vitic.* **2018**, *39*, 315–329. [CrossRef]
49. Erasmus, D.J.; Cliff, M.; Van Vuuren, H.J.J. Impact of yeast strain on the production of acetic acid, glycerol, and the sensory attributes of icewine. *Am. J. Enol. Vitic.* **2004**, *55*, 371–378.
50. Arroyo-López, F.N.; Romero-Gil, V.; Bautista-Gallego, J.; Rodríguez-Gómez, F.; Jiménez-Díaz, R.; García-García, P.; Querol, A.; Garrido-Fernández, A. Potential benefits of the application of yeast starters in table olive processing. *Front. Microbiol.* **2012**, *3*, 1–4. [CrossRef]
51. Luckett, C.R. The Influences of Texture and Mastication Pattern on Flavor Perception Across the Lifespan. Ph.D. Thesis, University of Arkansas, Fayetteville, Arkansas, 2016.
52. Fernández-Bolaños, J.; Rodríguez, R.; Saldaña, C.; Heredia, A.; Guilén, R.; Jiménez, A. Factors affecting the changes in texture of dressed ("aliñ adas") olives. *Eur. Food Res. Technol.* **2002**, *214*, 237–241. [CrossRef]
53. Ramírez, E.; Gandul-Rojas, B.; Romero, C.; Brenes, M.; Gallardo-Guerrero, L. Composition of pigments and colour changes in green table olives related to processing type. *Food Chem.* **2015**, *166*, 115–124. [CrossRef]
54. Mínguez-Mosquera, M.I.; Gandul-Rojas, B.; Mínguez-Mosquera, J. Mechanism and Kinetics of the Degradation of Chlorophylls during the Processing of Green Table Olives. *J. Agric. Food Chem.* **1994**, *42*, 1089–1095. [CrossRef]

foods

MDPI

Article

Do Best-Selected Strains Perform Table Olive Fermentation Better than Undefined Biodiverse Starters? A Comparative Study

Antonio Paba [1], Luigi Chessa [1], Elisabetta Daga [1], Marco Campus [1,2], Monica Bulla [1], Alberto Angioni [3], Piergiorgio Sedda [1] and Roberta Comunian [1,*]

1 AGRIS Sardegna, Agenzia regionale per la ricerca in agricoltura, Loc. Bonassai, km 18.600 S.S. 291, 07100 Sassari, Italy; apaba@agrisricerca.it (A.P.); lchessa@agrisricerca.it (L.C.); edaga@agrisricerca.it (E.D.); campus@portocontericerche.it (M.C.); monica.bulla@libero.it (M.B.); psedda@agrisricerca.it (P.S.)
2 Current affiliation: Porto Conte Ricerche S.r.l, S.P. 55 Porto Conte-Capo Caccia km 8,400 Loc. Tramariglio, 07041 Alghero (SS), Italy
3 Dipartimento di Scienze della Vita e dell'Ambiente, Università di Cagliari, Viale S. Ignazio da Laconi 13, 09123 Cagliari, Italy; aangioni@unica.it
* Correspondence: rcomunian@agrisricerca.it; Tel.: +39-079-2842-329; Fax: +39-079-389-450

Received: 17 December 2019; Accepted: 22 January 2020; Published: 28 January 2020

Abstract: Twenty-seven *Lactobacillus pentosus* strains, and the undefined starter for table olives from which they were isolated, were characterised for their technological properties: tolerance to low temperature, high salt concentration, alkaline pH, and olive leaf extract; acidifying ability; oleuropein degradation; hydrogen peroxide and lactic acid production. Two strains with appropriate technological properties were selected. Then, table olive fermentation in vats, with the original starter, the selected strains, and without starter (spontaneous fermentation) were compared. Starters affected some texture profile parameters. The undefined culture resulted in the most effective *Enterobacteriaceae* reduction, acidification and olive debittering, while the selected strains batch showed the lowest antioxidant activity. Our results show that the best candidate strains cannot guarantee better fermentation performance than the undefined biodiverse mix from which they originate.

Keywords: undefined biodiverse starters; autochtonous cultures; lactic acid bacteria; *Lactobacillus pentosus*; Tonda di Cagliari; table olive; phenolic compounds; oleuropein

1. Introduction

Table olives are the most widely diffused traditional fermented vegetable product in the Mediterranean area [1]. The process is performed with the purpose of reducing olives bitterness to a palatable level, to enhance sensory features, while ensuring safety of consumption via acidification and/or biopreservation [2]. Natural fermentation is carried out by soaking raw olives in brines (6%–10% NaCl), where environmental microflora colonizing olives, vats, and tools used in previous processes give rise to a spontaneous fermentation, driven mainly by lactic acid bacteria (LAB) and yeasts. To improve the onset of favourable physical–chemical conditions during the early process stages, brines from previous fermentations can be used as microbial inoculum for new batches, according to the back-slopping method [3,4]. Thus, in several productions, natural fermentation is replaced by the use of microbial starters, yeast- or LAB-based, to enhance the fermentation performances, speeding up the acidification of brines [2], preventing the proliferation of spoilage bacteria [5], or conferring probiotic characteristics to the product [6,7]. The microbial starters used for table olives can be made by few (or even one) species and strains, as in the case of the selected starter cultures, or can consist of an indefinite number of microorganisms; in this case, we refer to natural biodiverse starter cultures [2].

Selected starters, frequently used in industrial productions [8], control the fermentation process and standardise the end product [9] by rapid domination of the indigenous microflora of raw olives, but reduce microbial biodiversity and sensory complexity of fermented table olives [4,10]. The microbial strains forming the selected starters are chosen based on their ability to survive to brine and adverse environmental conditions, i.e., high pH and NaCl concentration, and low temperature [11], and on their ability to hydrolyse oleuropein, produce aromas, and counteract the development of spoilage microorganisms and pathogens (e.g., *Enterobacteriaceae*, *Clostridium*, *Pseudomonas*, *Staphylococcus*, and *Listeria*) [4,12,13]. On the contrary, the use of undefined biodiverse starters, composed by autochthonous microflora, better adapted to the raw olives than allochthonous ones [14], could be advantageous in terms of taste richness, linking the product with the territory of production, in case of PDO and IGP products [2]. Moreover, the undefined biodiverse starters, characterised by a large number of strains [15], are more resistant to phage attacks, which is strain-specific, and phage-insensitive strains can mutually compensate for the loss of metabolic pathways of the sensitive strains attacked [11].

Recently, Campus et al. [16] and Comunian et al. [17] reported a new technological approach using a semi-natural starter culture (SIE, selected inoculum enrichment) consisting of an undefined number of *Lactobacillus pentosus* strains obtained from a natural fermentation of table olives of the variety *Tonda di Cagliari*, a local cultivar from Sardinia, Italy [18,19]. The SIE undefined mix of autochthonous strains was more adapted to the raw olives and brine conditions than the allochthonous selected starter, showing better technological performances. Natural biodiverse starters could be advantageous over single or dual strains, since complex microbial communities have undergone natural selection, adapting to specific environmental conditions.

In this study, 27 LAB strains were characterized for their technological properties in order to select the best candidates to be used as starters for table olives processing. The aim of this study was to compare the fermentation of table olives of the variety *Tonda di Cagliari* in brines inoculated with the autochthonous and undefined biodiverse starter (SIE), a selected double-strain starter (DSS), and natural fermentation (NF) without a starter.

2. Materials and Methods

2.1. Experimental Plan

A biodiverse *L. pentosus* starter culture (SIE, selected inoculum enrichment) obtained from a previous successful fermentation [16] and 27 *L. pentosus* strains, previously molecularly biotyped [17], were characterised for their technological features: 11 strains were isolated from the SIE starter; 14 came from vats of table olives inoculated with SIE; 2 from vats of table olives under natural fermentation. Two strains with appropriate technological properties, belonging to the 11 SIE isolates, were selected to be used as the double-strain starter (DSS) in a new table olive experimental trial, in comparison with the original SIE starter culture and natural fermentation (NF).

2.2. Technological Characterisation

Cultures kept frozen at −80 °C were reactivated by streaking on De Man, Rogosa and Sharpe (MRS) agar plates, incubated at 30 °C for 24 h, in anaerobiosis. All the phenotypic tests, described in the following paragraphs, were performed in triplicate using a standard inoculum of 1.5×10^5 CFU/mL. In spectrophotometric assays (BioPhotometer plus, Eppendorf AG, Hamburg, Germany), bacterial growth was expressed as optical density at 600 nm (OD_{600}), and only cultures showing an $OD_{600} \geq 0.15$ were considered positive.

2.2.1. Tolerance to Low Temperatures, High Saline Concentrations, and Alkaline pH

The 27 strains and the SIE starter culture were tested for their tolerance to low temperatures in MRS broth, at 10 and 15 °C, after 3 and 7 days of incubation.

Tolerance to high saline concentrations was assessed in MRS broth supplemented with 8 or 10% NaCl and incubated at 30 °C for 72 h.

In order to test the tolerance to alkaline pH, the bacterial cultures were inoculated in half-strength MRS broth adjusted to pH 8 with NaOH 0.25 N (International System of units (SI)), and incubated at 30 °C for 48 h in anaerobiosis [20].

To test the tolerance of the cultures to low temperatures, high saline concentration and alkaline pH, the bacterial growth was evaluated spectrophotometrically.

2.2.2. Bacterial Growth and Acidification Ability

The cultures were inoculated in MRS broth and incubated at 30 °C for 24 h. Then, different aliquots were used for the pH measurement (pH meter pH510, Eutech Instruments, City, Country), and for the bacterial growth evaluation, both spectrophotometrically (OD_{600}) and by plate count (Log CFU/mL), in MRS agar, incubated at 30 °C for 72 h in anaerobiosis.

2.2.3. Tolerance to Olive Leaf Extract

To test the tolerance to olive leaf extract (OLE), 5 μL of each overnight culture, at 1.5×10^5 CFU/mL, were spotted on MRS agar plates supplemented with 10% (*w/v*) of OLE, and incubated at 30 °C for 72 h in anaerobiosis. OLE powder was obtained by dehydrating olive leaves at 105 °C for 24 h and then grinding with a homogenizer (Type-A10 Janke & Kunkel GmbH & Co. Kg Ika-Werk, Staufen, Germany). Strains developing colonies on the medium were considered tolerant of OLE. A negative control without OLE was included in the assay [21].

2.2.4. Use of Oleuropein as Substrate

Modified MRS broth in which glucose was replaced with 1% (*w/v*) oleuropein (Applichem GmbH, Darmstadt, Germany) as the sole carbon source, was used for testing the oleuropein degradation ability of the microbial isolates and the SIE culture. The test was performed following a modified protocol of Ghabbour et al. [21], inoculating the cultures in a final volume of 100 μL in micro-plates. Degradation ability was assessed by visual examination of microbial growth after 7 days of incubation at 30 °C. Microplates wells showing cellular precipitate (pellet) at the bottom were considered positive. Standard MRS medium broth inoculated with the cultures was used as positive control.

2.2.5. Hydrogen Peroxide Production

The ability to produce hydrogen peroxide was tested according to Marshall [22] modified by Berthier [23], using Peptonized agar medium (PTM) containing HRP (horseradish peroxidase) and ABTS (2,2′-azino-bis (3-ethylbenzothiazoline-6-sulphonic acid)) as chromogenic substrate. Five microliters of each culture were spotted onto the plates and then incubated at 30 °C for 48 h in anaerobiosis. At the end of the incubation, the plates were exposed to air for 120 min at 30 °C, and for an additional 180 min at room temperature. The peroxide production was highlighted by the colour change of the colonies, and the tested strains were assigned to five categories. In order to perform the statistical analysis, a number was arbitrarily assigned to each category as follows: colourless, non-producer (0); green halo, very weak producer (1); green, weak producer (2); light purple, producer (3); dark purple, strong producer (4).

2.2.6. Lactic Acid Production

The test was performed on 11 strains, chosen among the best acidifying strains (tested in Section 2.2.2), and the SIE starter culture. Quantification of lactic acid D and L produced was carried out using the D-Lactic acid/L-Lactic acid Kit UV-method (R-Biopharm AG, Darmstadt, Germany), according to the manufacturer's instructions. The results were expressed in g/L of total D/L-lactic acid produced.

2.3. Starter Culture Origin and Preparation

The SIE starter culture, D104 and D702 strains, chosen among the SIE isolates and joined in the DSS starter, were reactivated by inoculating 10 μL of the concentrated culture stored at −80 °C in MRS broth and incubating overnight at 30 °C. The cultures grown were inoculated at a 1% rate in fresh MRS broth and incubated under the same conditions as the day before. The cultures were centrifuged (Centrifuge SL40R, Thermo Fisher Scientific, Lagenselbold, Germany) at 4500 rpm at 2 °C for 15 min in 500 mL volume Bio-bottles (Thermo Fisher Scientific). After discarding the supernatant, the pellets were washed with 200 mL of saline solution (0.89% *w/v* NaCl), in order to eliminate medium residues, resuspended in cryoprotectant (gelatin 5%, Na-citrate 5%, monosodium glutamate 5%, sucrose 10%, pH 7), and kept frozen at −80 °C. Before use, the cell concentration of the SIE starter and the strains D104 and D702 were checked by plate count in order to prepare a suitable inoculum for the SIE and DSS brines.

2.4. Pilot Scale Fermentation Trials

Olives from the variety *Tonda di Cagliari* were mechanically collected from an irrigated olive orchard, located in the south of Sardinia (Italy), at the green-yellow ripe stage. Defective fruits were discarded and then calibrated olives (fruit diameter between 17 and 20 mm) were carefully washed in tap water under continuous stirring, allowing the dripping of the excess water. The olives were placed in sanitised plastic vats that had a capacity of 220 L, filled up with NaCl brine (130 kg of olives and 90 L of 7% NaCl brine, kept constant manually throughout the process). An experimental design with 3 replicates and 3 repetitions per treatment was used. Vats were inoculated with DSS or SIE starter cultures, in order to reach an inoculum with a final concentration of 1.5×10^6 CFU/mL in brine. Natural fermentation (NF) vats were prepared as control. Vats were transferred to an acclimatized room and kept at 25 °C throughout the experiment.

2.5. Physical-Chemical Analyses

Olive brines were analysed for pH and titratable acidity (expressed as grams of lactic acid per 100 mL brine) using standard laboratory methods. Volatile acidity (expressed in grams of lactic acid per 100 mL of brine) was carried out by steam distillation, as follows: 10 mL of brine was put in a 50 mL flask, adding 1 g of tartaric acid. Volatile acids were distilled under steam current using a distillation apparatus and decarbonized distilled water as steam feeding. The distillate (250 mL) was collected and titrated with NaOH 0.1 N, using phenolphthalein as the indicator.

Sodium chloride in brines was determined according to the Mohr method: 1 mL of brine was diluted with 50 mL of distilled water, titrated with $AgNO_3$ 0.1 N with K_2CrO_4 as the indicator. All chemicals were purchased from Sigma Aldrich (Milan, Italy). Samples were analysed after 0, 7, 15, 30, 60, 90, and 180 days.

2.6. Phenolic Analysis

Phenolic compounds extracts were obtained according to the IOC method for the determination of biophenols by HPLC in olive oils [24], with some minor changes. Three grams of homogenized olives were extracted twice with 15 mL of a methanol/ water (80/20 *v/v*) solution and 10 mL of hexane. Tubes were agitated for 20 min in a rotatory shaker, then the organic layer was separated with a separatory funnel. The two $MeOH/H_2O$ extracts were combined, filtered through a 0.45 μm PTFE syringe filter (Whatman Inc., Clinton, NJ, USA), and dried in a rotary evaporator Rotavapor® R-300 (Buchi, Flawil, Switzerland) at 30 °C. The residue was dissolved in 15 mL of ethyl acetate, adding 2 g of anhydrous $MgSO_4$ to remove the remaining water fraction. One millilitre of the ethyl acetate solution was gently dried under N_2 stream, recollected with 1 mL of methanol and injected in HPLC/DAD for the analysis.

A HPLC 1100 (Agilent Technologies, Milan, Italy) equipped with a DAD detector UV 6000 (Thermo Finnigan, Milan, Italy) was used. The column was a Varian Polaris C18 (5 μm, 300 A, 250 X 4.6 mm).

Analyses were carried out at 280 and 360 nm, in gradient elution. Solvents were phosphoric acid 0.22 M (A), acetonitrile (B), and methanol (C), and the gradient program (T= time, in minutes) was: T = 0 A 96%, B and C 2%; T = 40 A 50%, B and C 25%; T = 45 A 40%, B and C 30%; T = 60 A 0%, B and C 50%, hold: 10 min; post time: 15 min., flow: 1 mL/min. Calibration curves were prepared in the range 5–50 μg/mL of authentic analytical standards of tyrosol, 3-hydroxytirosol, benzoic acid, paracumaric acid, ferulic acid, quercitin, luteolin, oleuropein, verbascoside and apigenin (Sigma-Aldrich Inc., St. Louis, MO, USA), except elenolic acid, which was synthetised in the laboratory. Stock solutions of the analytes were prepared in methanol (1000 μg/mL). Intermediate stock standard solutions were prepared at 100 μg/mL in methanol by dilution of stock standard solutions. Working standard solutions were prepared in methanol and used for qualitative and quantitative analysis.

2.7. DPPH Scavenging Activity as Trolox Equivalent Antioxidant Capacity (TEAC)

Five grams of destoned olives were homogenized, added with 10 mL of methanol and vigorously stirred for 20 minutes, then centrifuged at 4000 rpm for 25 min. DPPH-free radical scavenging capacity of phenolic extracts was evaluated according to the following protocol: 200 μL of the extracts or standard (Trolox) was added to 3 mL methanol solution of DPPH radical. After 1 min of vigorous shaking by vortex, the reaction mixture was left to stand at room temperature, in the dark, for 60 min. After that, the absorbance for the sample was read using a Varian Cary 50 UV–vis spectrophotometer (Varian Inc., Middelburg, The Netherlands), at $\lambda = 517$ nm, optical path 10 mm. A negative control was taken after adding the DPPH solution to the respective extraction solvent. The free radical scavenging capacity was expressed in Trolox equivalents (TE), e.g., mmol TE/kg, and quantified against a calibration curve of Trolox ($r = 0.99$).

2.8. Texture Analyses

Texture profile analyses (TPA) were carried out with a TA-XT Plus texture analyser (Stable Microsystems, Surrey, UK) with a plugged 30 kg load cell, coupled with the Exponent software (ver. 6.1.3.0) for acquisition and processing. Analyses were carried out on 30 fruits for each replicate, for a total of 90 fruits for each experimental condition. Olives were put on the heavy-duty platform and compressed along the longitudinal side by 15% of their thickness with the P/40 aluminium cylinder. Test speed was set at 1 mm/sec, time between compressions was 2 sec, and trigger force was set at 0.05 N. The TPA parameters computed were hardness, cohesiveness, gumminess, chewiness and springiness, according to Szczesniak [25] and Friedman et al. [26].

2.9. Microbiological Analyses

Samples of uninoculated brines, used for all the experimental theses, were collected. Decimal serial dilutions in saline solution (0.89% *w/v* NaCl) were prepared and plated, in duplicate, on FH agar medium, incubated at 30 °C for 72 h in anaerobiosis, for mesophilic lactobacilli enumeration; MEA agar medium (Microbiol, Uta Cagliari) supplemented with 0.01% of chloramphenicol (Sigma-Aldrich), incubated at 25 °C in aerobiosis, for yeasts and moulds; VRBGA medium (Microbiol), incubated at 30 °C for 18–24 h in aerobiosis, for *Enterobacteriaceae*. Furthermore, olives before brining and olives after 7, 15, 30, 60, 90, and 180 days from brining were collected. Samples constituting 130 g of olives and 90 mL of saline solution for olives before brining, or fermentation brine, were collected and homogenized for 10 min by a BagMixer paddle blender (Interscience Corporation, Saint Nom, France). Microbial counts were performed in duplicate on the growth media and incubation conditions indicated above. Analyses were performed on three vats for each experimental thesis (SIE, DSS and NF) and expressed as average Log CFU/mL.

2.10. *Statistical Analysis*

One-way analysis of variance (ANOVA) for the evaluation of significance ($P < 0.05$) was performed on the whole data set. Differences between the individual means were compared by Tukey's HSD post hoc test, using the software SPSS Statistics (v. 21.0; IBM Corp., Armonk, NY, USA).

3. Results

3.1. *Technological Characterisation*

The technological characterisation was based on the tolerance to low temperature, high saline concentration and alkaline pH, OLE resistance, oleuropein degradation, and acidification ability. Moreover, hydrogen peroxide and lactic acid production were also investigated.

3.1.1. Tolerance to Low Temperatures, High Saline Concentrations, and Alkaline pH

None of the isolates or the SIE starter culture were able to grow at 10 °C (data not shown). The bacterial growth was observed only at 15 °C after 7 days of incubation, and no significant ($P < 0.05$) differences were generally observed among the strains and the SIE culture, with few exceptions (Table 1). Most of the cultures tolerated saline concentrations up to 8% NaCl (*w/v*). D102, D104, D702 and SBOD300 strains showed better adaptability to the brine conditions. Only D714, D723, FNI901, SBOF1002, and SBOF901 strains were not able to grow (Table 1). None of the isolates and the SIE starter culture tolerated 10% NaCl.

Table 1. Technological properties (growth at low temperature, high salinity and alkaline pH) of isolates and natural communities.

Culture	Growth at 15 °C 7 day	Growth NaCl 8% 3 day	Growth pH 8 48 h	Growth 30 °C 24 h		pH 24 h
	OD600	OD600	OD600	OD600	CFU/mL	UpH
D101	1.00 ± 1.86 ^{abc}	1.07 ± 0.90	4.56 ± 0.36 ^{abc}	6.09 ± 0.31	8.96 ± 0.28	4.29 ± 0.09
D102	1.05 ± 1.20 ^{abc}	2.04 ± 0.65	4.13 ± 0.23 ^{abc}	6.14 ± 0.59	8.66 ± 0.49	4.15 ± 0.02
D104	0.26 ± 0.33 ^{a}	2.11 ± 0.27	4.33 ± 0.11 ^{abc}	6.28 ± 0.82	9.07 ± 0.35	4.16 ± 0.05
D701	3.56 ± 1.90 ^{abc}	0.50 ± 0.87	4.07 ± 0.42 ^{abc}	6.12 ± 1.00	8.41 ± 0.33	4.18 ± 0.07
D702	0.36 ± 0.34 ^{a}	2.01 ± 0.49	4.30 ± 0.10 ^{abc}	6.27 ± 0.45	8.58 ± 0.50	4.15 ± 0.03
D705	4.25 ± 0.33 ^{abc}	0.99 ± 1.15	4.38 ± 0.05 ^{abc}	6.49 ± 0.78	8.81 ± 0.50	4.05 ± 0.03
D710	4.01 ± 0.77 ^{abc}	1.02 ± 1.43	3.95 ± 0.29 ^{abc}	6.51 ± 1.07	8.17 ± 0.85	4.07 ± 0.04
D713	3.38 ± 2.71 ^{abc}	0.80 ± 1.39	3.66 ± 0.32 ^{a}	6.58 ± 2.13	8.74 ± 0.37	4.17 ± 0.19
D714	4.02 ± 1.95 ^{abc}	0.00 ± 0.00	3.95 ± 0.32 ^{abc}	5.68 ± 2.61	8.42 ± 0.67	4.34 ± 0.33
D716	4.33 ± 1.67 ^{abc}	0.44 ± 0.75	4.20 ± 0.39 ^{abc}	6.12 ± 1.11	8.28 ± 0.52	4.21 ± 0.13
D723	4.67 ± 1.62 ^{abc}	0.00 ± 0.00	4.00 ± 0.64 ^{abc}	3.80 ± 2.89	7.73 ± 0.79	4.68 ± 0.59
D724	4.62 ± 0.67 ^{abc}	0.57 ± 0.98	4.48 ± 0.27 ^{abc}	7.60 ± 0.55	8.34 ± 0.56	4.13 ± 0.05
D725	4.12 ± 1.47 ^{abc}	0.43 ± 0.74	3.97 ± 0.64 ^{abc}	6.47 ± 2.27	8.33 ± 0.49	4.15 ± 0.11
D730	3.43 ± 1.32 ^{abc}	0.26 ± 0.45	4.34 ± 0.10 ^{abc}	6.24 ± 1.26	8.53 ± 0.57	4.15 ± 0.17
SIE	3.74 ± 2.16 ^{abc}	1.09 ± 0.78	4.45 ± 0.33 ^{abc}	5.26 ± 0.82	8.19 ± 0.78	4.42 ± 0.16
FNH900	2.80 ± 2.29 ^{abc}	0.67 ± 0.58	4.58 ± 0.18 ^{abc}	5.12 ± 1.70	8.46 ± 0.24	4.54 ± 0.36
FNI901	0.44 ± 0.78 ^{ab}	0.00 ± 0.00	4.01 ± 0.55 ^{abc}	7.16 ± 0.43	8.05 ± 0.14	4.12 ± 0.02
SBOD104	0.98 ± 0.91 ^{abc}	1.91 ± 0.58	4.45 ± 0.11 ^{abc}	5.78 ± 0.82	8.65 ± 0.50	4.32 ± 0.06
SBOD300	1.09 ± 1.12 ^{abc}	2.19 ± 0.29	4.47 ± 0.09 ^{abc}	6.22 ± 0.61	8.91 ± 0.58	4.21 ± 0.08
SBOD501	3.01 ± 2.18 ^{abc}	1.85 ± 0.45	4.04 ± 0.44 ^{abc}	5.70 ± 1.68	8.30 ± 0.31	4.29 ± 0.03
SBOD503	5.09 ± 0.35 ^{bc}	0.89 ± 1.14	4.00 ± 0.90 ^{abc}	4.85 ± 3.23	8.57 ± 0.98	4.59 ± 0.74
SBOE1000	4.60 ± 0.30 ^{abc}	0.35 ± 0.61	4.85 ± 0.15 ^{bc}	6.89 ± 1.12	8.36 ± 0.55	4.15 ± 0.30
SBOE502	5.35 ± 0.92 ^{c}	1.26 ± 0.79	4.57 ± 0.30 ^{abc}	6.66 ± 1.81	7.98 ± 0.21	4.23 ± 0.08
SBOE603	0.81 ± 1.53 ^{abc}	1.00 ± 1.31	4.11 ± 0.46 ^{abc}	6.19 ± 1.09	8.43 ± 0.45	4.14 ± 0.11
SBOE801	2.42 ± 1.90 ^{abc}	0.21 ± 0.32	4.78 ± 0.16 ^{abc}	4.50 ± 3.00	7.99 ± 1.53	4.30 ± 0.14
SBOE802	3.86 ± 2.68 ^{abc}	0.35 ± 0.15	5.05 ± 0.25 ^{c}	6.06 ± 1.25	8.32 ± 0.58	4.31 ± 0.05
SBOF1002	1.99 ± 2.10 ^{abc}	0.00 ± 0.00	4.59 ± 0.45 ^{abc}	4.15 ± 3.46	8.40 ± 0.64	4.21 ± 0.09
SBOF901	3.49 ± 1.74 ^{abc}	0.00 ± 0.00	3.86 ± 0.32 ^{ab}	4.38 ± 3.29	8.60 ± 0.26	4.26 ± 0.11

Technological test performed for microbial isolates and natural communities. Absorbance at 600 nm (OD_{600}), enumeration of CFU/mL, and pH measuring were evaluated after 24 h, 48 h, 3 days, or 7 days (mean values ± SD, $n = 3$). For each parameter, average values sharing the same superscript letters (if present) do not differ significantly ($P < 0.05$), according to Tukey's HSD post hoc test.

All the cultures were able to grow in alkaline MRS (pH 8) after 48 h, and significant ($P < 0.05$) differences were observed among a few of the isolated tested. In particular, the growth of D713 was significantly lower than that of SBOE1000 and SBOE802, whereas SBOF901 showed a significantly lower growth than SBOE802 (Table 1).

3.1.2. Bacterial Growth and Acidification Ability

The growth of the isolates and the SIE culture was tested at 30 °C, and it was measured both optically (OD_{600}) and by plate count agar (CFU/mL). The OD_{600} values ranged from 3.80 of D723 to 7.60 of D724, whereas the number of CFU/mL ranged from 7.73 of D723 to 9.07 of D104 (Table 1). No significant ($p < 0.05$) differences in microbial growth after 24 h of incubation were observed among the cultures using both detection methods.

The acidification performance after 24 h was also evaluated. The final pH ranged between 4.07 of D710 and 4.68 of D723, and, similarly to as observed for the bacterial growth, no significant ($P < 0.05$) differences among the cultures were calculated (Table 1).

3.1.3. Olive Leaf Extract Tolerance and Use of Oleuropein as Substrate

All the isolates and the SIE culture were tolerant to 10% of OLE and showed degradation of 1% oleuropein.

3.1.4. Hydrogen Peroxide Production

The isolates revealed different levels of hydrogen peroxide production, with significant ($P < 0.05$) differences among the cultures. SBOE1000 and SBOE801 showed the highest production, which was not significantly ($P < 0.05$) higher than the isolates D104 and D702 (subsequently joined in the DSS culture), and the SIE culture (Figure 1).

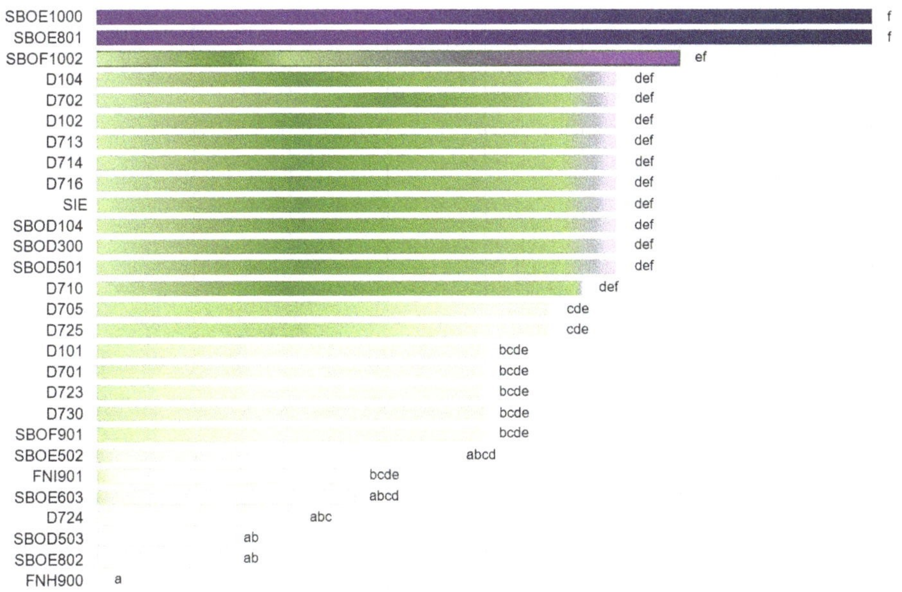

Figure 1. Hydrogen peroxide production of the characterised isolates and the semi-natural starter culture (SIE) starter culture. For each microbial culture tested, rows sharing the same letters do not differ significantly ($P < 0.05$), according to Tukey's HSD post hoc test.

3.1.5. Lactic Acid Production

The production of D, L, and total lactic acid revealed an interesting scenario among the bacterial cultures characterised. Significant ($P < 0.05$) differences in the amount of lactic acid produced were observed among the isolates, and generally, the SIE culture produced less lactic acid than most of the isolates (Table 2).

Table 2. Lactic acid production of selected bacterial isolates.

Culture	Lactic acid D− (g/L)	Lactic acid L+ (g/L)	Total Lactic acid (g/L)
D101	6.41 ± 0.82 ^{abcd}	2.53 ± 0.41 ^{abc}	9.55 ± 0.93 ^{ab}
D102	7.30 ± 0.97 ^{bcde}	3.09 ± 0.95 ^{abc}	10.40 ± 0.61 ^{ab}
D104	5.69 ± 0.39 ^{abc}	2.54 ± 0.25 ^{abc}	8.21 ± 0.45 ^{ab}
D702	7.94 ± 1.04 ^{cde}	2.82 ± 0.92 ^{abc}	11.08 ± 0.01 ^{cde}
D705	4.21 ± 0.02 ^{a}	3.43 ± 0.57 ^{abc}	7.54 ± 0.75 ^{cdef}
D710	7.03 ± 0.79 ^{abcde}	4.03 ± 0.94 ^{abc}	11.27 ± 0.44 ^{def}
D724	4.81 ± 1.04 ^{ab}	4.76 ± 0.21 ^{c}	11.19 ± 1.09 ^{def}
D730	7.08 ± 1.00 ^{bcde}	4.31 ± 0.48 ^{bc}	11.39 ± 0.52 ^{def}
SIE	4.66 ± 0.48 ^{ab}	2.18 ± 0.94 ^{ab}	8.48 ± 0.90 ^{ab}
FNI901	8.90 ± 0.95 ^{def}	1.94 ± 0.56 ^{a}	10.12 ± 0.15 ^{def}
SBOE1000	11.53 ± 0.23 ^{f}	2.18 ± 0.95 ^{ab}	12.71 ± 0.85 ^{ef}
SBOE603	9.46 ± 0.35e ^{f}	2.36 ± 0.44 ^{ab}	11.82 ± 0.74 ^{f}

Concentration (mean values ± SD, $n = 3$) of lactic acid D−, L+, and DL produced by selected bacterial isolates and natural communities. For each isomeric form of lactic acid, average values sharing the same superscript letters do not differ significantly ($P < 0.05$), according to Tukey's HSD post hoc test.

Based on the results obtained by the technological characterisation, two strains (D104 and D702) from the SIE undefined culture were selected and joined to make the double-strain starter (DSS) for table olive fermentation in vats. These two strains were among the best hydrogen peroxide producers and tolerated low temperature (i.e., 15 °C), high saline concentration (NaCl 8%), alkaline pH (8), and OLE (10%). Furthermore, their capacity to grow at the temperatures tested in this work (15 and 30 °C), the acidification ability, and the lactic acid production were comparable and not significantly different to the SIE culture.

3.2. Microbiological Analyses

Preliminary investigation on uninoculated brines and olives before brining revealed a very low yeast contamination (1.82 and 3.49 Log CFU/mL, respectively), while mesophilic lactobacilli were not detected. *Enterobacteriaceae* were found only in the olives (4.60 Log CFU/mL).

After 7 days from the inoculum in brine, mesophilic lactobacilli were below the level of detectability in NF samples, while reached 6.76 and 5.51 Log CFU/mL in SIE and DSS, respectively (Figure 2a). During the early stage of fermentation, higher counts were found in SIE than in DSS, showing better adaptability of the undefined starter SIE to brine conditions. Statistical differences ($P < 0.05$) among the three theses were found up to 15 days from brining. After 30 days from the inoculum, mesophilic lactobacilli counts were comparable in the three vats, remaining constant at around 6 Log until the end of the trial.

Yeast development was well controlled by the SIE starter culture (Figure 2b), as well as the *Enterobacteriaceae* (Figure 2c). In particular, yeasts, starting from about 3 Log CFU/mL in the three theses, slightly increased throughout the incubation period in SIE, whereas they were about 2 Log higher ($P < 0.05$) in DSS and NF at 15 and 30 days. At 60 days, yeasts reached similar levels in all the theses, then tended to decrease reaching a concentration between 3.56 Log CFU/mL (SIE) and 4.15 Log CFU/mL (DSS) at 180 days.

Enterobacteriaceae were about 5 Log CFU/mL after 7 days from brining in all the three theses. During the first 30 days of incubation, they decreased rapidly, not being detectable in SIE samples, while in NF and DSS *Enterobacteriaceae* were no more detectable from the 60th day.

Moulds were never found in all of the samples analysed.

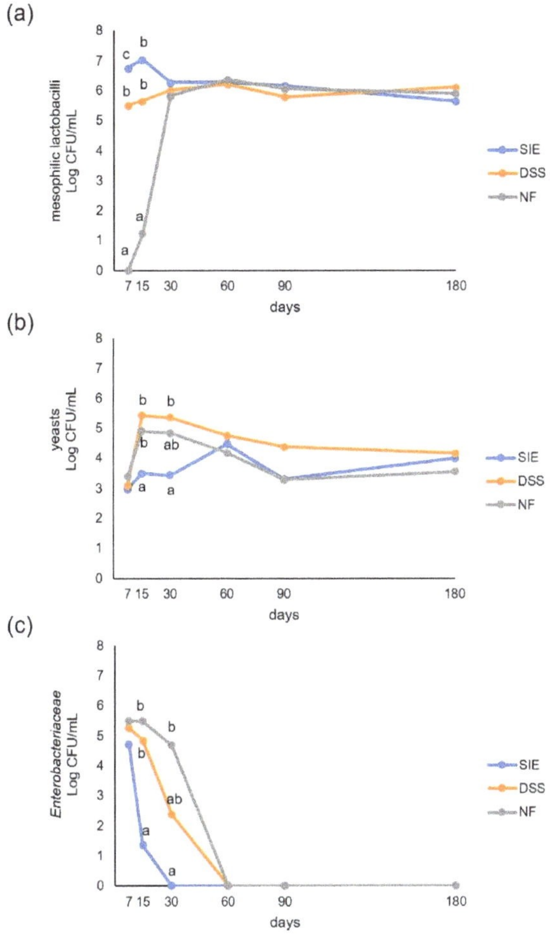

Figure 2. Microbial counts of viable mesophilic lactobacilli (**a**), yeasts (**b**), and *Enterobacteriaceae* (**c**) in vats inoculated with SIE and double-strain starter (DSS) starter cultures, and with natural fermentation (NF), evaluated after 7, 15, 30, 60, 90, and 180 days from the inoculum. For each microbial group and time-point of detection, counts, expressed as Log CFU/mL, sharing the same letters do not differ significantly ($P < 0.05$), according to Tukey's HSD post hoc test.

3.3. Physical-Chemical Analyses

No differences were observed in salinity among the three theses throughout the fermentation (Figure 3). Generally, DSS and NF showed not significant ($P < 0.05$) differences in titratable acidity and pH values. On the contrary, SIE showed significantly ($P < 0.05$) higher values during the evolution of titratable acidity. The monitoring of volatile acidity revealed significant differences between DSS and the other theses, which showed slightly higher values throughout the trial. A rapid fall in pH was observed in SIE, reaching values lower than 4 in 15 days, remaining almost constant until the end of observations (at 180 days, pH was 3.81), while DSS and NF never reached pH < 4 till the end of the trial (4.12 and 4.06, respectively).

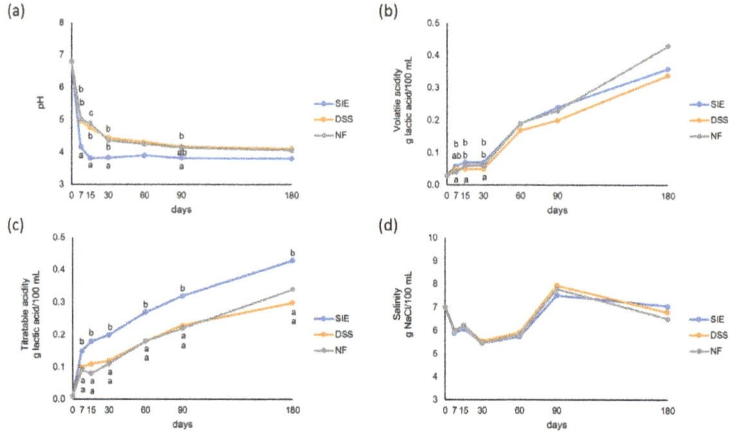

Figure 3. Physical–chemical parameters evolution during fermentation. (**a**) pH, (**b**) volatile acidity (g of lactic acid/100 mL), (**c**) titratable acidity (g of lactic acid/100 mL), and (**d**) salinity (*w/v*), measured immediately after the inoculum (0 d) and after 7, 15, 30, 60, 90, and 180 days. For each parameter and sampling time, values sharing the same superscript letters (if present) do not differ significantly ($P < 0.05$), according to Tukey's HSD post hoc test.

3.4. Phenolic Compounds Concentration and Antioxidant Activity as TEAC (Trolox Equivalent Antioxidant Capacity)

The HPLC analysis of phenols in the pulp in the different treatments showed 13 main compounds accounting for almost 90% of total phenols detected. For most of the individual phenols, significant ($P < 0.05$) differences between the concentrations were detected. Hydroxytyrosol was the most abundant in all samples, showing higher levels in SIE, according to the negligible values of oleuropein in these samples (Table 3), followed by verbascoside. Elenoic acid, 4-OH benzoic acid, paracumaric acid, quercetin dihydrate, and apigenin showed similar values in all treatments, while tyrosol, luteolin, luteolin 7-glucoside, and the unknown compound showed higher values in SIE samples, and comparable amounts in DSS and NF treatments.

Table 3. Phenolic compounds concentration (mg/kg ± SD) and TEAC activity.

Phenolic Compounds	SIE	DSS	NF
Elenolic acid	44.44 ± 8.46 ^{a}	31.76 ± 4.24 ^{a}	41.64 ± 7.80 ^{a}
OH tyrosol	264.22 ± 5.20 ^{b}	214.51 ± 9.87 ^{a}	217.08 ± 27.75 ^{a}
Tyrosol	34.74 ± 2.08 ^{b}	25.25 ± 1.99 ^{a}	25.27 ± 2.78 ^{a}
4 OH benzoic acid	21.46 ± 1.82 ^{a}	16.68 ± 2.64 ^{a}	19.79 ± 6.23 ^{a}
unknown	8.87 ± 0.47 ^{b}	5.12 ± 0.51 ^{a}	5.54 ± 1.04 ^{a}
Paracumaric acid	9.59 ± 1.29 ^{a}	11.69 ± 2.07 ^{a}	18.85 ± 3.35 ^{b}
Ferulic acid	6.11 ± 1.01 ^{ab}	4.97 ± 0.11 ^{b}	7.82 ± 1.09 ^{a}
Verbascoside	175.14 ± 16.57 ^{b}	124.57 ± 6.09 ^{a}	130.96 ± 19.31 ^{a}
Luteolin 7-glucoside	9.38 ± 2.21	n.d.	n.d.
Oleuropein	n.d.	17.05 ± 1.75 ^{a}	21.01 ± 3.64 ^{a}
Quercetin dihydrate	1.13 ± 0.28 ^{a}	2.41 ± 0.45 ^{a}	3.10 ± 0.55 ^{a}
Luteolin	30.50 ± 3.52 ^{b}	15.17 ± 1.25 ^{a}	15.59 ± 3.11 ^{a}
Apigenin	2.24 ± 0.23 ^{a}	1.98 ± 0.34 ^{a}	1.98 ± 0.41 ^{a}
Total phenolic compounds	3942.93 ± 478.78 ^{a}	3977.64 ± 612.15 ^{a}	4182.20 ± 213.90 ^{a}
TEAC	350.36 ± 33.82 ^{a}	339.95 ± 43.38 ^{a}	350.55 ± 63.12 ^{a}

Concentration of main phenolic compounds identified in pulp extracts and antioxidant activity as TEAC (Trolox equivalent antioxidant capacity). For each compound, average values (n = 3) sharing the same superscript letters do not differ significantly ($P < 0.05$) according to Tukey's HSD post hoc test. n.d.: not detected.

Oleuropein was not detectable in SIE samples while showed comparable amounts in DSS and NF in vitro antioxidant activity as TEAC was comparable among the theses although DSS showed the lowest values.

3.5. Texture Analyses

The TPA tests, carried out at the end of the fermentation, showed no differences ($P < 0.05$) among olives from the three theses in all texture parameters except for "gumminess" and "chewiness" (Table 4). "Gumminess" is "hardness × cohesiveness", thus this parameter refers to the "solidity" of the material and its resistance to deformation. "Chewiness" is "gumminess × elasticity". SIE samples showed significantly higher values of these parameters.

Table 4. Texture evaluation in olives at the end of fermentation.

TPA Parameters	SIE	DSS	NF
Hardness (g)	2397.31 ± 506.84^{a}	2185.96 ± 560.90^{a}	2209.41 ± 530.11^{a}
Springiness	0.64 ± 0.05^{a}	0.62 ± 0.06^{a}	0.62 ± 0.06^{a}
Cohesiveness	0.52 ± 0.05^{a}	0.50 ± 0.04^{a}	0.51 ± 0.05^{a}
Gumminess	1228.17 ± 236.15^{b}	1084.29 ± 251.77^{a}	1117.14 ± 239.66^{a}
Chewiness (g/mm)	782.25 ± 157.29^{b}	673.48 ± 168.16^{a}	690.67 ± 154.63^{a}
Resilience	0.27 ± 0.03^{a}	0.26 ± 0.03^{a}	0.26 ± 0.03^{a}

For each TPA parameters, average values (± SD, $n = 3$) sharing the same superscript letters do not differ significantly ($P < 0.05$), according to Tukey's HSD post hoc test.

4. Discussion

To answer the question raised in the title, two strains (D104 and D702), chosen among the best performers, isolated from the autochthonous SIE starter, were used as a double-strain starter (DSS) in a table olive fermentation trial, in comparison with the SIE starter culture and natural fermentation (NF). Overall, the SIE starter carried out the fermentation with better results than DSS and NF, even though D104 and D702 showed better performances in the technological characterisation tests. These strains showed among the best peroxide production and resistance to salt (i.e., 8% NaCl) performances, while the oleuropein hydrolysis and growth after 24 h, at all the temperatures tested, were comparable to the SIE culture, as well as the acidification ability. During the fermentation, SIE pushed more acidification, lowering the pH to a value <4.0, which is fundamental for the preservation of table olives since it prevents the proliferation of harmful and spoilage bacteria [5]. The pH drop observed during fermentation is due to the conversion of carbohydrates into organic acids, mainly lactic acid, by LAB fermentation. In addition, the hydrolysis of oleuropein, which is decomposed by endogenous and bacterial enzymes in sugars and simple phenols such as OH tyrosol and elenolic acid, may contribute to the pH fall and acidity rise [27]. The use of the starters (SIE and DSS) revealed a greater performance in controlling the evolution of spoilage bacteria and the development of favourable physical–chemical conditions during the fermentation compared to NF. The effectiveness of the starter culture addition was also observed in yeast control and the *Enterobacteriaceae* reduction, greater in the SIE vats, where, in the early fermentation phase, mesophilic lactobacilli were almost 1 and 6 Log CFU/mL higher than in DSS and NF, respectively. Interestingly, in NF, despite mesophilic lactobacilli slowly developed and reached the same level found in SIE and DSS only after 30 days of fermentation, it was observed that there was a pH trend similar to DSS, since a contribution to pH decrease could also come from the diffusion of organic acids from pulp to brine. The pH decreasing is involved in the prevention of spoilage microorganisms and pathogen contamination requested for table olive safety [28]. Indeed, *Enterobacteriaceae*, which could cause infections in humans and be responsible for table olive defects such as gas pockets formation, are the first microbial group able to grow during the early olive fermentation but are rapidly supplanted by LAB [29] through the decrease of pH [30]. Therefore, the use of the SIE starter could be a good hygiene practice in table olive processing, according to Campus et al. [16,31]. The faster disappearance of *Enterobacteriaceae*, as observed in the SIE thesis, has beneficial effects also on the table olive sensory quality [32].

Yeasts are involved in milder taste defects, excessive CO_2 production, and olive cells wall degradation [33,34]. However, yeasts can even improve the final product by the production of volatile compounds and the enhancement of LAB growth [34–36]. In this study, the vats inoculated with the SIE starter culture showed an almost constant yeast concentration throughout the fermentation, lower than in NF and in vats inoculated with the DSS starter. Due to its biodiversity, the SIE culture could have limited and better regulated yeast development during the fermentation.

The main phenomena responsible for changes in phenolic concentrations are the osmotic dehydration and the enzymatic activity exerted by endogenous and microbial enzymes. Olives are submerged in a hypertonic medium (brine), and plant tissues act as semipermeable membranes in relation to water movements when immersed in a hypertonic solution [37]. During the process, two major countercurrent flows take place simultaneously. The setting up of gradients across the product–medium interface leads to water flows from the product into the osmotic solution, whereas osmotic solute (NaCl) is transferred from the solution into the product. As a result, table olives increase in salt content during processing and lose sugars, phenols, acids, minerals, and vitamins into the solution [38]. The rate of diffusion varies according to the concentration and temperature of the osmotic solution, size and geometry of the material, solution to material mass ratio, and level of agitation of the solution [39]. The lower content of oleuropein in SIE is due to enzymatic hydrolysis carried out by inoculated lactic acid bacteria with β-glycosidase and esterase activity. Hydroxytyrosol, together with elenolic acid, derives from the hydrolysis of oleuropein by β-glycosidases and esterases, enzymes of endogenous and microbial origin. As reported by Cardoso et al. [40], hydroxytyrosol was the most abundant phenolic compound in MeOH extracts of olive pulp. Marsilio et al. [41] reported that in processed Greek-style table olives coming from var. *Ascolana tenera*, both naturally fermented and inoculated with a *Lactobacillus plantarum* based starter culture, oleuropein and hydroxytyrosol were the most abundant phenols.

The olives analysed in this study resulted overall in comparable texture, although the SIE samples showed a significantly higher resistance to deformation, as shown by the gumminess and chewiness parameter magnitudes. As reported in literature [42], changes in texture during natural fermentation of olives can be ascribed to hydrolysis of cell wall pectic polysaccharides, which results in loss of structural coherence of olive tissues, as observed by Servili et al. [43] with SEM techniques.

Recently, Bleve et al. [13] described a selection procedure for the production of mixed autochthonous starters for table olive fermentation. The autochthonous starters, isolated from the microbiota of raw olives, could have the advantage of being better adapted to the matrix to be processed than the allochthonous ones, with extended shelf-life [44] and better sensory quality of the final product [3,14,45]. Moreover, the use of biodiverse and complex microbial communities as starter cultures, instead of the mono- or two-selected strains frequently employed [11], is advantageous in terms of resistance against phage attacks and possible failure of the fermentation [5]. Phage infections are usually strain-specific and, in case of attack, in a biodiverse culture, the other phage-insensitive strains can survive and compensate for the lack of the sensitive-strains [2].

5. Conclusions

In this study, the SIE starter, an undefined mix of autochthonous *L. pentosus* strains, has been shown to be more efficient in brine acidification, leading to a safer product, supplanting spoilage bacteria earlier than the DSS starter and natural fermentation. Debittering was achieved in a shorter time. The hydrolysis of oleuropein into elenolic acid and hydroxytyrosol was more intense using the SIE starter, resulting in a higher amount of most of the phenolic compounds compared to the double-strain starter. Moreover, instrumental texture was not substantially affected by the use of microbial starters. Overall, the DSS did not reach the same performances of the SIE starter, showing behaviour similar to NF or in-between the two experimental theses.

Autochthonous complex microbial communities coming from the same environment of the raw material to be processed have more adaptability to harsh fermentation conditions, preserving safety and quality characteristics of naturally fermented olives faster, thus reducing production costs.

Author Contributions: Conceptualization, A.P., E.D. and R.C.; Methodology, E.D. and R.C.; Formal Analyses, M.C., A.P., E.D., M.B., A.A. and P.S.; Data Curation, A.P., L.C., M.C. and A.A.; Writing—Original Draft Preparation, M.C., A.P., L.C., and R.C.; Writing—Review & Editing, M.C., A.P., L.C., A.A. and R.C.; Supervision, R.C. All authors have read and agreed to the published version of the manuscript.

Funding: This work was supported by C.R.P: Regione Sardegna L.R. 7/2007, Project "S.A.R.T.OL.".

Acknowledgments: The authors thank: Roberto Zurru for funds acquisition and coordination of the S.A.R.T.OL. Project and Maria Carmen Fozzi for her contribution in microbiological analyses.

Conflicts of Interest: The authors declare no conflict of interest.

References

1. Ciafardini, G.; Zullo, B.A. Use of selected yeast starter cultures in industrial-scale processing of brined Taggiasca black table olives. *Food Microbiol.* **2019**, *84*, 103250. [CrossRef] [PubMed]
2. Campus, M.; Değirmencioğlu, N.; Comunian, R. Technologies and Trends to Improve Table Olive Quality and Safety. *Front. Microbiol.* **2018**, *9*, 617. [CrossRef] [PubMed]
3. Aponte, M.; Blaiotta, G.; La Croce, F.; Mazzaglia, A.; Farina, V.; Settanni, L.; Moschetti, G. Use of selected autochthonous lactic acid bacteria for Spanish-style table olive fermentation. *Food Microbiol.* **2012**, *30*, 8–16. [CrossRef] [PubMed]
4. Corsetti, A.; Perpetuini, G.; Schirone, M.; Tofalo, R.; Suzzi, G. Application of starter cultures to table olive fermentation: An overview on the experimental studies. *Front. Microbiol.* **2012**, *3*, 248. [CrossRef] [PubMed]
5. Lanza, B. Abnormal fermentations in table-olive processing: Microbial origin and sensory evaluation. *Front. Microbiol.* **2013**, *4*, 91. [CrossRef]
6. Bonatsou, S.; Benitez, A.; Rodriguez-Gomez, F.; Panagou, E.Z.; Arroyo-Lopez, F.N. Selection of yeasts with multifunctional features for application as starters in natural black table olive processing. *Food Microbiol.* **2015**, *46*, 66–73. [CrossRef]
7. Bonatsou, S.; Tassou, C.C.; Panagou, E.Z.; Nychas, G.E. Table Olive Fermentation Using Starter Cultures with Multifunctional Potential. *Microorganisms* **2017**, *5*, 30. [CrossRef]
8. Hurtado, A.; Reguant, C.; Bordons, A.; Rozes, N. Lactic acid bacteria from fermented table olives. *Food Microbiol.* **2012**, *31*, 1–8. [CrossRef]
9. Leroy, F.; De Vuyst, L. Lactic acid bacteria as functional starter cultures for the food fermentation industry. *Trends Food Sci. Tech.* **2004**, *15*, 67–78. [CrossRef]
10. Martorana, A.; Alfonzo, A.; Settanni, L.; Corona, O.; La Croce, F.; Caruso, T.; Moschetti, G.; Francesca, N. An innovative method to produce green table olives based on "pied de cuve" technology. *Food Microbiol.* **2015**, *50*, 126–140. [CrossRef]
11. Heperkan, D. Microbiota of table olive fermentations and criteria of selection for their use as starters. *Front. Microbiol.* **2013**, *4*, 143. [CrossRef]
12. Bevilacqua, A.; Altieri, C.; Corbo, M.R.; Sinigaglia, M.; Ouoba, L.I. Characterization of lactic acid bacteria isolated from Italian Bella di Cerignola table olives: selection of potential multifunctional starter cultures. *J. Food Sci.* **2010**, *75*, M536–M544. [CrossRef]
13. Bleve, G.; Tufariello, M.; Durante, M.; Grieco, F.; Ramires, F.A.; Mita, G.; Tasioula-Margari, M.; Logrieco, A.F. Physico-chemical characterization of natural fermentation process of Conservolea and Kalamata table olives and developement of a protocol for the pre-selection of fermentation starters. *Food Microbiol.* **2015**, *46*, 368–382. [CrossRef]
14. Di Cagno, R.; Surico, R.F.; Paradiso, A.; De Angelis, M.; Salmon, J.C.; Buchin, S.; De Gara, L.; Gobbetti, M. Effect of autochthonous lactic acid bacteria starters on health-promoting and sensory properties of tomato juices. *Int. J. Food Microbiol.* **2009**, *128*, 473–483. [CrossRef]
15. Bassi, D.; Puglisi, E.; Cocconcelli, P.S. Comparing natural and selected starter cultures in meat and cheese fermentations. *Curr. Opin. Food Sci.* **2015**, *2*, 118–122. [CrossRef]

16. Campus, M.; Sedda, P.; Cauli, E.; Piras, F.; Comunian, R.; Paba, A.; Daga, E.; Schirru, S.; Angioni, A.; Zurru, R.; et al. Evaluation of a single strain starter culture, a selected inoculum enrichment, and natural microflora in the processing of Tonda di Cagliari natural table olives: Impact on chemical, microbiological, sensory and texture quality. *LWT Food Sci. Tech.* **2015**, *64*, 671–677. [CrossRef]
17. Comunian, R.; Ferrocino, I.; Paba, A.; Daga, E.; Campus, M.; Di Salvo, R.; Cauli, E.; Piras, F.; Zurru, R.; Cocolin, L. Evolution of microbiota during spontaneous and inoculated Tonda di Cagliari table olives fermentation and impact on sensory characteristics. *LWT* **2017**, *84*, 64–72. [CrossRef]
18. Bandino, G.; Moro, C.; Mulas, M.; Sedda, P. Survey on olive genetic resources of sardinia. *Acta Hortic.* **1999**, *474*, 151–154. [CrossRef]
19. Bandino, G.; Sedda, P. Le Varietà di olivo della Sardegna. In *L'olio in Sardegna*; ILISSO, Ed.; ILISSO: Nuoro, Italy, 2013; pp. 171–222. ISBN 978-88-6202-309-2.
20. Sawatari, Y.; Yokota, A. Diversity and Mechanisms of Alkali Tolerance in Lactobacilli. *Appl. Environ. Microbiol.* **2007**, *73*, 3909. [CrossRef]
21. Ghabbour, N.; Lamzira, Z.; Thonart, P.; Cidalia, P.; Markaoui, M.; Asehraou, A. Selection of oleuropein-degrading lactic acid bacteria strains isolated from fermenting Moroccan green olives. *Grasas y Aceites* **2011**, *62*, 84–89. [CrossRef]
22. Marshall, V.M. A Note on Screening Hydrogen Peroxide-producing Lactic Acid Bacteria Using a Non-toxic Chromogen. *J. Appl. Bacteriol.* **1979**, *47*, 327–328. [CrossRef]
23. Berthier, F. On the screening of hydrogen peroxide-generating lactic acid bacteria. *Lett. Appl. Micbobiol.* **1993**, *16*, 150–153. [CrossRef]
24. International Olive Council. Determination of Biophenols in Olive Oils by HPLC. Available online: https://www.oelea.de/downloads/COI-T20-DOC-29-2009-DETERMINATION-OF-BIOPHENOLS-IN-OLIVE-OILS-BY-HPLC.pdf (accessed on 17 December 2019).
25. Szczesniak, A.S. Classification of Textural Characteristicsa. *J. Food Sci.* **1963**, *28*, 385–389. [CrossRef]
26. Friedman, H.H.; Whitney, J.E.; Szczesniak, A.S. The texturometer—A new instrument for objective texture measurement. *J. Food Sci.* **1963**, *28*, 390–396. [CrossRef]
27. Kiai, H.; Hafidi, A. Chemical composition changes in four green olive cultivars during spontaneous fermentation. *LWT Food Sci. Technol.* **2014**, *57*, 663–670. [CrossRef]
28. International Olive Council. Trade Standard on Table Olives. Available online: https://www.internationaloliveoil.org/what-we-do/chemistry-standardisation-unit/standards-and-methods/ (accessed on 17 December 2019).
29. De Angelis, M.; Campanella, D.; Cosmai, L.; Summo, C.; Rizzello, C.G.; Caponio, F. Microbiota and metabolome of un-started and started Greek-type fermentation of Bella di Cerignola table olives. *Food Microbiol.* **2015**, *52*, 18–30. [CrossRef]
30. Medina-Pradas, E.; Pérez-Díaz, I.M.; Garrido-Fernández, A.; Arroyo-López, F.N. Review of Vegetable Fermentations With Particular Emphasis on Processing Modifications, Microbial Ecology, and Spoilage. In *The Microbiological Quality of Food: Foodborne Spoilers*; Elsevier: Duxford, UK, 2017; pp. 211–236.
31. Campus, M.; Cauli, E.; Scano, E.; Piras, F.; Comunian, R.; Paba, A.; Daga, E.; Di Salvo, R.; Sedda, P.; Angioni, A.; et al. Towards Controlled Fermentation of Table Olives: LAB Starter Driven Process in an Automatic Pilot Processing Plant. *Food Bioprocess Tech.* **2017**, *10*, 1063–1073. [CrossRef]
32. Tofalo, R.; Schirone, M.; Perpetuini, G.; Angelozzi, G.; Suzzi, G.; Corsetti, A. Microbiological and chemical profiles of naturally fermented table olives and brines from different Italian cultivars. *Antonie van Leeuwenhoek* **2012**, *102*, 121–131. [CrossRef] [PubMed]
33. Arroyo-López, F.N.; Bautista-Gallego, J.; Rodríguez-Gómez, F.; Garrido-Fernández, A. Predictive microbiology and table olives. In *Current Research, Technology and Education Topics in Applied Microbiology and Microbial Biotechnology*; Formatex Research Center: Badajoz, Spain, 2010; pp. 1452–1461.
34. Hernandez, A.; Martin, A.; Aranda, E.; Perez-Nevado, F.; Cordoba, M.G. Identification and characterization of yeast isolated from the elaboration of seasoned green table olives. *Food Microbiol.* **2007**, *24*, 346–351. [CrossRef] [PubMed]
35. Arroyo-López, F.N.; Querol, A.; Bautista-Gallego, J.; Garrido-Fernández, A. Role of yeasts in table olive production. *Int. J. Food Microbiol.* **2008**, *128*, 189–196. [CrossRef]
36. Fernandez, A.G.; Adams, M.R.; Fernandez-Diez, M.J. *Table Olives*, 1st ed.; Springer: Heidelberg, Germany, 1997.

37. Lazarides, H. Reasons and possibilities to control solids uptake turing osmotic treatment of fruits and vegetables, osmotic dehydration and vacuum impregnation. In *Osmotic Dehydration and Vacuum Impregnation*; Taylor & Francis Group: London, UK, 2019; pp. 33–42.
38. Akbarian, M.; Ghasemkhani, N.; Moayedi, F. Osmotic dehydration of fruits in food industrial: A review. *Int. J. Biosci.* **2013**, *3*, 234–237. [CrossRef]
39. Maldonado, M.; Zuritz, C.; Miras, N. Influence of brine concentration on sugar and sodium chloride diffusion during the processing of the green olive variety Arauco. *Grasas y Aceites* **2008**, *59*, 267–273. [CrossRef]
40. Cardoso, S.M.; Guyot, S.; Marnet, N.; Lopes-da-Silva, J.A.; Renard, C.M.; Coimbra, M.A. Characterisation of phenolic extracts from olive pulp and olive pomace by electrospray mass spectrometry. *J. Sci. Food Agric.* **2005**, *85*, 21–32. [CrossRef]
41. Marsilio, V.; Seghetti, L.; Iannucci, E.; Russi, F.; Lanza, B.; Felicioni, M. Use of a lactic acid bacteria starter culture during green olive (*Olea europaea* L. cv *Ascolana tenera*) processing. *J. Sci. Food Agric.* **2005**, *85*, 1084–1090. [CrossRef]
42. Coimbra, M.A.; Waldron, K.W.; Delgadillo, I.; Selvendran, R.R. Effect of Processing on Cell Wall Polysaccharides of Green Table Olives. *J. Agric. Food Chem.* **1996**, *44*, 2394–2401. [CrossRef]
43. Servili, M.; Minnocci, A.; Veneziani, G.; Taticchi, A.; Urbani, S.; Esposto, S.; Sebastiani, L.; Valmorri, S.; Corsetti, A. Compositional and tissue modifications induced by the natural fermentation process in table olives. *J. Agric. Food Chem.* **2008**, *56*, 6389–6396. [CrossRef] [PubMed]
44. Di Cagno, R.; Coda, R.; De Angelis, M.; Gobbetti, M. Exploitation of vegetables and fruits through lactic acid fermentation. *Food Microbiol.* **2013**, *33*, 1–10. [CrossRef]
45. Bevilacqua, A.; de Stefano, F.; Augello, S.; Pignatiello, S.; Sinigaglia, M.; Corbo, M.R. Biotechnological innovations for table olives. *Int. J. Food Sci. Nutr.* **2015**, *66*, 127–131. [CrossRef]

foods

Article

Lactic Acid Bacteria and Yeast Inocula Modulate the Volatile Profile of Spanish-Style Green Table Olive Fermentations

Antonio Benítez-Cabello [1], Francisco Rodríguez-Gómez [1], M. Lourdes Morales [2,*], Antonio Garrido-Fernández [1], Rufino Jiménez-Díaz [1] and Francisco Noé Arroyo-López [1]

1 Departamento de Biotecnología de Alimentos, Instituto de la Grasa (CSIC), Ctra. Utrera km 1, Edificio 46, Campus Universitario Pablo de Olavide, 41013 Sevilla, Spain

2 Área de Nutrición y Bromatología, Dpto. Nutrición y Bromatología, Toxicología y Medicina Legal. Facultad de Farmacia, Universidad de Sevilla, C/P. García González, n°2, 41012 Sevilla, Spain

* Correspondence: mlmorales@us.es

Received: 4 July 2019; Accepted: 22 July 2019; Published: 24 July 2019

Abstract: In this work, Manzanilla Spanish-style green table olive fermentations were inoculated with *Lactobacillus pentosus* LPG1, *Lactobacillus pentosus* Lp13, *Lactobacillus plantarum* Lpl15, the yeast *Wickerhanomyces anomalus* Y12 and a mixed culture of all them. After fermentation (65 days), their volatile profiles in brines were determined by gas chromatography-mass spectrometry analysis. A total of 131 volatile compounds were found, but only 71 showed statistical differences between at least, two fermentation processes. The major chemical groups were alcohols (32), ketones (14), aldehydes (nine), and volatile phenols (nine). Results showed that inoculation with *Lactobacillus* strains, especially *L. pentosus* Lp13, reduced the formation of volatile compounds. On the contrary, inoculation with *W. anomalus* Y12 increased their concentrations with respect to the spontaneous process, mainly of 1-butanol, 2-phenylethyl acetate, ethanol, and 2-methyl-1-butanol. Furthermore, biplot and biclustering analyses segregated fermentations inoculated with Lp13 and Y12 from the rest of the processes. The use of sequential lactic acid bacteria and yeasts inocula, or their mixture, in Spanish-style green table olive fermentation could be advisable practice for producing differentiated and high-quality products with improved aromatic profile.

Keywords: table olives; starter cultures; GC-MC analysis; volatile composition

1. Introduction

Table olives are a fermented vegetable with a pronounced influence on the Mediterranean diet and culture. Nowadays, worldwide production exceeds 2.5 million tons/year [1]. Due to the presence of oleuropein, the fresh fruits are strongly bitter and, therefore, should be appropriately conditioned before consumption. The most common processing styles are: (a) alkali-treated green olives (Spanish-style); (b) ripe olives, obtained by oxidation in an alkaline medium (Californian style); and (c) directly brined olives (Greek style) [2].

Lactic acid bacteria (LAB) are the main beneficial microorganisms found in the fermentation of Spanish-style green table olives [3], but yeasts are also always present and provide exciting technological and probiotic features [4]. Together, they form stable biofilms on the olive surface [5–8] which are ingested by consumers. As a result, the interest in multifunctional starters with adequate technological properties and probiotic potential, as well as the efforts for finding synergy between these two groups of microorganism, has strongly increased. LAB stabilise the product through the production of lactic acid, which lows the pH, whereas the production of enzymes such as lipase and esterase may contribute to the biological hydrolysis of bitter compounds [7,9]. Simultaneously, yeasts improve organoleptic quality [4]. Then, their coexistence provides the table olives with an attractive sensory appeal in which the presence of volatile organic compounds (VOCs) plays an unquestionable role. Such substances are produced, in both the fruit matrix and brine, by the action of endogenous enzymes, such as lipoxygenases, or exogenous, released by the microorganisms.

There are several studies related to the determination of VOCs in table olives. Sabatini and Marsilio [10] studied the volatile profile in Spanish-style, Greek-style and Castelvetrano-style green olives of the Nocellara del Belice cultivar. The twenty-two VOCs formed during this olive fermentation were significantly affected by the type and time of processing. López-López et al. [11] studied the sensory profile and volatile composition of 24 samples of Spanish-style green table olives, identifying a total of 133 VOCs and finding a trend to separate samples according to sampling time whereas the segregation by olive cultivar was poor. However, none of these studies associated the presence of VOCs with the use of inoculum.

Recently, De Angelis et al. [12] identified 47 different VOCs during fermentation of directly brined Bella di Cerignola table olives, reporting differences between uninoculated and inoculated treatments with *Lactobacilli* and *Wickerhanomyces anomalus*. Tufariello et al. [13] studied the influence of the type of inoculation on VOCs production in directly black table olives belonging to two Italian (Leccino and Cellina di Nardò) and two Greek (Conservolea and Kalamàta) cultivars, using sequential inoculation of native yeasts and selected LAB starter. De Castro Sánchez et al. [14] reported the influence of the inoculation with *Lactobacillus* on the volatile composition of Manzanilla green olives. Among a considerable number of new VOCs, a remarkable amount of 4-ethylphenol was detected in inoculated olives compared to the uninoculated processes. The same group related the formation of some VOCs with the presence of *Propionibacterium* and *Clostridium* genera in spoilt Spanish-style green table olives [15]. Pino et al. [16] determined the influence of the inoculation with *Lactobacillus plantarum* and *Lactobacillus paracasei* on the VOCs composition of directly brined Sicilian table olives, finding differences between spontaneous and inoculated processes. All these studies show that the addition of starter cultures could have a marked influence on the VOC composition of fermented olives.

The working hypothesis of this study was to support that the VOCs profile of olive fermentation may be modulated by the addition of starter culture. For this purpose, gas chromatography-mass spectrometry (GC-MS) analysis was used for the analysis of VOCs and diverse multivariate statistical techniques were applied for studying the results.

2. Materials and Methods

2.1. Olive Fermentations

Fermentations were carried out in the 2017/2018 season using olives from Manzanilla variety, processed according to the Spanish-style in cylindrical fermentation vessels (9.5 kg olives/5 L liquid). To hydrolyse the oleuropein, fruits were treated with a solution containing 32.4 g/L NaOH, 21.9 g/L NaCl and 0.89 g/L $CaCl_2$ (97% purity), for 7 h, until NaOH penetrated 2/3 pulp. To remove the excess of alkali, fruits were washed in tap water for 5 h. Then, olives were placed in fermentation brines containing 120 g/L (w/v) NaCl, 1.3 g/L $CaCl_2$ and 0.012 L de HCl. After performing all these operations in the industry, the fermentation vessels were transported to the pilot plant of the Instituto de la Grasa (CSIC, Sevilla, Spain) for their inoculation, fermentation and analysis.

2.2. Experimental Design

Two strain of *L. pentosus* (LPG1, Lp13), one of *L. plantarum* (Lpl15) and the yeast strain *W. anomalus* Y12, all of them previously isolated from the biofilm of table olives, were used for single and co-inoculation experiments. Their selection was based on their technological and probiotic properties determined in previous studies [7]. The experimental design consisted of four individual inoculations of each organism (T1, for LPG1; T2, for Lp13; T3, for Lpl15; T4, for Y12), a combination of all them

(T5, for Y12+LPG1+Lp13+Lpl15), and a spontaneous process (T6). All experiments were performed in duplicate.

Previously to the inoculation, LAB strains were grown at 37 °C overnight on Man Rogosa and Sharpe (MRS) broth medium (Oxoid, Basingstoke, Hampshire, England) whereas the yeast was grown on YM broth (Difco, Le Pont de Claix, France) at 28 °C during 48 h. To favour the acclimation of inoculum, culture media were supplemented with 4% NaCl. Previous to inoculation, to remove the medium, cultures were washed and re-suspended in 0.9% sterile saline buffer. Inoculation was executed 1 day after brining for yeasts and at the 9th day for LAB to reach 5 $\log_{10}$ CFU/mL and 6 $\log_{10}$ CFU/mL in the cover brine, respectively. In the mixed treatment (T5), yeast and LAB were inoculated sequentially after the same periods and population levels, but using 1/3 for each LAB strain in the case of LAB.

2.3. Control Points of Fermentations

At the moment of LAB inoculation (9 days), and at the end of fermentation (65 days), brine were analysed to determine their main physicochemical parameters (pH, salt, free and combined acidity). LAB, yeast and *Enterobacteriaceae* populations were also determined in brine at the end of fermentation and, in the case of LAB, also before inoculation. These parameters were determined according to procedures described in Benítez-Cabello et al. [7]. Rep-PCR with GTG_5 primer and clustering analysis were used to determine LAB inoculum imposition at 19 days of fermentation when the LAB populations were at the highest level. For this purpose, 10 colonies from each treatment were randomly picked from the highest dilution and their fingerprinting compared with LPG1, Lp13, and Lpl15 profiles according to the protocol described in Benítez-Cabello et al. [7]. Each fermentation vessel was individually analysed.

2.4. Olive Brines' Sequential Extraction and GC-MS Analysis

At the end of fermentation, 100 mL of brines were removed from each treatment and stored at 4 °C until further analysis. The brines' volatile fraction was submitted to a sequential sorptive extraction with Twisters® (Gerstel, Müllheim an der Ruhr, Germany). The sequential extraction procedure was performed using two polydimethylsiloxane Twisters® in each sample, i.e., first in immersion (SBSE) and then in the headspace (HSSE) [17]. Six mL of the olive brine was placed in a 20 mL vial, and 1.8 gr of NaCl (30%) plus 8 μL of the internal standard 4-methyl-2-pentanol were added (1,044 mg/L final concentration). A special device called Twicester® was used. This device enables to position the Twister magnetically on the wall of a sample vial and, in this way, to keep it immersed and prevent it from brushing against the vial wall. Extraction by immersion was performed for 1 h, and the sample was stirred with a conventional magnetic stir bar (non-coated stir bar) at 200 rpm at room temperature during the extraction process. The headspace extraction was performed by placing a new Twister® in an open glass insert inside the vial and heating the sample in a water bath at 62 °C for 1 hour. In both cases, after extraction, the Twister® was removed with tweezers, rinsed with Milli-Q water, and dried with a lint-free tissue paper. Both Twisters® were then introduced into the same desorption tube and thermally simultaneously desorbed in a gas chromatograph/mass spectrometer (GC-MS).

Analyses were conducted using an Agilent 6890 GC system coupled up to an Agilent 5975 inert quadrupole mass spectrometer (Agilent, Santa Clara, CA, US) equipped with a Gerstel Thermo Desorption System (TDS2) and a Cooling Injector System CIS-4 PTV inlet (Gerstel, Müllheim an der Ruhr, Germany). The desorption temperature program was the following: the temperature was held at 35 °C for 0.1 min, was ramped at 60 °C /min to 250 °C and held for 5 min. The temperature of the CIS-4 PTV injector, with a Tenax TA inlet liner, was held at −35 °C using liquid nitrogen for the total desorption time and was then raised at 10 °C /s to 260 °C and held for 4 min. The solvent vent mode was used to transfer the sample to the analytical column. A J&W CPWax-57CB column with dimensions 50 m × 0.25 mm and a 0.20 μm film thickness (Agilent, Santa Clara, CA, US) was used, and the carrier gas was He at a flow rate of 1 mL/min. The oven temperature program was the following: the temperature was 35 °C for 4 min and was then raised to 220 °C at 2.5 °C/min (held 15 min). The quadrupole, source and transfer line temperatures were maintained at 150 °C, 230 °C and 280 °C, respectively. The electron ionization mass spectra in the full-scan mode were recorded at 70 eV with the electron energy in the range of 29 to 300 amu.

Compound identification was based on mass spectra matching using the standard NIST 98 library and the linear retention index (LRI) of authentic reference standards. LRIs were calculated by injecting an *n*-alkanes mixture (C_{10}–C_{40}) under identical conditions as the samples. We considered identified compound the one which mass spectrum and LRI value matched with those of standards, tentatively identified (TI) when mass spectrum matched with those from NIST mass spectral library and LRI value with literature LRI, compound with identification not confirmed when only the mass spectrum of compound matched with those from NIST library, as unknown compounds we include the compounds which mass spectrum reached a low value of probability of right identification in library search report.

2.5. Statistical Analyses

The values of relative peak area of the diverse VOCs found in the treatments were first subjected to analysis of variance (ANOVA) according to treatments. Only those who showed a significant difference between at least two fermentations conditions (Fisher's LSD post-hoc test) were used later for studying the influence of inoculation.

The contribution of the diverse inocula to the selected VOCs was also modelled by ANOVA, using treatments as explicative factors and VOCs as dependent variables (tolerance = 0.0001 and confidence interval for $p = 0.05$), with an = 0 constrain (that is, considering T6, the spontaneous process, as a standard or control). The treatments' contributions were assessed by the corresponding standardised coefficients of the explicative factors. When the contribution was positive, it was estimated that the treatment significantly contributed to the formation of the corresponding VOCs over the levels reached in the spontaneous. On the contrary, treatments with negative coefficients mean that they decreased the presence of the volatile below the level in the spontaneous process (T6).

The relationship between main microbial population or final physicochemical characteristics with the volatile profile was achieved by PLS-R, using a fast algorithm, automatic stop conditions, Jackknife (LOO) validation, as well as centred and reduction of variables. For the study, the final microbial population (LAB and yeast) and the physicochemical characteristics (pH, titratable and combined acidity) were used as independent and the VOCs as dependent variables. To notice that the physicochemical characteristics corresponded to the final conditions when the samples for the volatile profiles were taken, but they did not represent necessarily those in which the compounds were formed, although both may, in some way, summarize the overall fermentation process. The relationships between independent and dependent variables were measured by the respective standardized coefficients of the first for each VOC (the independent variable). Positive (negative) coefficients mean that the independent and dependent variables changed in the same (oppose) direction.

Besides, the relationships between treatments and volatile profile were also analysed by biplots and bicluster graphs. Biplots are an exciting tool to study simultaneously the relationship between cases and variables since they can represent both (scores and loadings) in the same plot. Both covariance (more appropriate to study the relationships among variables) and form (more useful for segregating cases) biplots were studied. Also, bicluster was suitable for simultaneously clustering observations and VOCs, therefore providing a map of their relationship.

The statistical analysis was achieved using XLSTAT v2018 (Addinsoft, Paris, France), for ANOVA and PLS analysis, and R package Multbiplot v 2018 [18], for biplot, clustering, and biclustering.

3. Results

3.1. Fermentation Process

Table 1 shows the main physicochemical characteristics of the brines at the moment of the LAB inoculation (9 days after olive brining). LAB and yeast strains were inoculated in a pH ranging from 6.19 to 6.33, titratable acidity of 0.09–0.14%, combined acidity of 0.12–0.15 Eq/L, and a salt concentration of 6.61–6.77%. At the 19th day of fermentation, LAB inoculum imposition was determined by molecular methods, finding that the frequency of isolation of the strains LPG1, Lp13 and Lpl15 in their inoculated treatments (T1, T2, and T3, respectively) were 100%. However, Lp13 also was detected (100% frequency) in the rest of the treatments (T4, T5, and T6), showing that this strain has a high ability for brine colonization. All treatments developed safe final pH values (<4.5). However, T4, inoculated only with the yeast Y12, had a particular performance since its pH was higher than the values observed for the rest of the inocula, although the final difference was only significant regarding T1. However, its titratable acidity was significantly lower than the other treatments. On the contrary, treatment inoculated with LPG1 was the most technologically efficient, reaching the significantly lowest pH value (see Table 1). NaCl concentration was similar in all treatments (7.47 ± 0.21% average).

Regarding LAB growth, they were not detected in any treatment before inoculation. At the VOCs sampling time, they have reached similar populations in all inoculated treatments, although their average value (6.94 $\log_{10}$ CFU/ml) in the spontaneous fermentation was the highest at the end of fermentation (65 days). On the contrary, no significant differences between treatments were found in the yeast population at the end of the process (5.99 ± 0.10 $\log_{10}$ CFU/mL average value). *Enterobacteriaceae* were never detected during the process.

3.2. ANOVA Analysis

A total of 131 VOCs were determined in the brines from the 12 fermentation trials (6 treatments in duplicate). Results were expressed as relative area values respect to the internal standard (see Table S1 in Supplementary Materials). The chemical group with the highest number of compounds was alcohols (32) followed by ketones (14), aldehydes (9) and volatile phenols (9). A first ANOVA screening of the VOCs according to treatments (Table S1) showed that the levels of only 71 compounds were significantly different between at least two fermentation processes. Therefore, 60 VOCs were produced regardless of the process and represent a common profile which, at least in the current fermentation conditions, could always be found and included both identified and not assigned formula components. Because in this study the interest was focused exclusively on those which presence could be attributed to the inocula, the compounds not significantly different among treatments were not considered for further analysis.

To investigate the relationships between the inoculated starter cultures and the initially significant VOCs, the ANOVA was again repeated with only these response variables, but using T6, the spontaneous process, as reference. As a result, the contribution of each treatment to the concentration of each volatile (versus the spontaneous) was evaluated through the standard coefficient of the respective models. Due to the large number of VOCs remaining in the study, only a few examples of treatments contributions to the formation of volatile will be illustrated graphically (Figure 1) while the rest are summarised (including only significant coefficients) in Table 2.

Table 1. Microbiological and physicochemical analysis of the treatments at the moment of LAB inoculation (I, 9 days) and at the end of fermentation (F, 65 days). Inoculum imposition was determined by rep-PCR with GTG_5 primer at 19 days of fermentation.

Treatment	Inoculum Strain	Inoculum Imposition	TA-I * (%)	TA-F (%)	pH-I	pH-F	CA-I ** (Eq/L)	CA-F (Eq/L)	Salt-I (%)	Salt-F (%)	LAB-F (Log_{10} cfu/mL)	Yeasts-F (Log_{10} cfu/mL)	*Enterobacteriaceae*-F (Log_{10} cfu/mL)
T1	LPG1	100%-LPG1	0.14 (±0.06) ^{a}	0.77 (±0.03) ^{a}	6.29 (±0.06) ^{a}	4.06 (±0.11) ^{a}	0.15 (±0.00) ^{a}	0.17 (±0.02) ^{a}	6.67 (±0.02) ^{a}	7.60 (±0.25) ^{a}	5.97 (±0.02) ^{a}	5.95 (±0.08) ^{a}	Nd
T2	Lp13	100%-Lp13	0.09 (±0.01) ^{a}	0.79 (±0.00) ^{a}	6.31 (±0.11) ^{a}	4.18 (±0.00) ^{a,b}	0.15 (±0.00) ^{a}	0.15 (0.00) ^{a}	6.44 (±0.37) ^{a}	7.30 (±0.09) ^{a}	5.98 (±0.25) ^{a}	5.93 (±0.10) ^{a}	Nd
T3	Lpl15	100%-Lpl15	0.18 (±0.11) ^{a}	0.78 (±0.04) ^{a}	6.19 (±0.16) ^{a}	4.21 (±0.02) ^{a,b}	0.15 (±0.01) ^{a}	0.16 (±0.01) ^{a}	6.61 (±0.02) ^{a}	7.21 (±0.03) ^{a}	6.22 (±0.06) ^{a}	6.11 (±0.04) ^{a}	Nd
T4	Y12	100%-Lp13	0.10 (±0.00) ^{a}	0.58 (±0.05) ^{b}	6.29 (±0.00) ^{a}	4.38 (±0.09) ^{b}	0.15 (±0.00) ^{a}	0.14 (±0.00) ^{a}	6.62 (±0.06) ^{a}	7.80 (±0.10) ^{a}	6.23 (±0.01) ^{a}	6.01 (±0.17) ^{a}	Nd
T5	Y12+Lp13+ LPG1+Lpl15	100%-Lp13	0.11 (±0.01) ^{a}	0.76 (±0.03) ^{a}	6.33 (±0.02) ^{b}	4.13 (±0.01) ^{a,b}	0.15 (±0.01) ^{a}	0.15 (±0.01) ^{a}	6.77 (±0.09) ^{a}	7.50 (±0.17) ^{a}	6.15 (±0.08) ^{a}	6.04 (±0.06) ^{a}	Nd
T6	Control	100%-Lp13	0.12 (±0.06) ^{a}	0.81 (±0.01) ^{a}	6.22 (±0.05) ^{a}	4.18 (±0.06) ^{a,b}	0.12 (±0.06) ^{a}	0.15 (±0.00) ^{a}	6.63 (±0.15) ^{a}	7.43 (±0.31) ^{a}	6.94 (±0.07) ^{b}	5.92 (±0.04) ^{a}	Nd

* TA. Titratable Acidity, ** CA. Combined Acidity, Nd. Not detected. LAB were absent before inoculation. Different superscript letter, within the same column, are significantly different ($p \leq 0.05$) according to post-hoc comparison test.

Table 2. Contribution of treatments (inoculation with different LAB and yeast species) to the production of the different VOCs found in brine at the end of the fermentation as assessed by their standardized effects. Only 71 significant compounds (from a total of 131) with differences between at least two treatments were used for these analyses. Contribution of treatments for the different VOCs was compared with respect to the spontaneous fermentation (T6). T1 stands for treatment inoculated with LPG1, T2 with Lp13, T3 with Lpl15, T4 with Y12, and T5 with Y12+LPG1+Lp13+Lpl15.

		Contributor (ANOVA) and Sign (Standaridised Coefficient)						PLS-R Analysis. Significant Coefficients		
Compound	Code	T1	T2	T3	T4	T5	Pooled SD	TA *	pH	CA **
Methyl acetate	A	-	-	-	0.710	0.703	0.155	−0.351 (±0.092)	-	−0.171 (±0.046)
Ethyl acetate	B	-	-	-	0.957	0.470	0.053	−0.464 (±0.199)	0.345 (±0.114)	−0.226 (±0.104)
***cis*-3-Hexenyl acetate**	C	-	−0.542	-	-	-	0.226	−0.187 (±0.061)	-	−0.091 (±0.028)
2-Phenylethyl acetate	D	0.596	-	-	0.970	0.625	0.024	-	-	-
3-Methylbutanoic acid	E	-	-	-	-	0.725	0.185	-	-	-
Methanol	F	−0.341	-	0.771	−0.242	-	0.112	-	-	-
Ethanol	G	-	-	-	0.750	-	0.147	−0.459 (±0.144)	0.341 (±0.135)	−0.223 (±0.072)
2-Butanol	H	0.688	-	-	0.377	0.764	0.132	-	-	-
2-Methyl-1-propanol	I	-	−0.819	-	-	−0.819	0.152	-	-	-
1-Butanol	J	0.501	0.413	0.351	1.208	0.753	0.048	-	0.308 (±0.103)	-
2-Methyl-1-butanol	K	-	−0.346	-	0.798	-	0.090	-	0.348 (±0.118)	−0.227 (±107)
3-Methyl-1-butanol	L	-	−0.430	-	0.738	-	0.101	−0.449 (±0.179)	0.334 (±0.121)	−0.219 (±0.088)
3-Methyl-3-buten-1-ol	M	0.684	−0.200	-	0.587	0.650	0.079	-	-	-

Table 2. *Cont.*

		Contributor (ANOVA) and Sign (Standaridised Coefficient)						PLS-R Analysis. Significant Coefficients		
1-Pentanol	N	-	-	-	0.816	0.436	0.123	−0.351 (±0.159)	-	−0.171 (±0.069)
***cis*-2-Penten-1-ol**	*O*	0.774	-	-	0.760	0.765	0.046	-	-	-
2-Methyl-2-buten-1-ol	P	0.613	-	-	0.631	0.502	0.160	-	-	-
1-Hexanol	Q	0.399	-	-	0.637	0.539	0.146	-	-	-
***cis*-3-Hexen-1-ol**	*R*	0.564	−0.301	-	0.464	0.604	0.108	-	-	-
2-Methyl-3-hexanol	S	0.611	-	-	0.881	0.389	0.142	-	-	-
1-Heptanol	T	-	−0.428	-	0.744	-	0.117	−0.394 (±0.160)	+0.293 (±0.126)	−0.192 (±0.071)
6-Hepten-1-ol	U	-	-	-	0.749	-	0.165	−0.421 (±0.164)	+0.313 (±0.133)	−0.205 (±0.081)
***cis*-5-Octen-1-ol**	*V*	-	−0.385	-	0.615	-	0.178	−0.376 (±0.132)	+0.280 (±0.120)	−0.183 (±0.062)
Benzyl alcohol	W	0.283	−0.282	-	0.788	0.433	0.090	-	+0.237 (±0.099)	-
2-Phenylethanol	X	0.558	−0.339	−0.122	0.665	0.466	0.038	-	-	-
2-Ethenyl-2-butenal	Y	-	-	0.370	−0.412	−0.412	0.147	-	−0.221 (±0.064)	-
Isoxylaldehyde	Z	-	-	0.670	-	-	0.212	-	-	-
Dimethyl Sulfoxide	AA	-	-	-	-	-	0.212	-	-	-
β-Damascenone	AB	−0.667	-	-	-	-	0.201	−0.348 (±0.147)	-	−0.169 (±0.068)
Ethyl lactate	AC	-	-	-	0.470	-	0.191	-	-	−0.111 (±0.045)
Ethyl 5,6-dimethylnicotinate	AD	-	-	0.129	0.981	-	0.045	-	+0.369 (±0.122)	-
Unknown ester (m/z 88)	AE	−0.732	-	-	−0.767	-	0.178	-	-	-
Furfuryl methyl ether	AF	-	−0.349	-	-	−1.031	0.103	-	-	-
Acetoin	AG	−0.358	-	−0.328	−0.319	0.567	0.143	-	-	-
6-Methyl-3,5-heptadien-2-one	AH	-	-	0.363	0.495	-	0.183	−0.344 (±0.120)	+0.256 (±0.084)	−0.167 (±0.066)
Purpurocatecho	AI	-0.151	−0.251	-	−0.887	−0.887	0.065	-	-0.259 (±0.077)	-
Iridomyrmecine	AJ	-	−0.496	-	0.324	−0.688	0.118	-	-	-
Methyl lactate	AK	-	-	-	−0.864	-	0.132	-	-0.328 (±0.108)	-
Methyl hydrocinnamate	AL	-	-	-	−0.673	-	0.217	-	-	-
Methyl 4(methylamino)benzoate	AM	-	-	-	0.786	-	0.136	−0.373 (±0.142)	+0.278 (±0.108)	−0.182 (±0.076)
3-Ethylpyridine	AN	−0.457	-	-	-	−0.526	0.182	-	-	-
4-Methylguaiacol	AO	0.347	−0.483	-	0.662	-	0.100	-	-	-
4-Ethylguiacol	AP	0.705	−0.263	−0.196	0.608	0.216	0.067	-	-	-
4-Ethylphenol	AP	-	-	0.632	−0.407	-	0.149	-	-	-
Isovanillic acid	AR	−0.239	−0.551	-	0.260	−0.739	0.111	-	+0.184 (±0.081)	-
Coumaran	AS	−1.093	−0.898	−0.571	−7.460	−0.903	0.109	-	-	-
5-tert-Butylpyrogallol	AT	-	−0.610	−0.455	0.781	−0.371	0.097	-	-	-
Methoxyeugenol	AU	−0.247	−0.430	-	0.366	−0.723	0.105	-	+0.220 (±0.092)	-
Vainillin	AV	-	-	−0.692	−0.799	−0.529	0.187	+0.320 (±0.075)	-	+0.156 (±0.036)

Table 2. *Cont.*

		Contributor (ANOVA) and Sign (Standaridised Coefficient)						PLS-R Analysis. Significant Coefficients		
α-Isophorone	AW	−0.481	−0.584	-	-	−0.513	0.200	-	-	-
α-Terpineol	AX	-	-	-	-	0.572	0.198	-	-	-
Geraniol	AY	0.712	−0.297	-	-	0.546	0.127	-	-	-
Unknown A (m/z 71-59)	AZ	-	-	-	-	−0.627	0.187	-	-	-
Unknown B (m/z 123-138-96)	BA	-	−0.402	-	0.590	-	0.167	-	+0.279 (±0.107)	-
Unknown C (m/z 83-112-97)	BB	0.270	-	-	0.974	-	0.098	-	+0.300 (±0.137)	-
Unknown D (m/z 55-93-108)	BC	0.151	-	-	1.005	0.506	0.037	-	+0.329 (±0.116)	-
Unknown E (m/z 111-198)	BD	-	−0.326	-	0.808	-	0.125	-	+0.309 (±0.115)	-
Unknown F (m/z 95-154-110)	BE	0.261	-	-	0.940	0.665	0.085	−0.376 (±0.159)	−0.183 (±0.072)	-
Unknown G (m/z 138)	BF	−0.463	−0.664	-	-	−0.716	0.153	-0.250 (±0.106)	−0.121 (±0.051)	-
Unknown H (m/z 113-81-153)	BG	0.548	−0.457	-	0.518	-	0.127	-	-	-
Unknown I (m/z 99-139-67-81)	BH	-	−0.413	-	−0.807	-	0.179	-	−0.296 (±0.129)	-
Unknown K (m/z 93-79)	BI	-	−0.414	0.289	-	−0.754	0.123	-	-	-
Unknown L (m/z 222-43-85-177)	BJ	−0.822	−0.612	-	−0.576	−0.776	0.218	-	-	-
Unknown M (m/z 138-120)	BK	−0.629	−0.799	-	-	−0.710	0.186	-	-	-
Unknown N (m/z 151-43)	BL	-	-	-	−0.894	-	0.157	-	−0.344 (±0.113)	-
Unknown O (m/z 95-110-138)	BM	−0.704	−0.859	-	−0.721	−0.370	0.195	-	-	-
Unknown P (m/z 138)	BN	−0.669	−0.644	-	-	−0.694	0.172	-	-	-
Unknown Q (m/z 102-55-69)	BO	−0.208	−0.194	-	0.866	0.235	0.059	−0.497 (±0.185)	+0.70 (±0.130)	−0.242 (±0.097)
Unknown S (m/z 167-121)	BP	0.364	−0.613	-	-	-	0.157	-	-	-
Unknown T (m/z 70-55-82)	BQ	-	-	-	−0.590	-	0.217	-	−0.245 (±0.065)	-
Unknown U (m/z 119-159-192)	BR	−0.457	-	-	−0.928	−0.528	0.181	-	-	-
Unknown W (m/z 121-136-161)	BS	-	−0.761	-	-	-	0.193	-	-	-

Notes: LAB and yeast columns were removed from the PLS-R information since these variables were never significant. * TA, titratable acidity; ** CA, combined acidity. In parenthesis standard errors.

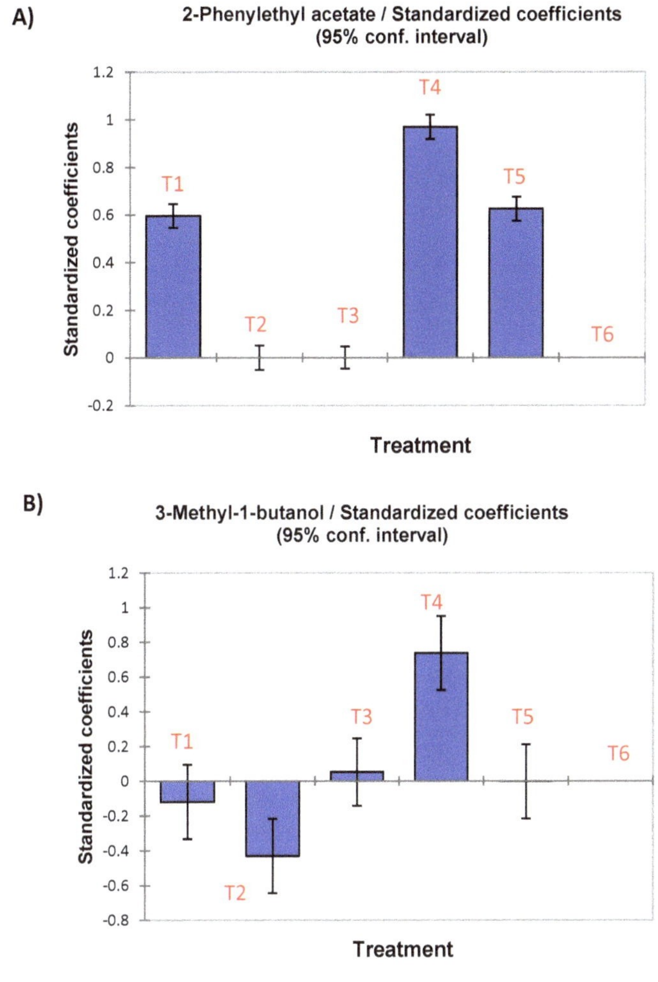

Figure 1. Standardised coefficients obtained after the ANOVA analysis (a_n = O, equivalent to stablish T6 treatment, spontaneous fermentation, as standard) for two selected VOCs: (**A**) 2-phenylethyl acetate, and (**B**) 3-methyl-1-butanol.

In the case of 2-phenylethyl acetate (Figure 1A), the treatments 4 and 5 promoted the formation of this compound due to the presence of the yeast in the inoculum. Interestingly, this compound was stimulated not only by the presence of the yeast but also LPG1 led to an important contribution and, therefore, the presence of this LAB did not interfere with its possible formation but even stimulated it. However, this effect was not observed in the treatments inoculated with Lp13 or Lpl15, since their 2-phenylethyl acetate contents were similar to those in the spontaneous process. Hence, the behaviour of LPG1 differs from Lp13 and Lp115 regarding the formation of 2-phenylethyl acetate. Similarly, 3-methyl-1-butanol was promoted by the presence of the yeast (Figure 1B); however, Lp13 had a marked negative effect on its formation, which is also reflected in T5, where the joint presence of Lp13 and Y12 has practically prevented its presence. The effects of inoculation treatments on these two volatile substances are reflected in Table 2 with a sign (+ means promotion or increase versus the spontaneous process whereas - indicates prevention or decrease) and the coefficient value (the large the value the most important the effect while the absence of data means not significant contribution). A similar methodology was also followed for the other 69 compounds. It should be noted that due to their standardization, the contributions are independent of the volatile concentrations. Besides, due to the large number of compounds, only an overview of the inoculation with the diverse LAB, yeast, and their mixtures can be commented on. For a detailed relationship for specific compounds, please see Table 2. Overall, inoculation with LGP1 (T1) reduced (negative sign) the production of several VOCs with respect to the control (T6, spontaneous fermentation), with methanol, β-damascenone, and other unknown volatiles among them. On the contrary, it promoted (by itself or by allowing its formation by the spontaneous yeasts, by chemical reaction, or a combination of transformations pathways) of many others like 2-phenylethyl acetate, 2-butanol, 1-butanol, 3-methyl-3-buten-1-ol, *cis*-2-penten-1-ol or 2-methyl-3-hexanol. Therefore, the analysis of VOCs was useful to study the influence of LPG1 presence on, at least, an aspect (VOCs) of the metabolomic related to the fermentation process.

Particularly interesting was the effect of the inoculation with Lp13 strain. In this case, almost all standardized coefficients had a negative sign (only that for 1–butanol was positive); that is, its presence had an important effect on the volatile composition reduction. On the contrary, the inoculation with the Lpl15 had an almost neutral effect on the formation of the VOCs since the significant coefficients were very reduced and had both positive and negative signs; however, it promoted the formation of methanol, isoxylaldehyde and 4-ethylphenol, while reducing that of coumarin, 5-*tert*-butylpyrogalol and vanillin.

A radically opposed behaviour was shown by yeast inoculated treatments. The inoculation with Y12 was determinant for increasing dramatically the concentration of most of the VOCs over the spontaneous process (T6) with only a few coefficients with a negative sign. Among the compounds which formation promoted inoculation with Y12 were 1-butanol, ethanol, methyl acetate, ethyl acetate, 2-phenylethyl acetate, or 2-methyl-1-butanol, to mention only a few of them; but, it depressed the levels of methanol, coumarin, and vanillin. Therefore, it was evident that the inoculation with only the yeast increased the amount of VOCs of the fermented olives. In most cases, this increase was inversely related to the production of free acidity, combined acidity, and the subsequent high pH (Table 2, PLS regression). This inverse relationship shows a competence between the productions of one or several compounds *vs* the others. Finally, when the yeast was inoculated together with the rest of LAB strains (T5), the volatile compounds content was more abundant than in the case of the spontaneous treatment, although some of the compounds found in the presence of the yeast like ethanol, 1–heptanol, or *cis*-5-octen-1-ol were reduced with respect to T4 and remain similar to T6 (spontaneous), possibly due to the competence of the LAB also present in T5. The effect on some VOCs like 2-methyl-1-butanol, 3-methyl-1-butanol, 1–heptanol, or *cis*-5-octen-1-ol could be directly associated with Lp13 presence, which did not promote their production. Therefore, the use of the only LAB in the starter cultures decreased or not affected the production of VOCs while the inoculation of only Y12 increased them. However, when both groups of microorganisms were mixed (T5), the yeast metabolites were affected (Table 2), revealing a competence between both groups.

3.3. PLS Analysis

The overall PLS-R model quality (one component) was reduced since Q^2cum explained low variances of both independent ($Q^2X = 0.404$) and dependent variables ($Q^2Y = 0.305$), although it may also be due, at least partially, to the noise introduced when working with numerous non-significant variables. Overall, the most influential variables in the model (Figure 2A) were titratable acidity, pH, and combined acidity (which in table olives are always strongly related to the first two) while LAB and yeasts counts were never significant, although the relationships could only be established with a reduced number of VOCs. An example of the coefficients is shown in Figure 2B, which corresponds to ethanol. The negative sign for titratable and combined acidities mean that high production of them (and their associated low pH), lead to low ethanol production (lactic acid fermentation predominated over yeast fermentation and in some way reduced the ethanol production). This opposed trend between these physicochemical characteristics and volatile composition was common to most compounds but, especially, to alcohols (Table 2). However, numerous compounds were also unaffected (2-phenylethyl acetate, 3-methylbutanoic acid or dimethyl sulfoxide), indicating a possible compatible, metabolic pathway or absence of competence for the nutrient/substrates between both LAB and yeasts (Table 2), with several of them being related to alcohols as well (e.g. 1-hexanol or *cis*-3-hexen-1-ol). Only in the case of vanillin, the production of lactic acid did not lead to a decrease in its formation.

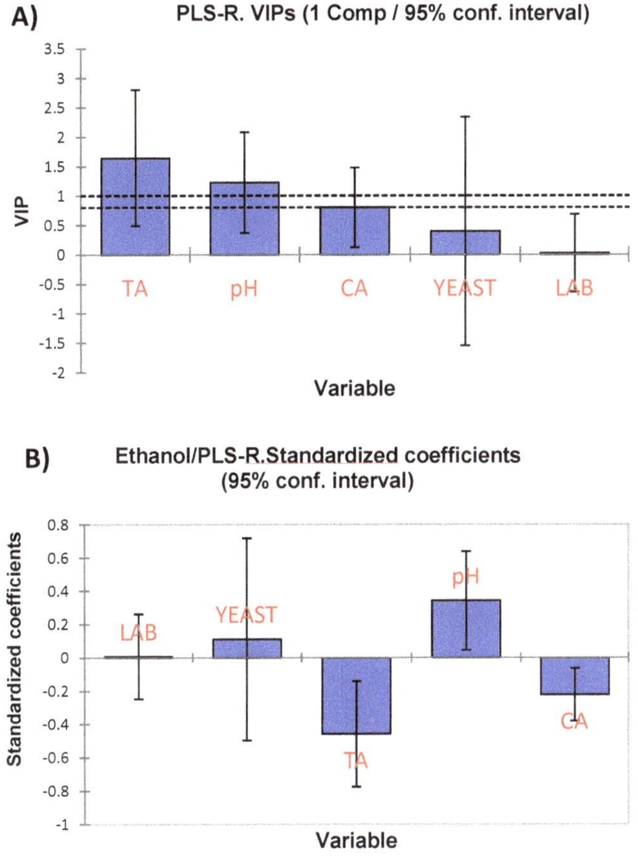

Figure 2. PLS-R analysis, using final physicochemical and microbiological characteristics as independent variables and VOCs as dependent. (**A**) Variable importance for the projection and (**B**) Significant standardised coefficients for the presence of ethanol.

3.4. *Biplots Analysis*

The analysis showed that two or three Factors accounted for 64.1 and 75.4% of the variance, respectively. Most of the treatments (cases) were well represented (big size names) onto the first two factors plane (Figure 3A) while T1 and T6 treatments were better associated, at least partially, to F2 or F3 (Figure 3B). Besides, the study showed that four clusters produced the best segregation among fermentations. In this case, the boundaries delimited by the Voronoi lines help to recognize the appropriate influential areas and to ascribe VOCs to fermentation clusters (Figure 3).

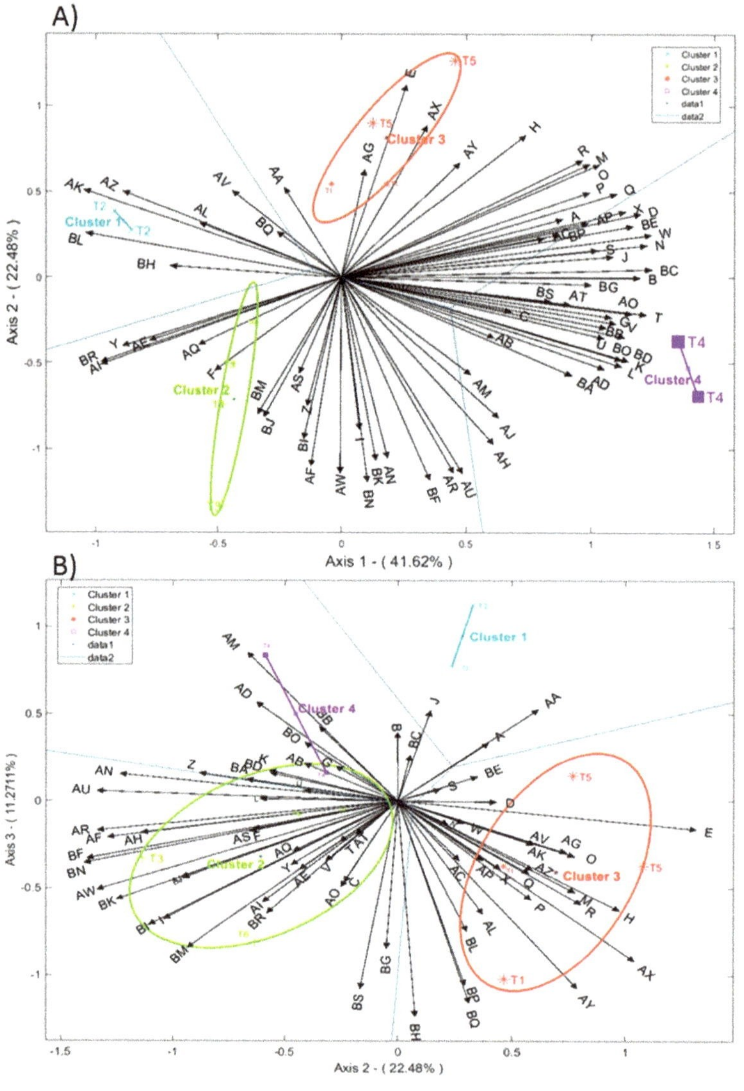

Figure 3. Biplot, including clustering and Voronoi borders, representing the projections of cases scores and variables loadings onto axis F1 vs F2 (**A**) and F2 vs F3 (**B**). Contributions of cases are proportional to the size of symbols and letters (see Table 2 for the meaning of symbols) while those of VOCs (see Table 2 for codes) are proportional to the length of their arrows.

As shown by the big sizes of their names, T2, T3 (one replicate), T4 and T5 are well represented on the F1/F2 plane. Similarly, variables with large arrows are better represented than those with the shortest lengths. The plot shows that T2 treatment (inoculated with Lp13) and T4 (Y12) represent two very different fermentation volatile profiles with T2 being characterized by a scarce volatile content (methyl lactate (AK in the graph), and unknowns N (BL) and A (AZ), in slightly lower proportions) while T4 was abundant in many of them (4-methylguaiacol (AO)), 1-heptanol (T), unknowns B (BA), D (BB), F (BE), Q (BO), E (BD), 2-methyl-1-butanol (K), 3-methyl-1-butanol (L), ethyl 5,6-dimethyl nicotinate (AD), 6-hepten-1-ol (U), *cis*-5-octen-1-ol (V), ethanol (G), ethyl acetate (B), 1-pentanol (N), benzyl alcohol (W), 2-phenylethyl acetate (D), and 2-phenylethanol (X)). The other two treatments show intermediate values of these two volatile profiles plus some representative compounds of their fermentations. Thus, T3 and T6 (in lower extension) were also characterized by unknown U (BR) and P (BN), purpurocatechol (AI), furfuryl methyl ether (AF), α-isophorone (AW), unknown M (BK), isovanillic acid (AR), or methoxyeugenol (AU). Besides, T5 (which in the ANOVA was also identified with abundant VOCs and it is also at a large distance from the origin) have a reduced number of characteristic compounds as would correspond to treatments inoculated with the mixture of microorganisms. In this case, only 3-methylbutanoic acid (E) could be the most representative while also may participate other VOCs like 2-butanol (H), *cis*-3-hexen-1-ol (R), 3-methyl-3-buten-1-ol (M), *cis*-2-penten-1-ol (O), 2-methyl-2-buten-1-ol (P), or 1-hexanol (Q) which are included in its Voronoi area. On the contrary, T1 (inoculated with LPG1) was not well represented and will be commented later.

In the plane F1/F3, T1 (one replicate), T5, T6 or T3 had very poor contributions. However, in F2/F3 (Figure 3B), T5 and T1 (one duplicate) were well represented, but the other replicate of T1 was still close to the centre, indicating that, overall, this replicate had a low representation. A similar situation was also observed for one replicate of T6 and T3. Therefore, in the F2/F3 plane, the volatile compounds best related to T3 and T6 (in lower proportion) were: 3-ethylpyridine (AN), methoxyeugenol (AU), 4-ethylguiacol (AP), furfuryl methyl ether (AF), unknowns G (BF), P (BN), M (BK), K (BI), O (BM), and α-isophorone (AW); however, as T3 is closely related to T6 this means that the inoculation with Lpl15 produce, in general, quite similar volatile compounds than T6 (spontaneous process). Besides, T1 and T5 may be related to 3-methylbutanoic acid (E), α-terpineol (AX), and geraniol (AY), with the last two compounds being better represented in this plane than in that of F1/F2, where their contributions were markedly lower.

Therefore, overall, the biplot showed that most of the VOCs were associated with the F1 axis (most positively and a few of them negatively). Therefore, this axis was the most influential for the treatment segregation, particularly between T4 (rich in many volatile compounds, inoculated with Y12) and T2 (Lp13, abundant in only a few components) while the rest of the treatments were more similar, particularly T1 (inoculated with LPG1) and T5 (with mixture of LAB and Y12), and limited regarding their contributions to volatile compounds. The reduction of volatile composition in T5 could have been caused by Lp13 who, in the ANOVA table, showed a clear negative effect on the formation of volatile compounds. On the contrary, the F2 axis was associated with a reduced number of VOCs both positively (3-methylbutanoic acid (E), α-terpineol (AX) and geraniol (AY)) and negatively (furfuryl methyl ether (AF), α-isophorone (AW), unknowns P (BN), G (BF), isovanillic acid (AR), or methoxyeugenol (AU)), which were linked to T5 and T1 and, T3 and T6, respectively. Therefore, F2 axis was also efficient for segregating these two groups, although one should always have in mind that, in half of the T1, T3 and T6 replicates, the presence of the volatile compound was not particularly relevant.

Regarding relationships among variables, they could be deduced from the angles of their respective arrows. It is evident that those pointing to the right in Figure 3A are related, and their production may be assigned to Y12. On the contrary, those looking towards the left (methyl lactate (AK), unknowns N (BL), U (BR), or purpurocatechol (AI) are strongly related among them and possibly linked to the metabolic pathways of the strain Lp13. Non-related to these VOCs might be those variables pointing up in the plot (Figure 3A) like 3-methylbutanoic acid (E)) and down as furfuryl methyl ether (AF), α-isophorone (AW), unknowns P (BN), G (BF), isovanillic acid (AR), methoxyeugenol (AU), or 6-methyl-3,5-heptadien-2-one (AH), with these two groups showing, in turn, opposed relationship among them. Besides, strong relationships may be observed in Figure 3B for volatile compounds pointing to the left (associated with T3 and T6) and right (linked to T5 and T1), but opposed between them.

3.5. Biclustering Analysis

An appropriate presentation of the whole relationship between VOCs and treatments may also be achieved by biclustering; that is, according to treatments and VOCs simultaneously (Figure 4).

The clustering of treatments also led to four clusters (indicated as b1-4) while the VOCs were grouped into four other big clusters (v1-4). The first (b1) was composed of only T2 treatment (inoculated with Lp13); the second (b2) consisted of T3 (Lpl15) and T6 (spontaneous); the third (b3) included T1 (LPG1) and T5 (mixture of Y12 and LAB strains), with a possible segregation between them; and the forth (b4) was devoted to only T4 (Y12). Therefore, this segregation was similar to that achieved in the biplot analysis where only Lp13 (T2) and Y12 (T4) led to individually differentiated VCOs profiles. Combining this segregation with the volatile compounds, it may be observed that the profile of T4 (Y12) was characterized by the high production of compounds such as methyl 4 (methylamino) benzoate (AM), ethyl 5,6-dimethylnicotinate (AD), unknowns B (BA), C (BB), D (BC), E (BD), F (BE), G (BF), H (BG), W (BS), Q (BO), and S (BP), ethanol (G), 6-hepten-1-ol (U), 2-methyl-1-butanol (K), 3-methyl-1-butanol (L), β-damascenone (AB), *cis*-3-hexenyl acetate (C), 5-tert-butylpyrogallol (AT), *cis*-5-octen-1-ol (V), 4-methylguaiacol (AO), 1-heptanol (T), 1-butanol (J), methyl acetate (A), 2-methyl-3-hexanol (S), 2-phenylethyl acetate (D), benzyl alcohol (W), 1-pentanol (N), ethyl lactate (AC), 4-ethylphenol (AP), 2-phenylethanol (X), 1-hexanol (Q), *cis*-2-penten-1-ol (O), 3-methyl-3-buten-1-ol (M), *cis*-3-hexen-1-ol (R), 2-methyl-2-buten-1-ol (P), 3-ethylpyridine (AN), methoxyeugenol (AU), isovanillic acid (AR), 6-methyl-3,5-heptadien-2-one (AH), and iridomyrmecine (AJ). On the opposite side is the T2 treatment (inoculated with Lp13), which is low or minimal in most of the compounds but high in dimethyl sulfoxide (AA), vanillin (AV), unknown A (AZ), N (BL), methyl lactate (AK), and 2-ethenyl-2-butenal (Y). The cluster consisting of T6 and T3 treatments is low or minimal in VOCs clustered in v1 while high or moderated in those included in v2. A more detailed specific relationship may be read directly from the graph (Figure 4).

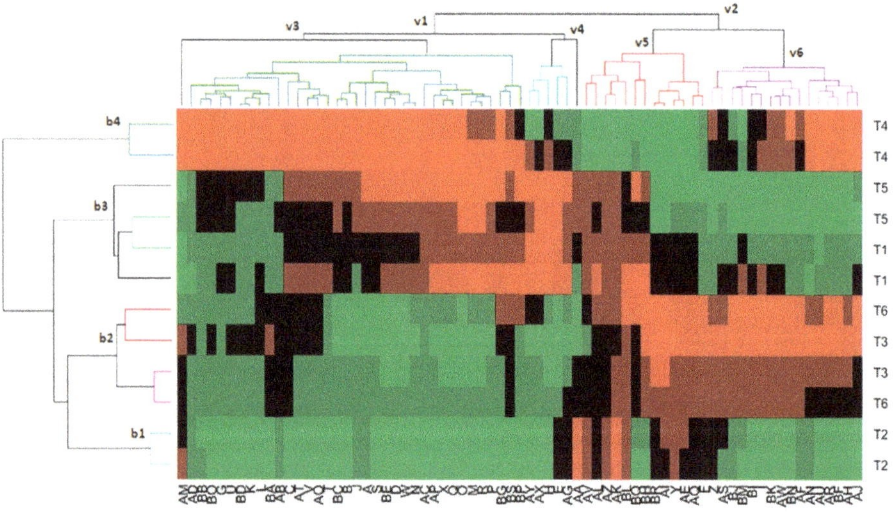

Figure 4. Bicluster plot of the relationship among treatments and VOCs. The presence of volatile is proportional to the colour scale, ranging from red (major) to green (low). See Table 2 for the meaning of the symbols for treatments and the codes of the VOCs. b1-b4 and v1-v4 refer to cluster from treatments and VOCs, respectively.

4. Discussion

In this work, a total of 131 VOCs, formed during Spanish-style table olive fermentation inoculated with diverse LAB and yeast native strains, have been determined using GC-MS analysis. Panagou and Tassou [19] studied through GC analysis the VOCs during the fermentation of Conservolea variety green olives inoculated with *L. pentosus* and *L. plantarum* strains, finding that ethanol, methanol, acetaldehyde, ethyl acetate, and isobutyric acid were the major VOCs identified during fermentation, some of them also found in this work. Recently, Cosmai et al. [20] applied SMPE/GC-MS analysis to study the VOCs of directly brined green table olives from *Bella di Cerignola variety* in treatments inoculated with *W. anomalus* and strains of *L. pentosus* and *L. plantarum*. They specially reported higher levels of lipoxygenase pathway-derived compounds as 1-hexanol or *cis*-3–hexen-1-ol in treatments inoculated with the yeast in which these compounds were overrepresented in treatments inoculated with *W. anomalus* Y12. In this paper, similar results were obtained for the last compound. Tufariello et al. [13] reported that the use of sequential inoculation of yeast and *Lactobacilli* species in directly brined olives affected VOCs. Alcohol and ester contents increased during starter-driven fermentations, but with significant differences among olive cultivars, and always in higher concentrations than in the corresponding spontaneous fermentations. No variation of hydrocarbons and terpenes was detected between spontaneous and starter-driven fermentations.

One of the most desirable objectives of designing an inoculum is to improve the organoleptic profile of olive fermentations, especially aroma [21]. In this work, very relevant differences between the VOC levels in the brine obtained after fermentation processes appear to depend on the microorganism used as inoculum, especially when yeasts are involved in the fermentation process. Hence, Sabatini et al. [22] observed that ethanol was produced in brine medium mostly by yeasts fermentation (alcoholic fermentation) and, in a lesser extent, during lactic acid fermentation (heterolactic fermentation). Our results are in agreement with them, and ethanol was produced mainly in the brine inoculated with yeast, doubling the amount found in fermentation processes carried out by *Lactobacillus* strains. Similar results were found for other alcohols closely related to alcoholic fermentation pathways such as isoamyl alcohols (2-methyl and 3-methyl-1-butanol), or 1-butanol. Sabatini and Marsilio [10] also detected by GC/MS analysis diverse VOCs, comprising alcohols, aldehydes, ketones, esters as well as acids, formed during olive fermentation of Spanish-style, Greek-style and Castelvetrano-style green olives of the Nocellara del Belice cultivar. Their results suggested that different processing technologies significantly affected the VOCs of samples, as well as the time of processing. Recently, Pino et al. [16] using GC-MS analysis found that the addition of *L. plantarum* and *L. paracasei* as starters significantly modified the volatile profile of directly brined Sicilian table olive fermentations. Specifically, compounds responsible for fruity and floral notes, such as methyl 2-methylbutanoate and phenylethyl alcohol, highly increased, while isoamyl alcohol and ethanol decreased compared to non-inoculated samples. The high content of alcohols in un-inoculated brine samples could be related to yeast metabolic activities, which was mainly dominated by *W. anomalus*, but this yeast species were also present during LAB inoculated fermentations.

Acetic acid esters are compounds formed by condensation between acetic acid and an alcohol. *W. anomalus* yeast has been reported to be an acetic acid ester producer [23]. The significant different high contents of ethyl acetate and 2-phenylethanol acetate are another relevant result of this work. This fact showed that, in table olive fermentation, this yeast might also develop its capability to produce such kinds of esters.

4-Ethylphenol, a compound with an unpleasant aroma, could be produced during lactic acid fermentation [24,25]. On the one hand, Randazzo et al. [26] studied the VOCs produced by different *Lactobacillus* strain inocula in brines of Nocellara Etnea table olive fermentations. Among strains compared, they studied the effect of a pure culture of one *L. plantarum* strain and other *L. pentosus* strain. They did not find a significant difference concerning the production of 4-ethylphenol. However, de Castro et al. [14] suggested that 4-ethylphenol formation is strain-dependent. Our results suggest that *L. plantarum* Lpl15 strain has a high capability for the production of this volatile phenol and support the idea of strain-dependent production.

5. Conclusions

The statistical approach used in the present work has allowed identifying the main modification in the volatile profile produced by inoculation with diverse starter cultures. Our study has demonstrated that the type of inoculum modulates the volatile composition of the final product significantly. The inclusion of yeast in the inoculum increases the production of VOCs while the presence of *Lactobacillus* alone, in general, decrease the concentrations of some compounds or keep them at the same levels than in the spontaneous process. This lack of impact on the VOCs by *Lactobacillus* strains may be explained because the emphasis when selecting starters was mainly focused on the acidification and the pH lowering characteristics. However, as the process is better known from the microbiological point of view, the introduction of genomic methodologies and the application of more accurate and sophisticated methods for the identification of metabolites formed during the process could make possible the design of inocula with wider and better-identified characteristics, including their aromatic profile. Therefore, to enhance the organoleptic characteristics of final products, the inclusion of yeasts in the inoculum appears as a promising alternative. By studying in detail, the relationships between the VOCs formed and the sensory characteristics, appropriate selection of yeast could be achieved. Besides, the relationships found in this work between starter cultures and VOCs may facilitate further studies on the numerous metabolic transformation occurring in table olive fermentations.

Supplementary Materials: The following are available online at http://www.mdpi.com/2304-8158/8/8/280/s1, Table S1: Table S1. Volatile composition determined by GC-MS analysis in the brines of the different treatments assayed at the end of fermentation. T1 stand for treatment inoculated with LPG1, T2 inoculated with Lp13, T3 inoculated with Lpl15, T4 inoculated with Y12, and T5 inoculated with Y12+LPG1+Lp13+Lpl15.

Author Contributions: Conceptualization: F.R.-G., R.J.-D., F.N.A.-L., and A.G.-F.; Methodology: A.B.-C., F.R.-G., and M.L.M.; Software: A.G.-F. and M.L.M.; Validation: F.R.-G. and M.L.M.; Formal Analysis: A.G.-F. and M.L.M.; Investigation: A.B.-C., F.R.-G., F.N.A.-L. and A.G.-F.; Resources: R.J.-D. and F.N.A.-L.; Data Curation: M.L.M. and A.G.-F.; Writing—Original Draft Preparation: A.B.-C., M.L.M., F.N.A.-L., and A.G.-F.; Writing—Review & Editing: M.L.M., A.G.-F., and F.N.A.-L.; Visualization: M.L.M. and F.N.A.-L.; Supervision: F.N.A.-L.; Project Administration: R.J.D. and F.N.A.-L.; Funding Acquisition: R.J.D. and FNA.-L.

Funding: The research was funded by the Spanish Government (Project OliFilm AGL-2013-48300-R: www.olifilm.science.com.es) AB-C thanks the Spanish Ministry of Economy and Competitiveness for their FPI grant.

Acknowledgments: AGF thanks the CSIC for his "Ad honorem" appointment.

Conflicts of Interest: The authors declare no conflict of interest.

References

1. International Olive Oil Council (IOC). *World Table Olive Figures*. 2019. Available online: http://www.internationaloliveoil.org/estaticos/view/132-world-table-olive-figures (accessed on 24 July 2019).
2. Garrido-Fernández, A.; Fernández-Díez, M.J.; Adams, R.M. *Table Olives Production and Processing*; Chapman & Hall: London, UK, 1997.
3. Hurtado, A.; Requant, C.; Bordons, A.; Rozes, N. Lactic acid bacteria from fermented olives. *Food Microbiol.* **2012**, *31*, 1–8. [CrossRef] [PubMed]
4. Arroyo-Lopez, F.N.; Romero-Gil, V.; Bautista-Gallego, J.; Rodriguez-Gomez, F.; Jimenez-Diaz, R.; Garcia-Garcia, P.; Querol, A.; Garrido-Fernandez, A. Yeasts in table olive processing: Desirable or spoilage microorganisms? *Int. J. Food Microbiol.* **2012**, *160*, 42–49. [CrossRef] [PubMed]
5. Arroyo-López, F.N.; Bautista-Gallego, J.; Domínguez-Manzano, J.; Romero-Gil, V.; Rodríguez-Gómez, F.; García-García, P.; Garrido-Fernández, A.; Jiménez-Díaz, R. Formation of lactic acid bacteria-yeasts communities on the olive surface during Spanish-style Manzanilla fermentations. *Food Microbiol.* **2012**, *32*, 295–301.
6. Benítez-Cabello, A.; Romero-Gil, V.; Rodríguez-Gómez, F.; Garrido-Fernández, A.; Jiménez-Díaz, R.; Arroyo-López, F.N. Evaluation and identification of poly-microbial biofilms on natural green Gordal table olives. *Antonie Van Leeuwenhoek* **2015**, *108*, 597–610. [CrossRef] [PubMed]
7. Benítez-Cabello, A.; Calero-Delgado, B.; Rodríguez-Gómez, F.; Garrido-Fernández, A.; Jiménez-Díaz, R.; Arroyo-López, F.N. Biodiversity and multifunctional features of lactic acid bacteria isolated from table olive biofilms. *Front. Microbiol.* **2019**, *10*, 836. [CrossRef] [PubMed]
8. Grounta, A.; Panagou, E.Z. Mono and dual species biofilm formation between Lactobacillus pentosus and Pichia membranifaciens on the surface of black olives under different sterile brine conditions. *Ann. Microbiol.* **2014**, *64*, 1757–1767. [CrossRef]

9. Behera, S.S.; Ray, R.C.; Zdolec, N. *Lactobacillus plantarum* with functional properties: An approach to increase safety and shelf-life of fermented foods. *BioMed Res. Int.* **2018**. [CrossRef]
10. Sabatini, N.; Marsilio, V. Volatile compounds in table olives (Olea Europaea, L., Nocellara del Belice cultivar). *Food Chem.* **2008**, *107*, 1522–1528. [CrossRef]
11. López-López, A.; Sánchez, H.; Cortés-Delgado, A.; de Castro, A.; Montaño, A. Relating sensory analysis with SPME-GC-MS data for Spanish-style green table olive aroma profiling. *LWT Food Sci. Technol.* **2018**, *89*, 725–734. [CrossRef]
12. De Angelis, M.; Campanella, D.; Cosmai, L.; Summo, C.; Rizzello, C.G.; Caponio, F. Microbiota and metabolome of un-started and started Greek-type fermentation of Bella di Cerignola table olives. *Food Microbiol.* **2015**, *52*, 18–30. [CrossRef]
13. Tufariello, M.; Durante, M.; Ramires, F.; Grieco, F.; Tommasi, L.; Perbellini, E.; Falco, V.; Tasioula-Margari, M.; Logrieco, A.F.; Mita, G.; et al. New process for production of fermented black table olives using selected autochthonous microbial resources. *Front. Microbiol.* **2015**, *6*, 1007. [CrossRef] [PubMed]
14. De Castro, A.; Sánchez, A.H.; Cortés-Delgado, A.; López-López, A.; Montaño, A. Effect of Spanish-style processing steps and inoculation with *Lactobacillus pentosus* starter culture on the volatile composition of cv. Manzanilla green olives. *Food Chem.* **2019**, *271*, 543–549. [CrossRef] [PubMed]
15. De Castro, A.; Sánchez, A.; López-López, A.; Cortés-Delgado, A.; Medina, E.; Montaño, A. Microbiota and metabolite profiling of spoiled Spanish-style green table olives. *Metabolites* **2018**, *8*, 73. [CrossRef] [PubMed]
16. Pino, A.; Vaccalluzzo, A.; Solieri, L.; Romeo, F.; Todaro, A.; Caggia, C.; Arroyo-López, F.N.; Bautista-Gallego, J.; Randazzo, C. Effect of sequential inoculum of beta-glucosidase positive and probiotic strains on brine fermentation to obtain low salt Sicilian table olives. *Front. Microbiol.* **2019**, *10*, 174. [CrossRef] [PubMed]
17. Ubeda, C.; Callejón, R.M.; Troncoso, A.M.; Peña-Neira, A.; Morales, M.L. Volatile profile characterisation of Chilean sparkling wines produced by traditional and Charmat methods via sequential stir bar sorptive extraction. *Food Chem.* **2016**, *207*, 261–271. [CrossRef] [PubMed]
18. Vicente Villalón, J.L. *MULTIPLOT. A Package for Multivariate Analysis Using Biplots*; Departamento de Estadística, Universidad de Salamanca: Salamanca, Spain, 2016; Available online: http://biplot.usal.es/ClassicalBiplot/index.htlm (accessed on 24 July 2019).
19. Panagou, E.Z.; Tassou, C.C. Changes in volatile compounds and related biochemical profile during controlled fermentation of cv. Conservolea green olives. *Food Microbiol.* **2006**, *23*, 738–746. [CrossRef] [PubMed]
20. Cosmai, L.; Campanella, D.; De Angelis, M.; Summo, C.; Paradiso, V.M.; Pasqualone, A.; Caponio, F. Use of starter cultures for table olives fermentation as possibility to improve the quality of thermally stabilized olive-based paste. *LWT Food Sci. Technol.* **2018**, *90*, 381–388. [CrossRef]
21. Bonatsou, S.; Tassou, C.C.; Panagou, E.Z.; Nychas, G.J.E. Table olive fermentation using starter cultures with multifunctional potential. *Microorganisms* **2017**, *5*, 30. [CrossRef] [PubMed]
22. Sabatini, N.; Perri, E.; Marsilio, V. An investigation on molecular partition of aroma compounds in fruit matrix and brine medium of fermented table olives. *Innov. Food Sci. Emerg. Technol.* **2009**, *10*, 621–626. [CrossRef]
23. Viana, F.; Gil, J.V.; Genoves, S.; Valles, S.; Manzanares, P. Rational selection of non-Saccharomyces wine yeasts for mixed starters based on ester formation and enological traits. *Food Microbiol.* **2008**, *25*, 778–785. [CrossRef]
24. Cavin, J.F.; Andioc, V.; Etievant, P.; Diviès, C. Ability to wine lactic acid bacteria to metabolize phenol carboxylic acids. *Am. J. Enol. Vitic.* **1993**, *44*, 76–80.
25. Couto, J.A.; Campos, F.M.; Figueiredo, A.R.; Hogg, T.A. Ability of lactic acid bacteria to produce volatile phenols. *Am. J. Enol. Viticul.* **2006**, *57*, 166–171.
26. Randazzo, C.L.; Todaro, A.; Pino, A.; Pitino, I.; Corona, O.; Caggia, C. Microbiota and metabolome during controlled and spontaneous fermentation of Nocellara Etnea table olives. *Food Microbiol.* **2017**, *65*, 136–148. [CrossRef] [PubMed]

foods

MDPI

Article

Volatile Composition, Sensory Profile and Consumer Acceptability of HydroSOStainable Table Olives

Lucía Sánchez-Rodríguez [1], Marina Cano-Lamadrid [1], Ángel A. Carbonell-Barrachina [1,*], Esther Sendra [2] and Francisca Hernández [3]

[1] Departamento Tecnología Agroalimentaria, Grupo Calidad y Seguridad Alimentaria, Escuela Politécnica Superior de Orihuela, Universidad Miguel Hernández de Elche, Carretera de Beniel, Km 3.2, 03312 Orihuela, Spain; lucia.sanchez@goumh.umh.es (L.S.-R.); marina.cano.umh@gmail.com (M.C.-L.)

[2] Departamento Tecnología Agroalimentaria, Grupo Industrialización de Productos de Origen Animal, Escuela Politécnica Superior de Orihuela, Universidad Miguel Hernández de Elche, Carretera de Beniel, Km 3.2, 03312 Orihuela, Spain; Esther.sendra@umh.es

[3] Departamento de Producción Vegetal y Microbiología, Grupo Producción Vegetal, Escuela Politécnica Superior de Orihuela, Universidad Miguel Hernández de Elche, Carretera de Beniel, km 3.2, 03312 Orihuela, Alicante, Spain; francisca.hernandez@umh.es

* Correspondence: angel.carbonell@umh.es

Received: 15 August 2019; Accepted: 8 October 2019; Published: 10 October 2019

Abstract: HydroSOStainable table olives (cultivar Manzanilla) are produced from olive trees grown under regulated deficit irrigation (RDI) strategies. Olives produced by RDI are known to have a higher content of some bioactive compounds (e.g. polyphenols), but no information about consumer acceptance (or liking) have been reported so far. In this study, the volatile composition, the sensory profile and the consumer opinion and willingness to pay (at three locations) for HydroSOStainable table olives produced from three RDI treatments and a control were studied. Volatile composition was affected by RDI, by increasing alcohols, ketones and phenolic compounds in some treatments, while others led to a decrease in esters and the content of organic acids. Descriptive sensory analysis (10 panelists) showed an increase of green-olive flavor with a decrease of bitterness in the HydroSOStainable samples. Consumers (study done with 100 consumers in 2-rural and 1-urban locations; n_{total} = 300), after being informed about the HydroSOStainable concept, preferred HydroSOStainable table olives to the conventional samples and were willing to pay a higher price for them (52% 1.35–1.75 € and 32% 1.75–2.50 € as compared to the regular price of 1.25 € for a 200 g bag). Finally, green-olive flavor, hardness, crunchiness, bitterness, sweetness and saltiness were defined as the attributes driving consumer acceptance of HydroSOStainable table olives.

Keywords: bitterness; consumer willingness to pay; descriptive sensory analysis; green-olive flavor; "Manzanilla" cultivar; pit hardening; regulated deficit irrigation

---

1. Introduction

Many irrigation treatments have been evaluated in different crops, including olive trees, due to an increasing interest in water-sustainable and environment-friendly products by modern consumers [1, 2]. "HydroSOStainable products" are defined for the first time by Noguera-Artiaga et al. [3] as fruits and vegetables cultivated under regulated deficit irrigation (RDI) treatments [3]. Furthermore, Corell et al. [4] have defined HydroSOStainable index for olive trees agronomic conditions. The main aim for application of these types of sustainable strategies is conservation of water (a hot topic in arid farming research) and improving the content of bioactive compounds in vegetables and fruits as a defense mechanism against water stress [5–7]. However, to date, the effects of RDI on the consumer acceptability of olives has not been evaluated.

During the last decade, several studies about the effect of RDI on table olives agronomical, chemical and functional characteristics have been published [5,8–13], but none of them included consumer insights. The use of moderate RDI (reducing water irrigation in a moderate way but without neglecting irrigation) in table olive orchards led to an enhanced antioxidant capacity and higher polyphenolic content [2,14,15]. Although in those studies, an improvement in the sensory attributes of trees growing under moderate RDI was reported by a trained sensory panel, no consumer acceptance study was conducted. Consumer studies are essential to adjust the sensory profile of food products to consumer demands and needs by adjusting irrigation treatments, to identify the main buying drivers, to develop successful marketing strategies, and to determine an acceptable price for HydroSOStainable table olives. Recently, an affective study carried out in HydroSOStainable almonds [16]; the main conclusion was that RDI strategies led to similar global acceptance than conventional treatments but being sustainable with the environment by saving irrigation water. In addition, consumers were willing to pay a higher price for HydroSOStainable almonds (~2 € kg^{-1} more), which could be an argument to convince farmers to implement these water-saving irrigation technologies. The same behavior was observed in a study with HydroSOStainable pistachios [3], in which authors concluded that consumers were willing to pay approximately 1 euro more per kg of HydroSOStainable pistachio as compared to control samples.

Consequently, the aim of the present study was to evaluate consumer insights about HydroSOStainable table olives produced using different technologies and to link consumer data with descriptive sensory analysis and the contents of the volatile compounds. For that purpose, table olives coming from three RDI treatments [moderate deficit irrigation (T1), severe deficit irrigation during short time (T2) and severe deficit irrigation during long time (T3), and a control were assayed at the field, and the following analyses were conducted: (i) volatile composition by gas-chromatography, (ii) descriptive sensory analysis by a trained panel, and (iii) affective opinion of consumers and their willingness to pay.

2. Materials and Methods

2.1. Plant Material and Experimental Design

Olives were collected on September 2017 from a farm, Doña Ana, which is located in Dos Hermanas (Seville, Spain) (37° 25′N, 5° 95′W). Olive trees (cultivar "Manzanilla") were approximately 32-year-old. Irrigation was performed during the night by drip, using lateral pipes per row of trees and four emitters per plant, split between the two rows (each delivering 2 L h^{-1}). A pressure chamber (PMS Instrument Company, Albany, OR, USA) was used to measured stem water potential at midday (Ψ_{stem}). Water stress integral (SI), calculated as Myers [17] was used to describe the cumulative effect of the water deficit [18]. Three different irrigation treatments and a control were carried out:

- control (T0), trees were fully irrigated, to avoid any water stress;
- moderate deficit irrigation (T1), the threshold value for water stress level (Ψ_{stem}) was set up at −2 MPa during pit hardening stage;
- severe deficit irrigation (short time) (T2), the threshold value for Ψ_{stem} was set up at −3 MPa during half period of pit hardening stage; and,
- severe deficit irrigation (long time) (T3), the threshold value for Ψ_{stem} was −3 MPa until the end of the period of pit hardening stage.

Table 1 shows the average of minimum stem water potential (min Ψ_{stem}) and SI values, together with the volume of applied water in each treatment.

Table 1. Minimum midday stem water potential (min Ψ_{stem}), water stress integral (SI) and water applied as affected by the irrigation treatment.

Sample	Min Ψ_{stem} (MPa)	SI (MPa × Day)	Water Applied (mm)
		ANOVA †	
	*	**	NS
		Multiple Range Tukey Test ‡	
T0	-2.16^{a}	17.5^{b}	274.3
T1	$-3.07^{b,c}$	$45.4^{a,b}$	294.9
T2	$-2.44^{a,b}$	$31.3^{a,b}$	347.7
T3	-3.69^{c}	69.2^{a}	105.1

† NS = not significant at $p > 0.05$. * and ** significant at $p < 0.05$, and 0.01, respectively. ‡ Values followed by the same letter within the same column were not significantly different ($p > 0.05$), according to Tukey's least significant difference test.

2.2. Spanish-style Processing

For each RDI treatment, four batches of fresh olives were processed. Each one was formed by 50 kg of raw olives that were mixed and transported to Cooperativa Nuestra Señora de las Virtudes (La Puebla de Cazalla, Seville, Spain). First, olives were submitted to lye treatment during 6–8 h with 1.3–2.6% (weight:volume) of NaOH. Then, olives were washed with water during 12 h for cleaning and they were put on 12% NaCl for fermentation (it began with 0.17 mol L^{-1} and finished with 0.09 mol L^{-1}). After 4 months of fermentation, table olives reached an equilibrium with brine (pH < 4.2, 8% NaCl, 0.8% lactic acid and residual alkalinity < 0.120 N).

2.3. Volatile Compounds

Volatile extraction was performed using headspace solid phase micro-extraction (HS-SPME). Analysis were carried out according to Cano-Lamadrid et al. [2]. Briefly, 5 g of olives mixed with 15 mL of ultrapure water and 1.5 g of NaCl were placed into a vial. The vial was put in a bath at 40 °C and, after equilibration, a 50/30 µm divinylbenzene/carboxen/polydimethylsiloxane fiber (2 cm, 24 ga, StableFlex) was manually exposed to the headspace during 50 min. Volatiles were desorbed from fiber into the Gas Chromatograph-Mass Spectrometry (GC-MS) for 3 min.

V+olatile compounds identification was performed in a gas chromatograph, Shimadzu GC-17A (Shimadzu Corporation, Kyoto, Japan), coupled with a Shimadzu mass spectrometer detector GC-MS QP-5050A. GC-MS was equipped with a Restek Rxi-1301 2016 column. Helium was used as carrier gas with same program previously reported by Cano-Lamadrid et al. [2]. Identification was based on: (i) retention indices, (ii) GC-MS retention times, and (iii) mass spectra matches in Wiley 09 MS library (Wiley, New York, NY, USA) and NIST14 (National Institute of Standards and Technology, Gaithersburg, MD, USA). Results for each of the volatile compounds were expressed as percentage of the total area.

2.4. Sensory Analysis

2.4.1. Descriptive Sensory Evaluation

Ten trained panelists (aged from 25–55 years) from the Food Quality and Safety research group (Miguel Hernández University of Elche, Alicante, Spain) carried out the descriptive sensory analysis of samples under study. Each panelist had more than 600 h of experience with a variety of products, mostly, vegetable or horticultural products. For the present study, the panel was trained during 3 sessions of 1 h each, where they worked on the International Olive Oil Council, IOOC [19] table olives lexicon and finally, the panel agreed on the useful lexicon for the samples: color (from yellow to green), saltiness, bitterness, sourness, sweetness, aftertaste, hardness, crunchiness and fibrousness, and off-flavors or negative attributes; if off-flavors were present panelists could choose among the options abnormal fermentation, musty, rancid, cooking effect, soapy, metallic, earthy, and winey-vinegary [19].

Odor-free disposable 100 mL plastic cups were used to serve samples to panelists at room temperature (~20 °C). Cups were half filled with table olives coded with random 3-digit numbers and covered. Distillated water and crackers were used to cleanse palates between samples. Three sessions were used for the descriptive sensory evaluation of samples (each sample was evaluated in triplicate). Panelists used a 0–10 scale (0: no intensity; and 10: extremely strong).

2.4.2. Consumer Acceptance

For affective sensory evaluation, 100 regular table olive consumers were invited from three locations: (i) L1: El Esparragal (Murcia, Spain); (ii) L2: Elche (Alicante, Spain); and, (iii) L3: Los Desamparados (Alicante, Spain). L1 and L3 were chosen to represent consumers from rural areas, while L2 was chosen to represent consumers from urban locations. Consumers were recruited by telephone from the database of SensoFood Solutions of Universidad Miguel Hernández de Elche. The eligibility criteria was that they consume, at least, three times per week table olives. Informed consent was obtained and it is available from the Principal Investigators of the project AGL2016-75794-C4-1-R, Prof. Carbonell-Barrachina. Demographic questions were added to the questionnaire. The consumer age range was 18–24 (13%), 25–35 (14%), 36–45 (19%), 45–55 (26%) and >55 (28%) with a 62:38 gender ratio (women:men). Forty-six percent of consumers participating in this study were full-time workers, 17% part-time, 17% were students and 20% were unemployed. Consumers were also asked about their interest on food labels, and 79% answered that pay attention to product labels, especially, for Spanish-products (64%), healthy products (57%) and sustainable products (25%).

The study was carried out using SensoFood Solutions individual booths (Inverso Estudio Creativo, Murcia, Spain) in all locations to isolate participants and ensure that they worked individually, with a randomized block design and using 3-digits codes for each sample. Samples were served following the same way as for descriptive sensory evaluation. Questionnaires were prepared using 9-point hedonic scale (1 = dislike extremely, 5 = neither like nor dislike, and 9 = like extremely) for color, flavor, bitterness, saltiness, sourness, hardness, crunchiness, fibrousness, aftertaste and overall. Just About Right (JAR) scale (1 = low intensity, and 9 = high intensity) was also used to score intensity attributes (flavor, bitterness, saltiness, sourness and aftertaste) to later evaluate how samples could be improve using penalty analysis. Additionally, preference test was done to rank irrigation treatments under study where consumers had to order table olive samples from dislike to like and later, Friedman test was carried out to interpret data.

All panelists (descriptive test) and consumers (affective tests) gave their informed consent for inclusion before they participated in the study. Universidad Miguel Hernández de Elche automatically exempts "general taste tests", including descriptive sensory tests from needing ethical approval, based on European Union guidelines. However, the study was conducted in accordance with the Declaration of Helsinki, and the protocol was approved by the Ethics Committee of the Escuela Politécnica Superior de Orihuela, Universidad Miguel Hernández de Elche (project AGL2016-75794-C4-1-R).

2.4.3. Consumer Willingness to Pay

Consumer were first informed about HydroSOStainability concept by a leaflet and answering their questions. Then, two samples of table olives were provided to them. Commercial Spanish-style "Manzanilla" table olives were purchased from Mercadona supermarket (Mercadona is one of the most popular food supermarkets in the Mediterranean area of Spain). These table olives were labeled as "conventional" as opposed to olives labeled "HydroSOStainable", with its logo (Figure 1); in this way, the same product was presented to the consumers but with and without the HydroSOStainability logo. Each sample ("conventional" or "HydroSOStainable") was presented to the consumer together with its corresponding questionnaire. Firstly, consumer evaluated "conventional" table olives green-olive flavor, saltiness, hardness and overall liking, and secondly, HydroSOStainable table olives green-olive flavor, saltiness, hardness overall liking and willingness to pay. They were given a price for conventional table olives of 1.35 € per 200 g (Mercadona price) and 4 options to pay for HydroSOStainable table olives: ≤1.35 € (distributor brand), range 1.35–1.75 € (known brand prices), range 1.75–2.50 € (known brand prices), and >2.50 € (gourmet table olives).

This study was done in the same three locations than the affective sensory evaluation but using 100 consumers in each site (some of them were the same than in the affective sensory evaluation).

Figure 1. HydroSOStainable logo. (**A**): English version. (**B**): Spanish version.

2.5. Statistical Analysis

Two or three-way analysis of variance (ANOVA) followed by Tukey's multiple range test were the chosen statistical tests. To assess panel performance, a 3-way ANOVA (factor 1: irrigation treatment; factor 2: panel session; and, factor 3: panelist) was carried out in the descriptive sensory evaluation. For affective sensory data, 2-way ANOVA was used (factor 1: irrigation treatment; and, factor 2: location). Additionally, penalty analysis was carried out with JAR data from the affective test to study how samples could be improved, and partial least squares regression (PLS) was also performed to correlate consumer overall liking with the volatile compounds and descriptive sensory attributes. All statistics were performed using XLSTAT Premium 2016 (Addinsoft, New York, NY, USA). Finally, data from the JAR analysis (Penalty analysis) were graphically represented.

3. Results and Discussion

3.1. Irrigation

Table 1 summarizes the information regarding the water stress achieved by the olive trees during 2017 season, by using 2 parameters (minimum midday stem water potential (min Ψ_{stem}) and water stress integral (SI)). Statistical differences were found among three RDI treatments and control in both parameters studied, Min Ψ_{stem} and SI. In fact, T3 was the treatment presenting the highest SI value (69.2 MPa × day) as well as the highest min Ψ_{stem} (−3.69 MPa) and this strong stress was basically due to the fact that the smallest volume of water was applied (105.1 mm). T1 and T2 occupied an intermediate position, reflecting a moderate water stress level as compared to T0 (control), which trees suffered the lowest stress. T1 and T2 were not statistically different although the stress applied was different (harder for T2) because of time of application, so applying moderate stress during log time and severe stress during short time caused similar stress on trees. These results followed a similar trend to those from previous seasons (2015 and 2016), as reported by Sánchez-Rodríguez et al. [18].

3.2. Volatile Compounds

Thirty-eight volatile compounds were identified in the table olives and their content for each irrigation treatment are shown in Table 2. Esters were the predominant volatiles in control table olives (38.48%), although their content decreased as RDI was more severe. On the contrary, terpenes were the predominant chemical family on HydroSOStainable table olives (T1–T3), with T2 olives (severe deficit irrigation, short time) having the highest content (47.39%). Organic acids were also in a high proportion (>10%) in all table olives, except T2 (2.95%). Besides, T2 showed the highest percentage of ketones (14.47%), while phenolic compounds and alcohols having similar contents in T1 and T3 samples but higher than those of T0 and T2.

There are some volatile compounds that showed the same trend in all RDI table olives, such as ethyl acetate, isoamyl acetate, *cis*-3-hexen-1-ol, 1-hexanol and γ-terpineol, that increased when water stress was applied, and, therefore, HydroSOStainable table olives would have, at least theoretically, stronger pineapple, banana, pear, green, woody and lilac notes than control samples. On the other hand, other compounds showed a decreased content when RDI treatments were applied (2-butanol, propanoic acid, ethyl cyclohexanecarboxylate and cyclohexanecarboxylic acid, butyl ester). Apart from these general trends, T1 experienced an increase on the contents of ethanol, dimethylsulfide (green, sulfurous), acetic acid (vinegar), ethyl propionate (fruity, pineapple), n-propyl acetate (celery), propyl propionate (oily, fruity), propyl butanoate and p-cresol (green, woody). With respect to T2, dimethylsulfide, propyl butanoate, D-limonene (citrus, lemon), p-cymene (citrus), γ-Terpinene (herbaceous, citrus), ethyl propanoate (fruity, melon, peach) and 6-methyl-5-hepten-2-one (herbaceous, oily) as compared to the control table olives, while 2-butanol, acetic acid and p-cresol were not found on these samples. Finally, T3 olives had an increased content of ethyl heptanoate, guaiacol (woody, smoky) and cyclohexanecarboxylic acid (fatty, fruity) but a decreased content on 2-butanol, propyl propionate and p-cresol always as compared to control samples. The sensory descriptors were obtained from relevant olive related references, including GC-olfactometry studies [2,20].

A previous study with "Manzanilla" Spanish-style table olives processed in the same way than in the current research, but under different irrigation conditions also showed statistically significant differences in a high number of volatile compounds [2]. For instance, it was found that acids and straight chain hydrocarbons increased their concentration simultaneously with the stress while aldehydes and phenol compounds decreased. These results did not agree with those found in the current research but it could be due to different irrigation conditions, among other agronomic differences such as soil characteristic or climate conditions. Brahmi, et al. [21] also found differences among volatile compounds as affected by the irrigation strategies on "Koroneiki" cultivar grown under Tunisian conditions. The content of some alcohols decreased, but others increased as it was found in the present work. In the same way, it was found that some aldehydes decreased.

Table 2. Retention indexes, sensory descriptors and percentage of total area of volatile compounds found in table olives as affected by the irrigation treatment.

Compounds	Chemical Family	Ions	RI		Descriptors §	ANOVA †	Content (%)			
		m/z	Exp.	Lit.			T0	T1	T2	T3
Ethanol	Alcohol	45	659			**	0.663 ^{b,‡}	1.135 ^{a}	0.604 ^{b}	0.998 ^{a,b}
Dimethylsulfide	Sulfur compound	62/47	679		Green, sulfurous	*	0.221 ^{c}	0.552 ^{b}	1.063 ^{a}	0.285 ^{c}
Ethyl acetate	Ester	45/61/70/88	703		Pineapple	**	1.243 ^{c}	1.856 ^{b}	2.319 ^{a}	2.115 ^{a,b}
2-Butanol	Alcohol	45	704			*	0.690 ^{a}	0.430 ^{a,b}	nd ^{c}	0.285 ^{b}
Acetic acid	Acid	45/60	724		Vinegar	***	11.86 ^{b}	14.11 ^{a}	nd ^{c}	11.03 ^{b}
Ethyl propionate	Ester	57	746	726	Fruity, pineapple	*	0.953 ^{b,c}	1.764 ^{a}	1.377 ^{b}	0.737 ^{c}
n-Propyl acetate	Ester	61/73	749	728	Celery	*	1.105 ^{b,c}	2.040 ^{a}	1.353 ^{b}	0.927 ^{c}
Propanoic acid	Acid	74/45	771		Dairy, acidic	*	0.925 ^{a}	0.614 ^{b}	0.217 ^{c}	0.238 ^{c}
2,4-dimethylhexane	Hydrocarbon	85/57/71	793			NS	0.580	1.135	0.773 ^{b}	0.523
Ethyl butanoate	Ester	71	812	802		NS	0.221	0.706	0.411	0.333
Propyl propionate	Ester	57/75	820	810	Oily, fruity	*	1.022 ^{b}	1.595 ^{a}	1.208 ^{b}	0.713 ^{c}
Butyl acetate	Ester	56/73	827	812	Fruity, greenish	NS	0.041	0.184	0.121	0.166
Ethyl lactate	Ester	45	846	813	Butter, fruity	NS	0.083	0.230	0.121	0.095
Ethyl 2-methyl butanoate	Ester	57/102/85	861	846		NS	0.124	0.368	0.242	0.190
Ethyl 3-methyl butanoate	Ester	88/57	865	859		NS	0.124	0.199	0.145	0.166
Isoamyl acetate	Ester	55/70	895	878	Banana, pear	*	0.041 ^{c}	0.138 ^{a}	0.072 ^{b}	0.048 ^{a}
cis 3-Hexen-1-ol	Alcohol	67/55/82	899	902	Green	***	0.097 ^{c}	0.245 ^{a}	0.121 ^{b}	0.119 ^{b}
1-Hexanol	Alcohol	56/69	907	912	Green, woody	**	0.069 ^{c}	0.153 ^{a}	0.097 ^{b}	0.143 ^{a}
Propyl butanoate	Ester	71/89/55	914	896		*	0.152 ^{c}	0.629 ^{a}	0.362 ^{b}	0.119 ^{c}
β-Myrcene	Terpene	93/69	997	992	Fruity, vegetable	***	0.801	1.089	1.594	1.426
Ethyl hexanoate	Ester	88	1016	1001		NS	1.229	2.086	2.126	1.949
D-Limonene	Terpene	68/93	1041	1044	Citrus, lemon	***	20.97 ^{b}	20.92 ^{b}	34.44 ^{a}	21.17 ^{b}
p-Cymene	Terpene	119/134/91	1044	1030	Citrus	**	3.148 ^{c}	3.896 ^{b,c}	6.449 ^{a}	4.705 ^{b}
γ-Terpinene	Terpene	93/91/136	1069	1076	Herbaceous, citrus	**	2.223 ^{b}	2.470 ^{b}	3.913 ^{a}	2.733 ^{a,b}
Methyl cyclohexanecarboxylate	Ester	55/87	1093	1056	Berry, creamy	NS	5.633	2.807	1.957	3.446
Ethyl heptanoate	Ester	88/115/60	1117	1095	Fruity, melon, peach	***	0.690 ^{b}	0.890 ^{b}	2.101 ^{a}	2.163 ^{a}
Guaiacol	Phenolic compound	109/124/81	1148	1114	Woody, smoky	***	0.318 ^{b}	0.322 ^{b}	0.725 ^{b}	18.560 ^{a}
Ethyl cyclohexanecarboxylate	Ester	55/83/101	1163	1170		***	25.81 ^{a}	8.943 ^{c}	10.72 ^{b}	2.614 ^{d}
p-Cresol	Phenolic compound	107	1180		Green, woody	***	2.844 ^{b}	12.62 ^{a}	nd ^{c}	0.285 ^{c}
2-Phenethylalcohol	Alcohol	91/107	1184	1159	Honey, rose	*	0.207	0.675	0.411	1.355
Cyclohexanecarboxylic acid	Acid	56/73/45/82	1197	1157	Fatty, fruity	**	0.801 ^{b}	0.123 ^{b}	nd ^{b}	10.91 ^{a}
6-Methyl-5-hepten-2-one	Ketone	55/108/69/91	1207		Herbaceous, oily	**	3.907 ^{b,c}	6.412 ^{b}	14.469 ^{a}	0.974 ^{c}
γ-Terpineol	Terpene	59/93/121/136	1243	1224	Lilac	*	0.400 ^{c}	0.660 ^{b}	0.990 ^{a,b}	1.972 ^{a}
1,4-Dimethoxy-benzene	Phenolic compound	123/138/95	1254		Fatty	**	2.968 ^{c}	5.093 ^{a}	5.217 ^{a}	4.111 ^{b}
Cyclohexanecarboxylic acid, butil ester	Acid	129/83/55/111	1266			*	6.227 ^{a}	1.411 ^{c}	2.729 ^{b}	1.854 ^{c}
4-Ethylphenol	Phenolic compound	107/122/77	1271		Alcohol, medicinal	NS	0.870	1.104	1.546	0.547
Ethyl dihydrocinnamate		104/91	1396	1390		NS	0.469	0.383	nd	nd
β-Bisabolene	Terpene	69/93	1525	1517		NS	0.262	nd	nd	nd
Σ Alcohols						*	1.726 ^{b}	2.638 ^{a}	1.233 ^{b}	2.900 ^{a}
Σ Sulfur compounds						NS	0.221	0.552	1.063	0.285
Σ Esters						**	38.48 ^{a}	24.44 ^{b}	24.64 ^{b}	15.78 ^{c}
Σ Ketones						**	3.907 ^{b,c}	6.412 ^{b}	14.47 ^{a}	0.974 ^{c}
Σ Terpenes						***	27.81 ^{c}	29.04 ^{b,c}	47.39 ^{a}	32.01 ^{b}
Σ Acids						*	19.81 ^{a}	16.26 ^{a}	2.95 ^{b}	24.03 ^{a}
Σ Phenolic compounds						***	7.000 ^{b}	19.14 ^{a}	7.488 ^{b}	23.50 ^{a}
Σ Hydrocarbons						NS	0.580	1.135	0.773	0.523

† NS = not significant at $p > 0.05$. *, ** and *** significant at $p < 0.05$, 0.01, and 0.001, respectively. ‡ Values followed by the same letter within the same row were not significantly different ($p > 0.05$), according to Tukey's least significant difference test. § Cano-Lamadrid et al. [2], Angerosa et al. [20], SAFC [22].R.I.: retention index; Exp.: experimental; Lit.: literature; nd: not detected.

3.3. Descriptive Sensory Analysis

Descriptive sensory analysis by trained panel (0–10 scale) of table olives under study was carried out and results are shown in Table 3. Saltiness, sweetness and fibrousness had mean values (for all treatments under study) of 5.4, 2.2 and 0.5, respectively; no statistically significant (ANOVA, $p < 0.05$) differences were found for these attributes and mean values are reported. With respect to color, T0 olives presented the highest color intensity (6.5), while T1 had the lowest intensity (5.4), and therefore the most yellowish color. T2 and T3 showed intermediate positions and thus, they presented intermediate colors between yellow and green. As far as the green-olive flavor is concerned, T1 table olives had the highest intensity (6.9), with T3 having the lowest score (6.2), and T0 and T2 having being in the middle. Bitterness decreased its intensity (up to 3 points) as the water stress increased. The T3 olives were the sourest ones (4.5 points higher than control) and at the same time had the longest aftertaste (2.2 points higher than control), but they simultaneously had the lowest intensity of hardness and crunchiness (3.5 and 1.7, respectively). Finally, it is important to mention that no off-flavors were found in any of the table olive under study.

Previous studies had also found changes on the intensity of key sensory descriptors as an effect of irrigation regimes on table olives. For instance, Cano-Lamadrid et al. [2] and Cano-Lamadrid et al. [13] showed the effect of two RDI treatments on the descriptive sensory profile of "Manzanilla" Spanish-style table olives. In those studies, saltiness, green-olive flavor, aftertaste, bitterness and hardness were affected by irrigation. It was found that moderate stress caused an increase of ~5% on the intensity value of the green-olive flavor attribute; result which agreed well with the trend just reported on the current research. However, results on bitterness and aftertaste showed an increase in trees grown under moderate stress [2] while in the current experiment a decreased intensity of bitterness and aftertaste (as compared to the control sample) at moderate level, while an increased aftertaste intensity was observed at severe stress. With respect to bitterness, a similar result was found on "Ascolana" olives [5], in which the bitter character decreased with the irrigation regime. The same trend was also found for hardness [5], which agreed with the low hardness of the T3 samples in the present work. In the case of "Nocellara del Belice" cultivar produced following Greek style [13], an increase on green-olive aroma, sourness, sweetness and crispness were reported under moderate water stress.

3.4. Consumer Acceptance

Affective sensory evaluation was carried out at three locations, although no statistical differences were found among data obtained; thus, the mean values of nine descriptors and the corresponding overall liking of consumers at the three locations is shown in Table 4. Table olives showed a high overall acceptability by consumers (mean of 6.3 in a scale up to a maximum score of 9). The rest of attributes under study (color, 6.5; flavor, 6.4; bitterness, 6.0, saltiness, 6.1; sourness, 6.0; hardness, 6.6; crunchiness, 6.6; fibrousness, 6.5; and aftertaste, 6.2) also received high values (1–9 scale) of consumer satisfaction degree.

Consumer preference for table olives was analyzed using the Friedman test. No statistical significant differences ($p < 0.05$) were found among preferences for control (T0) and HydroSOStainable table olives (T1–T3). Thus, this experimental finding confirmed that HydroSOStainable olives were as least as preferred as those coming from fully irrigated trees (T0), but saving water and being more sustainable; this sustainability makes these olives attractive for consumption [23].

From the best of our knowledge, only one affective sensory evaluation had been previously conducted for table olives coming for RDI treatments [2]. In this study, "Manzanilla" Spanish-style table olives under moderate deficit irrigation (but with different treatments than in the current research) were the preferred ones by consumers because of their flavor, crunchiness and aftertaste.

Table 3. Descriptive sensory attributes of table olives as affected by the irrigation treatment. Scale used ranged from 0 = no intensity to 10 = extremely strong intensity.

	Appearance	Flavor							Texture		
Sample	**Color**	**Green-Olive Flavor**	**Saltiness**	**Bitterness**	**Sourness**	**Sweetness**	**Aftertaste**	**Off-Flavor**	**Hardness**	**Crunchiness**	**Fibrousness**
	ANOVA †										
	**	*	NS	*	***	NS	*	NS	***	***	NS
	Multiple Range Tukey Test ‡										
T0	6.5 a,‡	6.5 a,b	5.9	5.8 a	2.4 b	2.9	5.9 a,b	0.0	7.8 a	7.3 a	0.3
T1	5.4 b	6.9 a	5.0	3.8 a,b	3.0 b	2.1	5.9 a,b	0.0	6.6 a	5.6 a	0.8
T2	5.9 a,b	6.4 a,b	5.9	4.0 a,b	2.6 b	2.2	5.6 b	0.0	7.2 a	6.1 a	0.3
T3	5.7 a,b	6.2 b	4.9	2.8 b	6.9 a	1.7	8.1 a	0.0	3.5 b	1.7 b	0.4

† NS = not significant at $p > 0.05$. *, **, and *** significant at $p < 0.05$, 0.01, and 0.001, respectively. ‡ Values followed by the same letter within the same column were not significantly different ($p > 0.05$), according to Tukey's least significant difference test.

Table 4. Affective sensory analysis (at 3 locations in Spain) of table olives as affected by irrigation treatment.

	Color	**Flavor**	**Bitterness**	**Saltiness**	**Sourness**	**Hardness**	**Crunchiness**	**Fibrousness**	**Aftertaste**	**Overall Liking**
	ANOVA †									
	NS	NS	NS	NS	NS	NS	NS	NS	NS	NS
	Multiple Range Tukey Test									
T0	6.2	6.6	6.3	6.2	6.3	7.0	6.7	6.5	6.6	6.5
T1	6.7	6.6	6.3	6.0	6.0	6.7	6.6	6.6	6.2	6.4
T2	6.5	6.3	5.7	6.3	5.8	6.5	6.6	6.5	6.2	6.4
T3	6.5	5.9	5.7	5.9	5.8	6.3	6.5	6.5	5.9	5.7

† NS = not significant at $p > 0.05$.

3.5. Driving Sensory Attributes

PLS Regression analysis was carried out to established drivers of liking for HydroSOStainable table olives (Figure 2). Two PLS maps were constructed to correlate the consumer overall liking (affective sensory analysis) with volatile compounds (total volatile contents for each chemical family) (Figure 2A) and with descriptive sensory attributes (trained panelists) (Figure 2B). Only attributes showing statistical differences among samples (ANOVA $p < 0.05$) were used to construct maps.

In the positive part of the x-axis (right side of the graph) volatiles associated with overall liking of consumers were acids, alcohols and phenolic compounds while in the negative part of the x-axis, ketones and terpenes can be found (Figure 2A). Although these volatile families are in opposite places on the map, consumer overall liking were not concentrate in any specific part of the map as a high dispersion on the map could be found; thus, it was not stated that no a clear relationship between overall consumer liking (affective sensory analysis) and volatile compounds was observed. Therefore, volatiles could not be considered as good driving sensory attributes for the acceptability of HydroSOStainable table olives.

Regarding map B (Figure 2B), consumer satisfaction (affective sensory analysis) was correlated with some positive attributes (descriptive sensory analysis by trained panel) of table olives such as green-olive flavor, hardness, crunchiness and bitterness, as it can be observed a high concentration of consumer overall liking in the right side of the map, where these descriptors are positioned. Consequently, these descriptors should be use as drivers to understand future consumer acceptance of HydroSOStainable table olives.

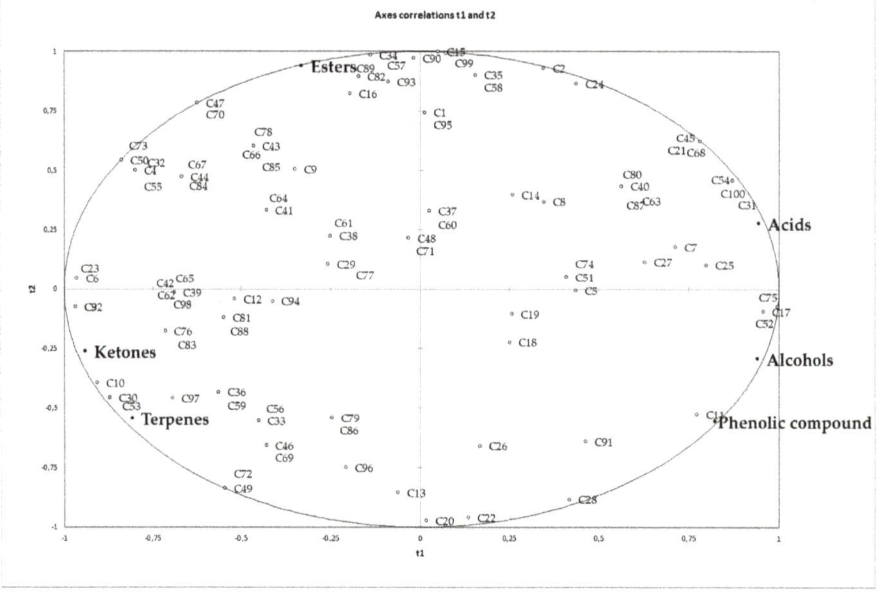

(A)

Figure 2. *Cont.*

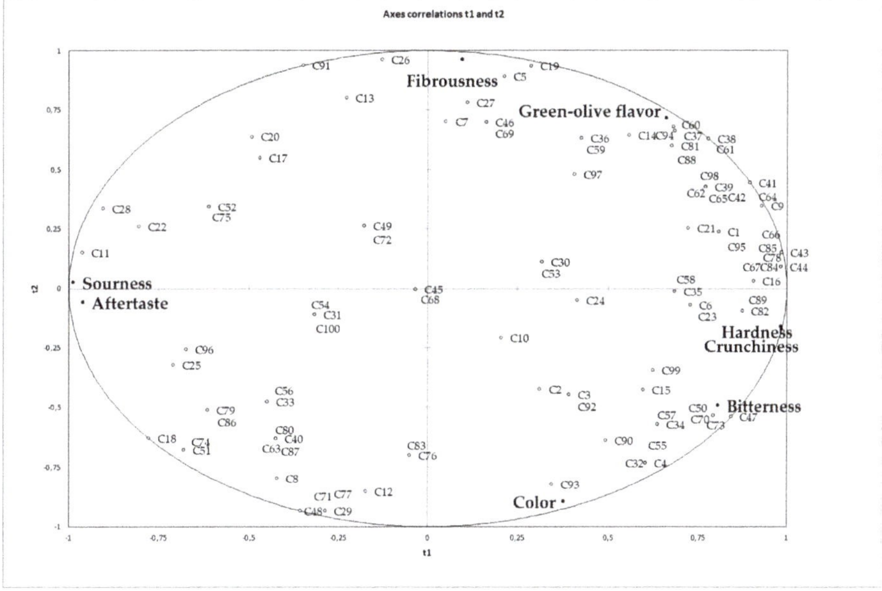

(B)

Figure 2. Partial least squares regression (PLS) of (**A**) volatile compounds (chemical families sum) (X axis: t2) and overall consumer liking (Y axis: t1) (unfiled circles: consumer (C + number of consumer); filled circle: volatile compound); and, (**B**) descriptive sensory attributes (X axis) and overall consumer liking (Y axis) (unfiled circles: consumer (C + number of consumer; filled circle: descriptor).

3.6. Consumer Willingness to Pay

Table 5 shows the results of overall liking and satisfaction degree study done regarding consumer willingness to pay for table olives at three locations. Green-olive flavor, saltiness, hardness and consumer overall liking were evaluated as the most important attributes valued by consumers to further understanding on their perception of HydroSOStainable logo. This logo (Figure 1), caused a clear effect on consumer overall liking and green-olive flavor perception, making HydroSOStainable samples to increase their values in 1.1 and 1.3 units, respectively, as compared to the control olives. Concerning the location, for green-olive flavor attribute, consumers in L1 punctuated olives with the highest score (7.7) while L2 with the lowest (7.0), but the opposite occurred for overall liking, where L2 scored with the highest satisfaction degree (7.3). Regarding the interaction logo and location, the highest scores of the green-olive flavor attribute were found in L1 and L3 samples with the HydroSOStainability logo, and the lowest values was found in the L3 table olives without the HydroSOStainability logo. It is important to consider that L2 consumers (Elche, Alicante, Spain), corresponding to people living in an urban location, scored the highest for the overall liking without any need for the hydroSOStainability logo. No significant statistical differences were found for the effects of logo, location and their interaction on table olives saltiness and hardness.

Table 5. Overall liking and satisfaction degree on flavor, saltiness and hardness of Table Olives affected by logo effect and location.

		Green-olive Flavor	Saltiness	Hardness	Overall Liking
		ANOVA Test †			
Logo effect		***	NS	NS	*
Location		***	NS	NS	*
Logo effect vs Location		***	NS	NS	*
		Multiple Range Tukey Test Logo effect			
	Conventional	6.7 ^{b,‡}	6.4	6.6	6.5 ^{b}
	HydroSOStainable logo	8.0 ^{a}	7.4	7.0	7.4 ^{a}
		Multiple Range Tukey Test Location			
	L1	7.7 ^{a}	6.6	6.9	6.9 ^{b}
Location	L2	7.0 ^{b}	7.1	7.2	7.3 ^{a}
	L3	7.3 ^{a,b}	7.0	6.3	6 ^{b}
		Multiple Range Tukey Test Logo effect vs. Location			
	L1	7.1 ^{a,b}	5.9	6.5	6.3 ^{a,b}
Conventional	L2	7.0 ^{a,b}	6.6	7.3	7.6 ^{a}
	L3	5.9 ^{c}	6.7	5.9	5.6 ^{b}
	L1	8.3 ^{a}	7.2	7.3	7.5 ^{a}
HydroSOStainable logo	L2	6.9 ^{b}	7.7	7.0	7.1 ^{a,b}
	L3	8.7 ^{a}	7.2	6.8	7.7 ^{a}

† NS = not significant at $p > 0.05$. *, and ***, significant at $p < 0.05$, and 0.001, respectively. ‡ Values followed by the same letter within the same column and factor (treatment and location) were not significantly different ($p > 0.05$), according to Tukey's least significant difference test.

Regarding willingness to pay, 88% of the participants in the study were willing to pay more than the usual price (1.35 € per 200 g) when they were informed about HydroSOStainable benefits. Concretely, 52% were willing to pay a price in the range 1.35–1.75 €, 32% 1.75-2.50 € and only 4% were willing to pay more than 2.50 €.

Previous study done with HydroSOStainable pistachios [3] also reported an increase of willingness to pay. In that case, the study was conducted in Galicia (northern Spain) and the Valencian Community (representing Mediterranean area of Spain) and consumers from Galicia willing to pay more than those from the Valencian Community; although all consumers agreed that the price for this product should be higher than for the conventional ones. A similar situation was reported by Lipan et al. [16], where Spanish and Romanian consumers were willing to pay more for HydroSOStainable almonds.

3.7. Penalty Analysis

Apart from the above described overall liking and satisfaction degree for specific sensory attributes, several JAR questions (flavor, bitterness, saltiness, sourness and aftertaste) were asked along the consumer study (affective sensory evaluation) with the purpose of analyzing the possible intensity attributes to be improved. Penalty analysis was conducted [24] an easier understanding of the relationship between JAR scores and consumer satisfaction degree scores. Figure 3 shows the proportion of consumer opinion plots against the mean penalty score. The attributes susceptible of improvement were those, which had the greatest negative impact on the sample liking for at least 20% of consumers and caused a drop of at least 1 point for liking. Results of the penalty analysis indicated that the studied deficit irrigation treatments (T1, T2 and T3) were not penalized by presenting low or high intensities of the studied attributes (Figure 2B–D). According to Spanish consumers, no improvement was necessary in these olive samples.

Previous research about overall consumer liking of HydroSOStainable almonds [16] results indicated that only the bitterness could be improved (decreasing it) when "sustained" deficit irrigation treatment was applied (deficit irrigation during whole season); however, when using RDI, HydroSOStainable almonds did not show any attribute to be improved, as it was found here for HydroSOStainable table olives, so this treatments were the best for consumer acceptance as their quality was as high as control table olives.

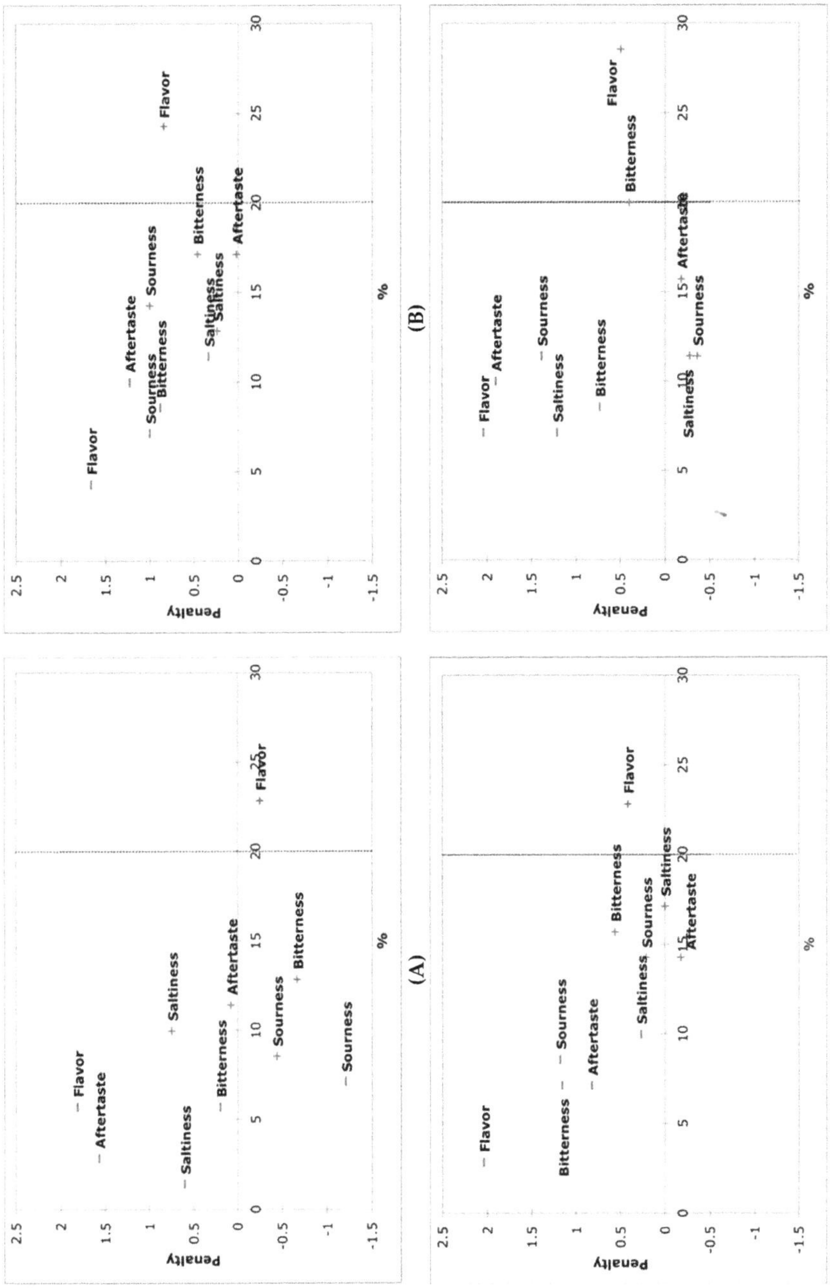

Figure 3. Penalty analysis of samples (**A**) = T0; (**B**) = T1; (**C**) = T2; (**D**) = T3. "Too low intensity" is indicated with "−" and "too high intensity" is indicated with "+".

4. Conclusions

This is the first study about consumer acceptance and willingness to pay for table olives under RDI treatments (HydroSOStainable table olives). Results indicated that RDI produced changes on volatile composition and on the intensity of several sensory descriptors. Green-olive flavor, hardness, crunchiness and bitterness seem to be the driving sensory attributes controlling consumer acceptance for HydroSOStainable table olives, although further studies are needed to fully prove this statement. Consumers preferred table olives with the HydroSOStainability logo and their satisfaction level was higher for the green-olive flavor and overall liking as compared to those of the conventional samples (without this logo). A high percentage of consumers were willing to pay a higher price for HydroSOStainable table olives. Information obtained in this research should be useful for developing the best irrigation strategy to produce table olives with the highest water saving, and the best sensory characteristics for consumers. For instance, T1 (moderate deficit irrigation where Ψ_{stem} was −2 MPa during pit hardening stage) and T2 (severe deficit irrigation during short time where Ψ_{stem} was −3 MPa during half period of pit hardening stage) strategies optimized for desirable sensory characteristics, such as green-olive flavor, hardness and crunchiness.

Author Contributions: Data curation, L.S.-R. and M.C.-L.; Formal analysis, L.S.-R. and M.C.-L.; Funding acquisition, A.C.-B.; Investigation, L.S.-R. and M.C.-L.; Methodology, E.S. and F.H.; Project administration, A.C.-B.; Resources, A.C.-B.; Supervision, E.S. and F.H.; Writing—original draft, L.S.-R.; Writing—review & editing, A.C.-B., E.S. and F.H.

Funding: The study has been funded (Spanish Ministry of Economy, Industry and Competitiveness) through a coordinated research project (hydroSOS mark) including the Universidad Miguel Hernández de Elche (AGL2016-75794-C4-1-R, hydroSOS foods) and the Universidad de Sevilla (AGL2016-75794-C4-4-R) (AEI/FEDER, UE). Author Marina Cano-Lamadrid was funded by a FPU grant from the Spanish Ministry of Education (FPU15/02158).

Acknowledgments: Authors are thankful to consumers participating on the study. Thank you to Junta Municipal de El Esparragal and Farmacia Iborra for their help with consumer studies.

Conflicts of Interest: The authors declare no conflict of interest.

References

1. Wei, S.; Ang, T.; Jancenelle, V.E. Willingness to pay more for green products: The interplay of consumer characteristics and customer participation. *J. Retail. Consum. Serv.* **2018**, *45*, 230–238. [CrossRef]
2. Cano-Lamadrid, M.; Girón, I.F.; Pleite, R.; Burló, F.; Corell, M.; Moriana, A.; Carbonell-Barrachina, A.A. Quality attributes of table olives as affected by regulated deficit irrigation. *LWT Food Sci. Technol.* **2015**, *62*, 19–26. [CrossRef]
3. Noguera-Artiaga, L.; Lipan, L.; Vázquez-Araújo, L.; Barber, X.; Pérez-López, D.; Carbonell-Barrachina, Á.A. Opinion of Spanish Consumers on Hydrosustainable Pistachios. *J. Food Sci.* **2016**, *81*, S2559–S2565. [CrossRef] [PubMed]
4. Corell, M.; Martín-Palomo, M.J.; Sánchez-Bravo, P.; Carrillo, T.; Collado, J.; Hernández-García, F.; Girón, I.; Andreu, L.; Galindo, A.; López-Moreno, Y.E.; et al. Evaluation of growers´effort to improve the sustainability of olive orchards: Development of the hydroSOStainable index. *Sci. Hortic.* **2019**, *257*, 108661. [CrossRef]
5. Marsilio, V.; d'Andria, R.; Lanza, B.; Russi, F.; Iannucci, E.; Lavini, A.; Morelli, G. Effect of irrigation and lactic acid on the phenolic fraction, fermentation and sensory characteristics of olive (*Olea europaea* L. cv. Ascolana tenera) fruits. *J. Sci. Food Agric.* **2006**, *86*, 1005–1013. [CrossRef]
6. D'Andria, R.; Lavini, A.; Morelli, G.; Sebastiani, L.; Tognetti, R. Physiological and productive responses of *Olea europaea* L. cultivars Frantoio and Leccino to a regulated deficit irrigation regime. *Plant. Biosyst.* **2009**, *143*, 222–231. [CrossRef]
7. Gómez-Rico, A.; Salvador, M.D.; Fregapane, G. Virgin olive oil and fruit minor constituents as affected by irrigation management based on SWP and TDF as compared to *ETc* in medium-density young olive orchards (*Olea europaea* L. cv. Cornicabra and Morisca). *Food Res. Int.* **2009**, *42*, 1067–1076. [CrossRef]

8. Baccouri, O.; Guerfel, M.; Bonoli-Carbognin, M.; Cerretani, L.; Bendini, A.; Zarrouk, M.; Daoud, D. Influence of irrigation and site of cultivation on qualitative and sensory characteristics of a Tunisian minor olive variety (cv. Marsaline). *Riv. Ital. Sostanze Grasse* **2009**, *86*, 173–180.
9. Collado-González, J.; Moriana, A.; Girón, I.F.; Corell, M.; Medina, S.; Durand, T.; Guy, A.; Galano, J.-M.; Valero, E.; Garrigues, T.; et al. The phytoprostane content in green table olives is influenced by Spanish-style processing and regulated deficit irrigation. *LWT Food Sci. Technol.* **2015**, *64*, 997–1003. [CrossRef]
10. Corell, M.; Martín-Palomo, M.J.; Pérez-López, D.; Centeno, A.; Girón, I.; Moreno, F.; Torrecillas, A.; Moriana, A. Approach for using trunk growth rate (TGR) in the irrigation scheduling of table olive orchards. *Agric. Water Manag.* **2017**, *192*, 12–20. [CrossRef]
11. Corell, M.; Pérez-López, D.; Martín-Palomo, M.J.; Centeno, A.; Girón, I.; Galindo, A.; Moreno, M.M.; Moreno, C.; Memmi, H.; Torrecillas, A.; et al. Comparison of the water potential baseline in different locations. Usefulness for irrigation scheduling of olive orchards. *Agric. Water Manag.* **2016**, *177*, 308–316. [CrossRef]
12. Kaya, Ü.; Öztürk Güngör, F.; Çamoğlu, G.; Akkuzu, E.; Aşik, Ş.; Köseoğlu, O. Effect of Deficit Irrigation Regimes on Yield and Fruit Quality of Olive Trees (cv. Memecik) on the Aegean Coast of Turkey. *Irrig. Drain.* **2017**, *66*, 820–827. [CrossRef]
13. Martorana, A.; Miceli, C.; Alfonzo, A.; Settanni, L.; Gaglio, R.; Caruso, T.; Moschetti, G.; Francesca, N. Effects of irrigation treatments on the quality of table olives produced with the Greek-style process. *Ann. Microbiol.* **2016**, *67*, 37–48. [CrossRef]
14. Cano-Lamadrid, M.; Hernández, F.; Corell, M.; Burló, F.; Legua, P.; Moriana, A.; Carbonell-Barrachina, Á.A. Antioxidant capacity, fatty acids profile, and descriptive sensory analysis of table olives as affected by deficit irrigation. *J. Sci. Food Agric.* **2017**, *97*, 444–451. [CrossRef]
15. Sánchez-Rodríguez, L.; Lipan, L.; Andreu, L.; Martín-Palomo, M.J.; Carbonell-Barrachina, Á.A.; Hernández, F.; Sendra, E. Effect of regulated deficit irrigation on the quality of raw and table olives. *Agric. Water Manag.* **2019**, *221*, 415–421. [CrossRef]
16. Lipan, L.; Cano-Lamadrid, M.; Corell, M.; Sendra, E.; Hernandez, F.; Stan, L.; Vodnar, D.C.; Vazquez-Araujo, L.; Carbonell-Barrachina, A.A. Sensory Profile and Acceptability of HydroSOStainable Almonds. *Foods* **2019**, *8*, 64. [CrossRef]
17. Myers, B.J. Water stress integral-a link between short-term stress and long-term growth. *Tree Physiol.* **1988**, *4*, 315–323. [CrossRef]
18. Sánchez-Rodríguez, L.; Corell, M.; Hernández, F.; Sendra, E.; Moriana, A.; Carbonell-Barrachina, Á.A. Effect of Spanish-style processing on the quality attributes of HydroSOStainable green olives. *J. Sci. Food Agric.* **2019**, *99*, 1804–1811. [CrossRef]
19. International Olive Oil Council (IOOC). *Method for the Sensory Analysis of Table Olives*; International Olive Oil Council: Madrid, Spain, 2011.
20. Angerosa, F.; Servili, M.; Selvaggini, R.; Taticchi, A.; Esposto, S.; Montedoro, G. Volatile compounds in virgin olive oil: Occurrence and their relationship with the quality. *J. Chromatogr. A* **2004**, *1054*, 17–31. [CrossRef]
21. Brahmi, F.; Chehab, H.; Flamini, G.; Dhibi, M.; Issaoui, M.; Mastouri, M.; Hammami, M. Effects of irrigation regimes on fatty acid composition, antioxidant and antifungal properties of volatiles from fruits of Koroneiki cultivar grown under Tunisian conditions. *Pak. J. Biol. Sci.* **2013**, *16*, 1469–1478. [CrossRef]
22. SAFC; Sigma-Aldrich. *Flavors & Fragances*; Sigma-Aldrich: Madrid, Spain, 2014.
23. Bollani, L.; Bonadonna, A.; Peira, G. The Millennials' Concept of Sustainability in the Food Sector. *Sustainability* **2019**, *11*, 2984. [CrossRef]
24. Narayanan, P.; Chinnasamy, B.; Jin, L.; Clark, S. Use of just-about-right scales and penalty analysis to determine appropriate concentrations of stevia sweeteners for vanilla yogurt. *J. Dairy Sci.* **2014**, *97*, 3262–3272. [CrossRef]

foods

MDPI

Article

Panel and Panelist Performance in the Sensory Evaluation of Black Ripe Olives from Spanish Manzanilla and Hojiblanca Cultivars

Antonio López-López *, Antonio Higinio Sánchez-Gómez, Alfredo Montaño, Amparo Cortés-Delgado and Antonio Garrido-Fernández

Food Biotechnology Department, Instituto de la Grasa (CSIC), Campus Universitario Pablo de Olavide, Edificio 46, Ctra. Utrera km 1, 41013 Sevilla, Spain; ahiginio@ig.csic.es (A.H.S.-G.); amontano@cica.es (A.M.); acortes@cica.es (A.C.-D.); garfer@cica.es (A.G.-F.)

* Correspondence: all@cica.es; Tel.: +34-9-5461-1550

Received: 17 October 2019; Accepted: 6 November 2019; Published: 8 November 2019

Abstract: There is vast experience in the application of sensory analysis to green Spanish-style olives, but ripe black olives ($\approx 1 \times 10^6$ kg for 2016/2017) have received scarce attention and panelists have less experience on the evaluation of this presentation. Therefore, the study of their performance during the assessment of this presentation is critical. Using previously developed lexicon, ripe olives from Manzanilla and Hojiblanca cultivars from different origins were sensory analysed according to the Quantitative Descriptive Analysis (QDA). The panel (eight men and six women) was trained, and the QDA tests were performed following similar recommendations than for green olives. The data were examined while using SensoMineR v.1.07, programmed in R, which provides a diversity of easy to interpret graphical outputs. The repeatability and reproducibility of panel and panelists were good for product characterisation. However, the panel performance investigation was essential in detecting details of panel work (detection of panelists with low discriminant power, those that have interpreted the scale in a different way than the whole panel, the identification of panelists who required training in several/specific descriptors, or those with low discriminant power). Besides, the study identified the descriptors of hard evaluation (skin green, vinegar, bitterness, or natural fruity/floral).

Keywords: panel performance; panelist; black ripe table olives; sensory descriptors; sensory profile

1. Introduction

World table olive production was around 2.6×10^6 tones in season 2016/2017 according to the last consolidated balance of the International Olive Oil Council [1]. Approximately, 40% of them were processed as black ripe table olives (Californian style). This style was first developed in the USA, which is still one of the most relevant contributors with current production of about 80×10^3 tons [1], but other countries, like Spain, Greece, Turkey, or Egypt, are progressively increasing their productions. Black ripe table olive processing includes a phase of storage, which is usually accomplished by immersing the fruits in brine or acidified solution, followed by a darkening step, which consists of the application of one (or several) lye treatments and subsequent immersion in tap water to remove the excess of alkali. During this oxidation phase, air is also bubbled through the suspension to accelerate browning. The colour is then fixed by a ferrous gluconate solution, after which the olives are packed and the cans sterilised [2]. The products usually offer a rather plain organoleptic profile, which has been a favourable condition for its introduction in new markets, due to their numerous treatments in aqueous solutions. In fact, according to the Trade Standards Applying to Table Olives [3], the only requisites for these olives are sensory characteristics and texture in agreement with their processing system.

Along the last decade, the International Olive Council developed a method for the sensory evaluation of table olives. However, it was mainly focused on green Spanish-style, since most of the descriptors included in the evaluation sheet are exclusively related to this product (e.g., abnormal fermentation, acidity, or bitterness) [4]. However, methods for the evaluation and classification of black ripe olives were developed in California, where this processing has a long tradition [5].

On the other hand, Quantitative Descriptive Analysis (QDA) is widely used for studying the sensory profile of diverse foods ([6–8], among many others). Recently, researchers have applied QDA to a list of 33 descriptors for the sensory comparison of American black ripe table olives with respect to those that are imported from other countries (Spain, Egypt, or Morocco) [9]. Similar descriptors were used to study the sensory profile of black ripe table olives from Spanish Manzanilla and Hojiblanca cultivars and successfully distinguishing among cultivars, farming origins, and storage period [10]. López-López et al. [11] have developed an entirely new lexicon for the application of QDA to Spanish-style green table olives; the results showed relevant differences between cultivars and origins. Therefore, the use of the QDA to black ripe table olives from the most important Spanish cultivar devoted to this elaboration is relevant.

Traditionally, the sensory analysis of table olive, regardless of style, has been mainly devoted to the characterization of products [12–16], but the panelists and panel performances were rarely studied in detail. However, along the last two decades, different authors have developed methodologies for evaluating the reliability of the panel [17–22]. Its application to the panel performance, discrimination power of descriptors of diverse green and black ripe table olives, following the COI/OT/MO No. 1/Rev. 2 methodology, has been recently published [17]. Nevertheless, the performance of a panel and panelists that were devoted to the sensory analysis of black ripe table olives using QDA has never been studied.

This work aims for the application of Quantitative Descriptive Analysis to black ripe table olives from Spanish Manzanilla and Hojiblanca cultivars, focusing interest on the panel and panelist performances as a tool for improving their training and reliability.

2. Materials and Methods

2.1. *Olives and Their Processing*

The olives were of the Manzanilla and Hojiblanca cultivars, harvested at green maturation stage in October 2016. Their origins were: Aljarafe (Sevilla) and Lora de Estepa (Sevilla) for *Manzanilla*, and Lora de Estepa (Sevilla), and Alameda (Málaga) for *Hojiblanca*. The samples were identified as MAL, ML, HL, and HA, according to cultivar (initial letter) and growing area (remaining letter/s).

Just harvested olives from each cultivar and origin were directly brined in 25 L (15 kg olives) PVC (polyvinyl chloride) fermenters in an acidified (2.4% acetic acid) solution. After three months of storage, the fruits were subjected to the darkening process. For this purpose, horizontal stainless steel cylindrical containers (0.4 m diameter, 0.7 m length) were used. The fruits were treated with a 3% lye solution until the alkali reached the pit. After removing the alkali, the olives were washed to low the pH up to 8.0 units. During both operations, an oxygen-saturated ambient was maintained in the suspension by bubbling air through a perforated tube lying along the bottom of the oxidation vessels. Subsequently, the black colour developed was fixed, while using a 0.1% ferrous gluconate solution with pH adjusted to 4.5 to prevent the precipitation of the element as hydroxide. Afterwards, the darkened olives were introduced in glass jars (145 g of olives), together with 170 mL of 3.5% NaCl cover solution, which also contained 0.2 g ferrous gluconate/L and had the pH adjusted to 4.5 with acetic acid. Finally, the jars were closed and sterilised at 130 °C for 20 min [23].

The sensory analysis of the above-prepared black ripe olives was achieved after storage at room temperature for 30 (to allow complete olive flesh/brine equilibrium) and 210 days (estimated maximum normal period of the product in the shelves before reposition). The new codes were those previously mentioned, plus 1 (one-month storage) and 2 (seven-month storage), respectively. Therefore, the symbols of the final samples: were: MAL1, MAL2, ML1, ML2, HL1, HL2, HA1, and HA2, which indicated the successive letters and figures cultivar, growing area, and the storage period, respectively.

A panel composed of eight men and six women, making a total of 14 panelists (40 years' average age) performed the analysis. They all belonged to the Instituto de la Grasa staff and had vast experience on sensory studies due to their participation in the development of the Sensory Analysis Method for Table Olives [4] and the permanent involvement in diverse IG table olive sensory projects (e.g., [10,11]). Before the tests, the panelists were trained for one h twice a week for two months to familiarise them with the QDA techniques and the black ripe olive descriptors, while using industrially processed Spanish cultivars black ripe olives. The presentation of the samples was always made in the standard glasses [24], which were coded with three randomly chosen digits. After each test, the mouth was washed with tap water, freely available in each booth. Therefore, the panelists were progressively familiarised with the product, the sensory descriptors that were included in the evaluation sheet, informal tentative evaluations, and, finally, allowed for practicioning with the unstructured scale (1, complete absence; 11, strongest perception) of the evaluation sheet for another month. After these periods, they were considered ready for the evaluation of the real samples because of the previous expertise of the panelists in sensory testing. The assessed descriptors included appearance (skin red, skin green, skin sheen, flesh red, flesh yellow, and flesh green), aroma (briny, mushroom, earth/soil, oak/barrel, nutty, artificial fruity/floral, natural fruity/floral, vinegary, alcohol, fishy smell/ocean, and cheese smell), taste (sourness, bitterness, and saltiness), flavor (ripeness, buttery, metallic, rancid, soapy smell/medicinal, and gassy smell), and texture/mouthfeel (firmness, fibrousness, moisture release, mouth coating, chewiness, astringency, and residual). Their definitions and references may be found elsewhere [10].

For performing the tests, the black ripe olive samples were presented to panelists at an ambient temperature (20 ± 1 °C) and in a panel room that was equipped with individual booths under incandescent white lighting and free from any odors. The panelists were asked to mark the intensity of the different descriptors in the evaluation sheets. The scores of the attributes were measured with the exactitude of one decimal point and the results tabulated.

2.2. Data Analysis

The data were mainly studied while using the SensoMineR v.1.07 software (Agrocampus Ouest, Rennes, France) [25], a package that was designed and programmed in R language [26]. It is characterized by combining classical sensory statistical methods as well as others directly conceived in the developers' laboratory. In this way, SensoMineR provides a synthesis of the results of the usual analysis of variance (ANOVA) models, as well as a diversity of easy to interpret graphical outputs. Notably, the package includes several options for the panel evaluation, such as multivariate analysis and the generation of virtual panels, by bootstrapping techniques, which allow for the estimation of the corresponding confidence limits. XLSTAT [27] was also applied in specific analysis and tests.

3. Results and Discussion

The matrix of data was constituted by the following variables: sample-storage period (just sample from now on), panelist, session, and the 33 descriptors making a total of 36 columns. Additionally, sample, panelist, and session had 8, 14, and 3 levels, respectively, making a total of 336 rows. Therefore, the overall number of cells was 12,096. The generated database was already used for product characterization [10], but, in this work, the analysis is focused on the panel and panelists performance as an exercise for improving their evaluation and training.

3.1. Overview of Results

After checking the dataset for possible outliers and typing errors, they were also subjected to a first overview (frequency histograms and boxplots), which indicates that several descriptors received low scores and they were hardly noticed; however, others were perceived by the panelists, distributed along the scale, and allowed for discrimination among samples (data not shown). Further details can be found elsewhere [10].

3.2. Panel Performance

The techniques that are available for panel and panelists performance are numerous, with ANOVA and multivariate analysis being the most common. Kermit and Lengard Almli [16] presented univariate and multivariate data analysis methods to assess the individual and group performances in a sensory panel. Notably, Husson et al., [25] developed the SensoMineR, which includes several innovative tools with this objective.

3.2.1. Effect of Sample (Power of Discrimination)

The evaluation of the panel performance is an essential premise not only for obtaining reliable results on sensory analysis, but also for improving the selection of panelists and their training. In this work, the *panelperf* instruction from SensoMineR, with the appropriate models and the corresponding analysis of variance, was used. The ANOVA was fitted to the following full model:

Score = sample + panelist + session + sample panelist + sample session + panelist session

where score stands for the expected evaluation value, while sample, panelist, and session for the predictive variables, with the effect of storage being included as levels of the variable sample. The panelist and the session were both studied as random effects, but the sample was considered to be fixed [28].

The results regarding performance (Table 1) showed that the panel was able to discriminate the samples based on skin green, flesh green, skin sheen, flesh red, firmness, fibrousness, flesh yellow, skin red, vinegary, moisture release, fishy smell/ocean, and saltiness. Good segregation among the samples or products by panelists is systematically reported in numerous publications ([6,17,28–30], among others).

3.2.2. Effect of Panelist

The significant effect of the panelist, with very low *p*-values, regardless of descriptors, indicates a different interpretation of the scales. Such an effect is not desirable, but it is usually observed. However, its presence does not represent any inconvenience for achieving appropriate conclusions, since the panelists' variance can be eliminated thanks to the ANOVA analysis and by centring the data with respect to panelists [31]. The assessors' performance will be studied in detail later.

3.2.3. Effect of Session

The effect of the session was not significant for any descriptor (Table 1), which indicates an overall good panelist performance over time (the samples were assessed in the same way from one session to another), which is an appropriated and desired situation. Subsequently, no further comments regarding this aspect are also required.

Table 1. Overall panel performance as assessed by analysis of variance (ANOVA) sorted by sample p-values, including main effects, and interactions. Panelist, session, and their interactions were considered as random, while the sample was studied as fixed factor/variable.

Sensory Attribute	Sample	Panelist	Session	Sample·Panelist	Sample·Session	Panelist·Session	Median
Skin green	**2.792×10^{-10}**	**9.069×10^{-26}**	2.177×10^{-1}	**2.404×10^{-8}**	6.211×10^{-1}	9.423×10^{-2}	**4.712×10^{-2}**
Flesh green	**1.720×10^{-9}**	**2.186×10^{-15}**	2.377×10^{-1}	**2.602×10^{-4}**	7.443×10^{-1}	6.387×10^{-1}	1.190×10^{-1}
Skin sheen	**1.900×10^{-6}**	**1.742×10^{-28}**	2.497×10^{-1}	**1.011×10^{-4}**	6.362×10^{-1}	2.689×10^{-1}	1.249×10^{-1}
Flesh red	**8.603×10^{-6}**	**2.654×10^{-34}**	6.326×10^{-1}	**1.796×10^{-9}**	5.180×10^{-1}	**3.250×10^{-3}**	**1.629×10^{-3}**
Firmness	**1.033×10^{-4}**	**4.141×10^{-35}**	3.320×10^{-1}	**1.881×10^{-3}**	9.960×10^{-2}	**3.305×10^{-2}**	**1.747×10^{-2}**
Fibrousness	**9.292×10^{-4}**	**2.088×10^{-39}**	4.165×10^{-1}	**1.397×10^{-3}**	4.263×10^{-1}	**9.392×10^{-3}**	**5.394×10^{-3}**
Flesh yellow	**1.328×10^{-2}**	**7.265×10^{-20}**	2.752×10^{-1}	**1.752×10^{-4}**	2.046×10^{-1}	2.332×10^{-1}	1.090×10^{-1}
Skin red	**1.760×10^{-2}**	**3.535×10^{-51}**	6.585×10^{-1}	**1.511×10^{-10}**	2.695×10^{-1}	1.342×10^{-1}	7.590×10^{-2}
Vinegary	**1.833×10^{-2}**	**3.683×10^{-30}**	1.613×10^{-1}	**1.794×10^{-3}**	6.612×10^{-2}	**2.908×10^{-2}**	**2.370×10^{-2}**
Moisture release	**2.978×10^{-2}**	**4.515×10^{-33}**	9.058×10^{-1}	**9.558×10^{-6}**	5.027×10^{-1}	6.828×10^{-2}	**4.903×10^{-2}**
Fishy smell/Ocean	**3.060×10^{-2}**	**1.312×10^{-12}**	3.342×10^{-1}	3.912×10^{-1}	7.507×10^{-1}	7.249×10^{-1}	3.627×10^{-1}
Saltiness	**3.117×10^{-2}**	**1.680×10^{-48}**	3.191×10^{-1}	**2.575×10^{-3}**	**3.940×10^{-3}**	**1.068×10^{-3}**	**3.258×10^{-3}**
Astringency	2.232×10^{-1}	**1.599×10^{-45}**	9.491×10^{-1}	**6.062×10^{-14}**	6.593×10^{-1}	1.447×10^{-1}	1.839×10^{-1}
Ripeness	2.614×10^{-1}	**1.586×10^{-44}**	8.095×10^{-1}	**4.834×10^{-5}**	6.614×10^{-1}	**6.012×10^{-3}**	1.337×10^{-1}
Soapy smell/Medicinal	2.636×10^{-1}	**2.478×10^{-51}**	4.467×10^{-1}	4.560×10^{-1}	6.792×10^{-1}	**2.451×10^{-2}**	3.552×10^{-1}
Bitterness	2.710×10^{-1}	**4.434×10^{-38}**	2.556×10^{-1}	5.075×10^{-1}	1.940×10^{-1}	**2.364×10^{-3}**	2.248×10^{-1}
Chewiness	3.334×10^{-1}	**1.538×10^{-39}**	7.989×10^{-1}	**4.862×10^{-11}**	2.964×10^{-1}	**2.947×10^{-3}**	1.497×10^{-1}
Briny	3.567×10^{-1}	**1.708×10^{-37}**	7.944×10^{-1}	**4.671×10^{-6}**	8.067×10^{-1}	**2.152×10^{-2}**	1.891×10^{-1}
Natural fruity/Floral	4.102×10^{-1}	**2.781×10^{-30}**	1.840×10^{-1}	**4.931×10^{-3}**	6.946×10^{-1}	3.302×10^{-1}	2.571×10^{-1}
Rancid	4.867×10^{-1}	**1.815×10^{-41}**	3.093×10^{-1}	2.110×10^{-1}	6.873×10^{-1}	**2.669×10^{-2}**	2.601×10^{-1}
Nutty	4.892×10^{-1}	**4.653×10^{-26}**	3.041×10^{-1}	**3.637×10^{-3}**	2.203×10^{-1}	**1.108×10^{-2}**	1.157×10^{-1}
Buttery	5.223×10^{-1}	**7.225×10^{-42}**	3.749×10^{-1}	**1.292×10^{-9}**	4.572×10^{-1}	**6.488×10^{-3}**	1.907×10^{-1}
Oak barrel	5.496×10^{-1}	**4.336×10^{-44}**	3.501×10^{-1}	**7.740×10^{-3}**	9.681×10^{-2}	**2.796×10^{-6}**	5.227×10^{-2}
Metallic	5.778×10^{-1}	**1.374×10^{-24}**	1.115×10^{-1}	1.010×10^{-1}	**5.508×10^{-6}**	8.859×10^{-1}	1.062×10^{-1}
Alcohol	6.690×10^{-1}	**9.806×10^{-60}**	8.765×10^{-2}	2.337×10^{-1}	7.464×10^{-1}	1.730×10^{-1}	2.033×10^{-1}
Mushroom	6.795×10^{-1}	**5.867×10^{-26}**	4.712×10^{-1}	**3.910×10^{-6}**	1.256×10^{-1}	**6.242×10^{-7}**	6.280×10^{-2}
Mouth coating	6.925×10^{-1}	**2.358×10^{-56}**	7.926×10^{-1}	**5.014×10^{-17}**	2.719×10^{-1}	**1.033×10^{-3}**	1.365×10^{-1}
Sourness	7.219×10^{-1}	**1.227×10^{-24}**	6.145×10^{-1}	**1.206×10^{-2}**	8.290×10^{-1}	**8.804×10^{-4}**	3.133×10^{-1}
Earthy/Soil	7.335×10^{-1}	**1.713×10^{-28}**	2.620×10^{-1}	2.050×10^{-1}	8.907×10^{-2}	3.771×10^{-1}	2.335×10^{-1}
Artificial fruity/Floral	8.387×10^{-1}	**8.279×10^{-37}**	2.036×10^{-1}	**1.018×10^{-4}**	9.372×10^{-1}	8.575×10^{-1}	5.211×10^{-1}
Residual	8.937×10^{-1}	**4.326×10^{-43}**	1.212×10^{-1}	**1.784×10^{-9}**	6.901×10^{-1}	9.118×10^{-2}	1.062×10^{-1}
Cheesy smell	9.075×10^{-1}	**2.015×10^{-27}**	6.441×10^{-1}	**8.088×10^{-3}**	3.692×10^{-1}	**2.652×10^{-4}**	1.886×10^{-1}
Gassy smell	9.389×10^{-1}	**3.652×10^{-33}**	5.786×10^{-1}	1.249×10^{-1}	9.484×10^{-1}	7.670×10^{-1}	6.728×10^{-1}

Note: Significant values at $p \leq 0.05$ are indicated in bold.

3.2.4. Sample·Panelist Interaction

In the case of a total consensus among the members of the panel to assess the descriptors in all samples, their effects should not be significant. However, in this work, there were numerous significant cases (Table 1). The evaluation of the interaction is usually measured by the coefficients of the ANOVA, defined as the difference between the expected mean score by all panelists and that given by a specific one. It is tedious to reproduce their meaning in all descriptors, so only the case of skin red and flesh red are shown as examples (Figure 1). The effect might be significant because of two circumstances: (i) the panelists do no rank the samples in the same order and (ii) they do no use the scale in the same way. Both situations were found in this work. Examples of different ranks were observed, among other descriptors, for skin red, panelist 1 gave the highest score to HA1, but panelist 2 ranked it as the second one from the bottom; a similar behaviour occurred for flesh red regarding panelist 5 with respect to panelist 6 (Figure 1).

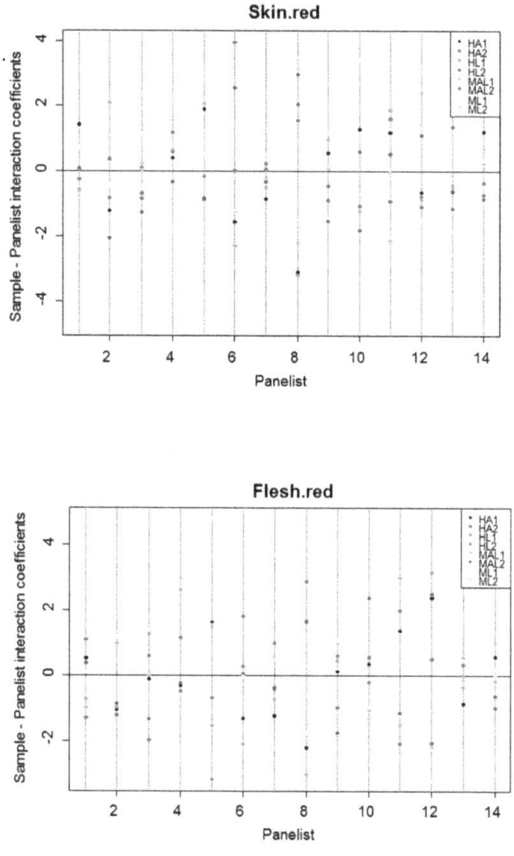

Figure 1. Panel performance. Sample·panelist interaction coefficients for selected descriptors (skin red and flesh red).

On the other side, for skin red, panelist 1 used a narrower scale than panelist 6; the same trend can be observed for flesh red by panelist 1 and panelist 12 (Figure 1). Therefore, to improve panel performance, it will be required further additional training in the scoring of some attributes and the amplitude of their scales.

The corresponding coefficients of each panelist in the ANOVA model were assessed by the identification of the panelists who mainly contributed to the interaction [19]. With this aim, the difference between the expected score and that given by a concrete panelist, overall sessions and samples, represent how far a specific panelist scores the sample differently to the product mean of the whole panel. No significant differences were usually observed (panelists had, in general, good reproducibility), but some peculiarities were noticed. For example, panelist A12 scored skin green (Figure 2A) sensibly higher than any other panelist; subsequently, he was critical in the significance of this interaction. Additionally, panelist A3 tends to scoring skin red, skin sheen, and flesh red above the panel average (Figure 2A).

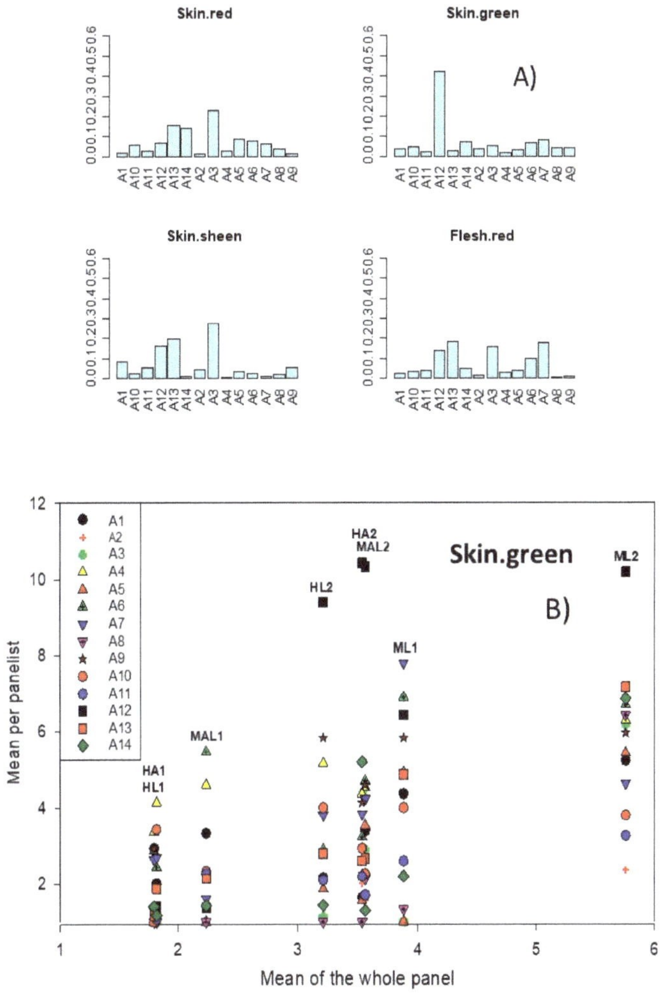

Figure 2. Panel performance. Sample·panelist interaction as assessed by (**A**) the panelist's contributions (coefficients) for selected descriptors (skin red, skin green, skin sheen, and flesh red), and (**B**) means of panelists over the whole panel according to samples.

Another way of observing the sample·panelist interaction and measuring the panelists' reproducibility is by plotting the mean per panelist over the mean on the whole panel according to samples. In agreement with previous comments, some panelists gave high scores to several descriptors and, in this line, panelist A12 overscored skin green in samples HL2, HA2, MAL2, and ML2 (Figure 2B). These high scores were due to a tendency of this panelist to evaluate several descriptors (flesh yellow and briny, data not shown) higher than other panel members. Similarly, outstanding scores were observed for panelist A5 in vinegary, alcohol, and sourness, and for panelist A8 in mouth coating, chewiness, stringency, and residual (data not are shown). However, most of the panelists differently scored only one descriptor like A4 in grassy smell, A10 in cheesy smell, A3 in a buttery, or A6 in rancid, to mention a few cases. Therefore, no panelist systematically contributed to the interaction, but the above-mentioned results could indicate that the panel performance would be improved by the further training of some panel members (A12, A5, and A8, on several descriptors or A4, A10, A3, or A6, only regarding specific ones). Kermit and Lengard Almli [19] also found several assessors who showed poor performance in some attributes, such as mealiness or fruity flavor.

3.2.5. Sample·Session Interaction

These interactions refer to the variation of the mean of each sample from one session to another and they should not be confused with the session effect, which applies to the mean of all samples between sessions. In the study (Table 1), the sample·session interaction was only significant in two cases: saltines (which was an important descriptor for sample discrimination) and metallic (Table 1). In saltiness, the significant interaction was mainly produced because of the different scoring for samples HA2, HL1, HA1, MAL1, and MAL2 in session S1 (Figure 3), while, in the case of metallic, the significant interaction is due to the abnormally high score of MAL1 in session S1 (Figure 3).

3.2.6. Panelist·Session Interaction

If significant, it means that one or more panelists do not similarly grade for all of the products from one session to another. There were several significant panelist·session interactions. Among the descriptors that contributed to discrimination, mushroom, oak barrel, cheesy smell, sourness, chewiness, bitterness, and saltiness had significant interactions (Table 1). The contribution of panelists to this interaction might also be evaluated by their respective coefficients, estimated as above-commented. Figure 4 shows examples.

Among the panelists that most contributed to the differences in scores between sessions according to descriptors, were: A13 for skin red, flesh red, and flesh green. Regarding other descriptors, A12 actively contributed to vinegar or A5 to natural fruity/floral, alcohol, and earthy soil (data not shown). However, most of the panelists had homogeneous contributions in most of the descriptors (skin green, skin sheen, flesh yellow, or briny, Figure 4). Moreover, no panelist showed a systematic trend for all descriptors, except a few of them, like A12 for skin sheen and flesh red or A7 for mushroom (Figure 4). Subsequently, the interaction was mainly due to the contribution of a reduced number of panelists (frequently only one) with limited influence on the panel repeatability.

The panelist·session interaction might also be presented as a plot of the mean per session over the mean on the whole sessions, according to panelists (Figure 5). Ideally, they should follow a line, regardless of sessions. In general, the panelists followed a similar trend over sessions (Figure 5 for some descriptors) with only punctual exceptions, like panelist A6 for rancid. Other cases were related to panelists A4, A12, and A8 for bitterness due to the abnormally low scores given by them (data not shown).

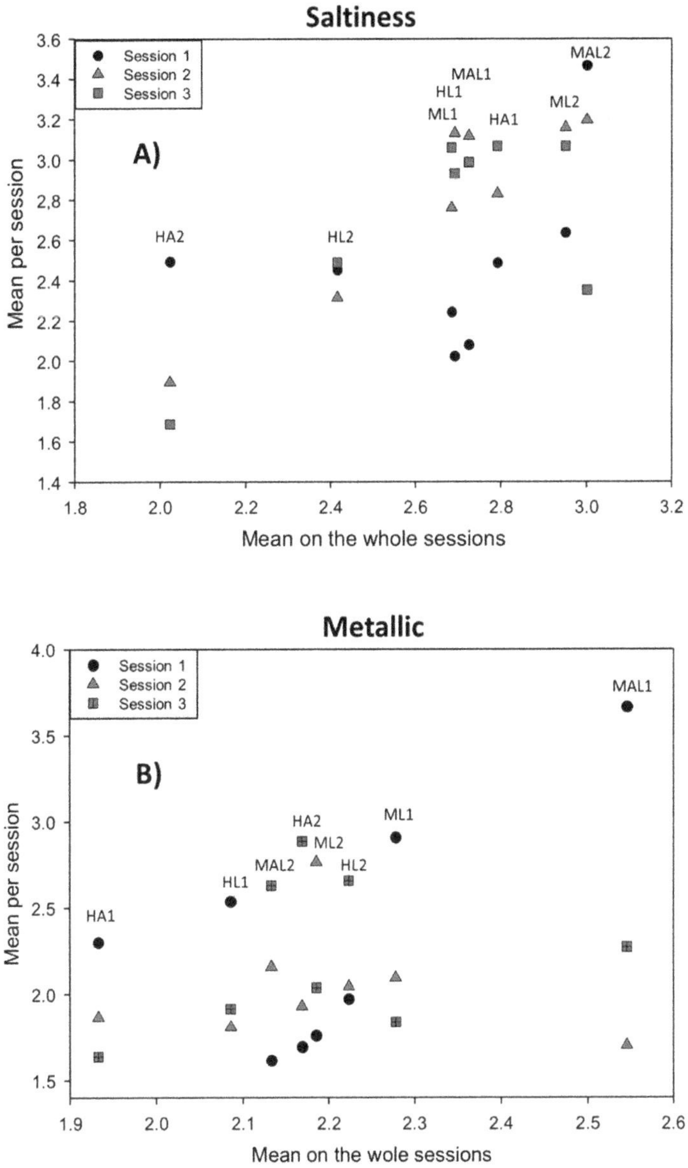

Figure 3. Panel performance. Sample-session interaction. Mean per session of panelists, according to samples, over the sample means of the whole sessions for significant descriptors: (**A**) saltiness, and (**B**) metallic.

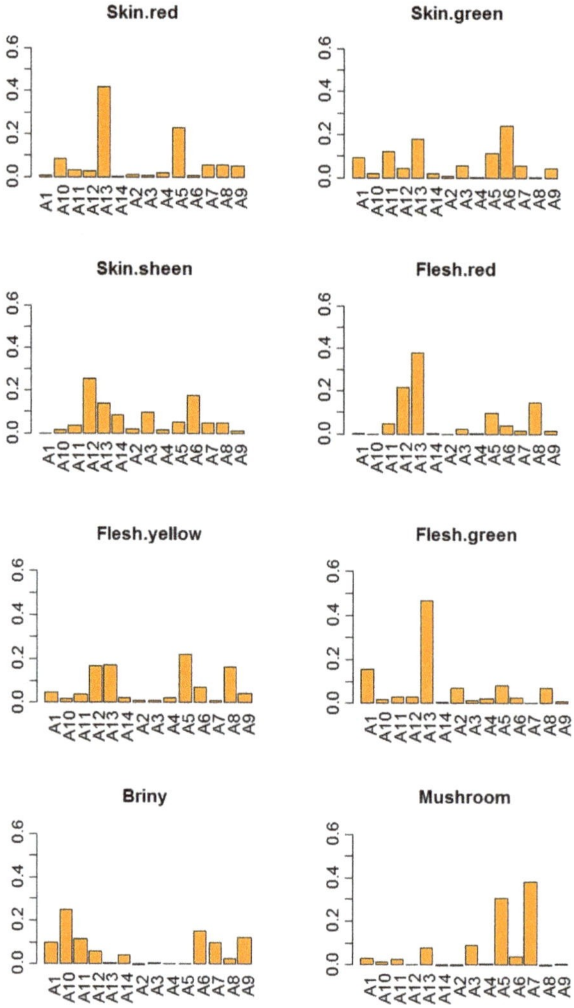

Figure 4. Panel performance. Panelist·session interaction. Contribution (coefficients) of panelists to the interaction for selected descriptors (skin red, skin green, skin sheen, flesh red, flesh yellow, flesh green, briny, and mushroom).

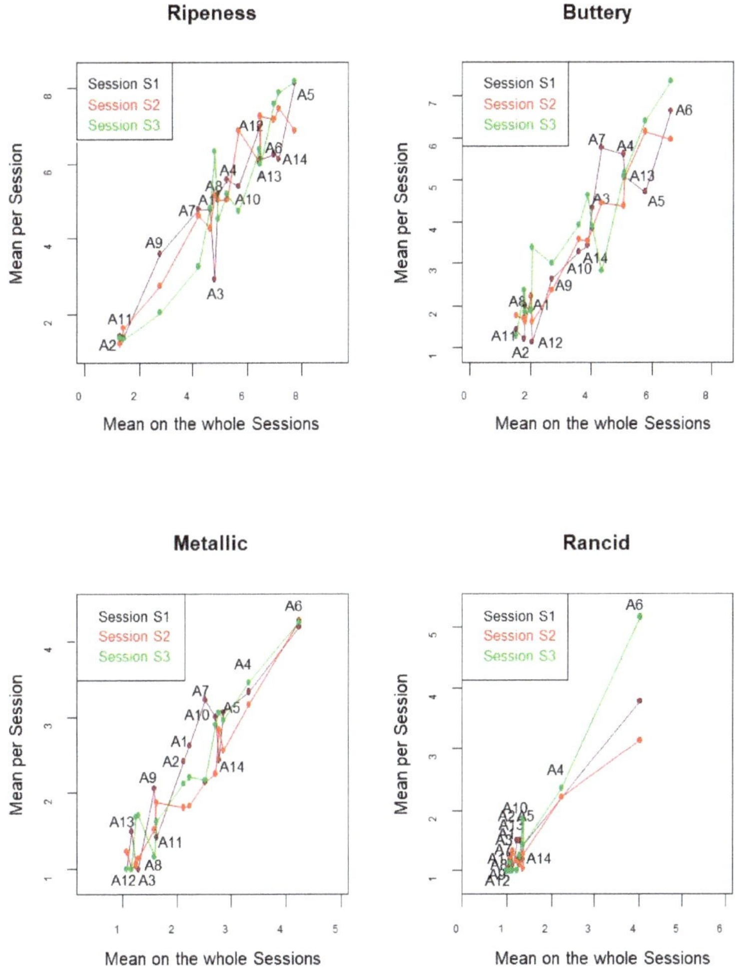

Figure 5. Panel performance. Panelist-session interaction. Means per session according to panelists over means of the whole sessions for selected descriptors (ripeness, buttery, metallic, and rancid).

Finally, the plot of the different coefficients over sessions is the most common evaluation of the panelist·session interaction (Figure 6, for flesh red as an example). In this case, the problems that could be observed are, again, of different ranking in successive sessions or different amplitude of scale over sessions. In Figure 6, panelist A13 assigned an excessive high score in the first session, while in the second session the score was low. Additionally, the amplitude of the scale for this descriptor was wider-spread in the first session than in the second. In saltiness, the situation was different, A12 had a very low contribution (coefficient) but the scale amplitude was similar among sessions; in firmness and fibrousness, panelist A13 was the only who had an excessive high score and, subsequently, a high contribution to the interaction, while, on the contrary, had low contribution on saltiness. Therefore, the analyses in detail of this interaction allowed for detecting some weakness of panel performance and lack of coherence in some panelist. Then, personalized training would be advisable.

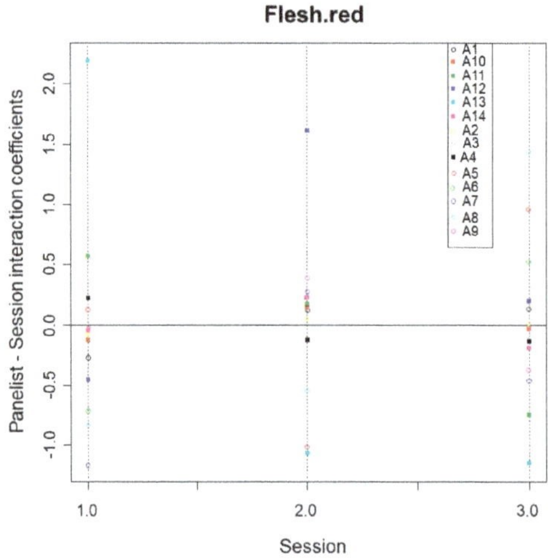

Figure 6. Panel performance. Panelist-session interaction. Detail of the coefficients through the three sessions for the flesh red descriptor.

3.3. *Panelist Performance*

When a panelist can discriminate among samples and is well repeatable and reproducible (that is, score the same product consistently and agrees with the rest of the panel), it is considered to be reliable according to Rossi [18]. There are several techniques for evaluating these panelist's performance parameters. Tomic et al. [20] develop a series of graphs for easy visualisation of the sensory profiling data for performance. Kermit and Lengard Almli [19] mentioned consonance analysis with PCA, full ANOVA model and notation, assessor sensitivity, assessor reproducibility, or agreement test as appropriate to evaluate the assessor and panel performance. Lanza and Amoruso [17] mention the repeatability index (RI_t) and deviation index (DI_t) to evaluate how assessors perform against themselves over time and their performance with respect to the whole panel, respectively. In this work, the diverse tools that were proposed by Husson et al. [31] for studying the panelist work will be particularly followed.

3.3.1. Discrimination Power of Each Panelist

The individual efficiency of panelists was evaluated with the model: score = sample + session. The *p*-values (Table 2) that are associated with the F-test of the sample effect on each panelist are, then, the appropriate parameter to measure this discrimination power. Their values, with rows and columns being sorted by the median estimated over them (Table 2), showed that most of the panelists were able to discriminate the black ripe table olive samples based on several of the descriptors that were developed by Lee et al. [9] and used later by López-López et al. [10]. Their efficiencies, in decreasing order, were: A14, A4, A2, A3, A6, A5, A8, A1, A12, A13, and A7, while only A11, A10, and A9 had not any discriminant power (Table 2). Skin green was the only descriptor that received an overall significant median; however, mouth coating, flesh red, briny, flesh green, or skin red were among the attributes most differently perceived in the samples (Table 2). On the contrary, soapy smell/medicinal, fishy smell, cheesy smell, alcohol, or metallic were among the most similarly perceived; however, this does not necessarily mean that the panelists were not able to differentiate samples, but that they were present in very low intensity or even completely absent (Table 2). There is controversy in the possible *p*-value that could be used as a cut off-level to consider one panelist acceptable. Stone et al. [32] proposed $p \geq 0.5$, but the problem was that there were so many *p*-values below 0.5 when evaluating tea that almost any laboratory would retain them. Powers [33] pointed out that the real question was establishing the number of attributes with significant performance being necessary for a judge to be an acceptable assessor. However, no agreement on this aspect was achieved. In this work, in general, the panelists were not systematically excellent in all descriptors, but most of them were good at some descriptors (significant *p*-value), and their overall performance was reasonable; however, the behaviour of panelists A11, A10, and A9 should be, according to these results, candidates for possible further training or even removal from the panel if their performance will not sufficiently improve. Kermit and Lengard Almli [19] also identified an assessor with further need for training in attributes pea flavor, sweetness, fruity, and off flavor.

3.3.2. Panelist Repeatability

The panelists' repeatability is the ability to consistently score the same product for a given attribute [18] and was evaluated by the standard deviation (SD) of the measurements of a descriptor from each panelist on each sample. It was considered that, when the residual of the ANOVA model for each panelist and descriptor (Table 3) was ≤ 1.96 ($p \leq 0.95$), the panelist scored the samples in a narrow range through the successive sessions and only panelists with residuals that were above this limit scored differently between sessions. In this work, there were no panelists who systematically graded the descriptors differently from one session to another (SD ≥ 1.96, in bold); however, several of them showed residuals above the limits for one to various descriptors, but not at a large distance. Therefore, in general, the panelists showed acceptable repeatability.

3.3.3. Panelist Reproducibility

The panelist agreement with the panel, as associated to reproducibility [18], was assessed by the correlation between the panelists' scores and the adjusted means of the panel (estimated by the ANOVA model) according to descriptors.

The procedure is similar to that used by Nyambaka et al. [30] to study the sensory changes in dehydrated cowpea leaves. The data are presented in a table, in which both panelists (in the column) and descriptors (in rows) are sorted from the highest to the lowest marginal median (Table 4). The panelists' agreement with the panel (significant correlation, in black) were, in descending order of their medians, A6, A8, A14, A5, A1, A7, A13, A10, A9, A3, A2, A4, A12, and A11, while the negative correlation (in black and italic) was distributed more or less evenly, indicating opposed agreement with the panel (divergent behaviour). The inconsistence of some panelists when evaluating cowpea leaves was attributed to particular preferences of assessors [30] and could also be possible in table olives for some attributes, like firmness or fibrousness.

Table 2. Discriminant power of panelists as assessed by the *p*-value of the F test.

	A14	A4	A2	A3	A6	A5	A8	A1	A12	A13	A7	A11	A10	A9	Median
Skin green	**0.0007**	**0.0119**	**0.0011**	**0.0010**	**0.0016**	0.0687	**0.0002**	0.1243	5.2×10^{-7}	**0.0402**	0.0655	0.5177	0.6086	0.1772	**0.0261**
Mouth coating	**0.0491**	**0.0188**	0.2445	**0.0023**	**0.0162**	**0.0063**	**4.8×10^{-7}**	0.0927	0.7896	**0.0042**	0.2880	0.4278	0.2185	0.8250	0.0709
Flesh red	**0.0103**	**0.0294**	0.448	**0.0019**	**3.9×10^{-6}**	**0.0073**	0.6432	0.3536	**0.0409**	0.1384	**2.4×10^{-5}**	0.1661	0.4706	0.1965	0.0897
Briny	0.0912	**0.0490**	**0.0074**	0.3023	0.0138	0.1940	0.0679	0.5585	**0.0370**	0.1132	0.4739	0.2358	0.1762	**0.0419**	0.1022
Flesh green	0.0564	**0.0049**	0.0885	0.0792	0.0001	0.2135	0.1192	0.1599	**0.0240**	0.1850	5.3×10^{-7}	0.3497	0.1588	0.1216	0.1039
Skin red	**6.7×10^{-6}**	0.1042	**0.0225**	**1.0×10^{-6}**	**0.0063**	0.5103	0.1245	0.3750	**0.0251**	**0.0046**	0.2455	0.0883	0.4822	0.2204	0.0963
Residual	0.1602	**0.0005**	0.4853	0.1151	**0.0480**	0.0976	**4.9×10^{-10}**	0.0514	0.5184	0.6976	0.3556	0.2728	0.7313	0.0727	0.1376
Oak barrel	0.1427	**0.0006**	0.9411	0.1950	0.2170	0.8693	0.4858	**0.0194**	0.1548	0.2937	0.4778	0.1655	0.1373	0.1540	0.1802
Fibrousness	0.1065	**0.0220**	**6.4×10^{-5}**	0.3311	0.1991	0.3859	**0.0055**	0.1004	**0.0410**	0.5118	0.0525	0.6702	0.4188	0.6076	0.1528
Ripeness	0.1602	**0.0005**	0.4853	0.4853	**0.0480**	0.0976	**4.9×10^{-10}**	0.0514	0.5184	0.6976	0.3556	0.2728	0.7313	0.0727	0.1376
Firmness	0.1345	**0.0402**	**0.0444**	**0.0022**	0.1130	0.7759	0.1552	0.4107	0.2639	0.1660	0.0009	0.3727	0.2481	0.8867	0.1606
Skin sheen	0.2700	0.2860	**0.0323**	**0.0003**	**0.3672**	0.0109	0.0269	**7.7×10^{-5}**	0.1841	**0.0009**	0.6877	0.1887	0.1489	0.8764	0.1665
Flesh yellow	0.1231	**0.0116**	0.081	0.1643	0.7031	0.2925	0.0878	0.5186	**0.0442**	0.5177	**6.6×10^{-5}**	0.3582	0.0754	0.2687	0.1437
Bitterness	**0.0100**	0.6474	**0.0056**	0.2079	0.5114	0.7813	0.2312	0.0501	0.2547	0.2341	0.9994	0.2390	0.1831	0.5097	0.2365
Astringency	0.5481	**0.0240**	**0.0374**	0.2538	0.6887	0.2064	**1.5×10^{-8}**	**0.0159**	0.1436	0.7140	0.0508	0.9816	0.6175	0.4225	0.2301
Moisture release	0.1299	0.2333	0.2658	**0.0004**	**0.0052**	0.6328	**0.0480**	0.3207	0.0604	0.0330	0.102	0.5576	0.8638	0.2983	0.1816
Nutty	**0.0392**	0.0793	0.3598	0.2546	0.1677	0.0824	0.2438	0.2683	0.4706	0.8304	0.4658	0.1908	0.3928	0.1083	0.2492
Buttery	**7.6×10^{-7}**	0.2047	0.2191	**6.1×10^{-6}**	0.6094	0.1268	0.8115	**0.0356**	0.2662	0.1992	0.0634	0.5658	0.3017	0.9222	0.2119
Mushroom	0.8962	**0.0003**	0.6540	**0.0006**	0.6046	0.3799	0.7030	**0.0476**	0.4706	0.5158	**0.0077**	0.1854	0.5196	0.136	0.4253
Chewiness	**0.0003**	0.4448	0.7152	**0.0156**	0.2636	0.6584	**6.5×10^{-9}**	**0.0003**	**0.0422**	0.2922	0.1441	0.6937	0.9195	0.6294	0.2779
Sourness	0.2963	0.6765	0.6530	0.4706	0.0155	0.1844	0.4783	0.2397	0.3977	0.2461	0.3597	0.8106	0.0778	0.7095	0.3787
Saltiness	0.5401	0.6428	0.4611	0.1292	0.0715	0.2307	**0.0075**	0.0147	0.1889	0.9875	0.7995	0.7577	0.4706	0.1714	0.3459
Vinegary	0.1315	0.6339	0.3024	0.3830	0.2282	0.1108	0.4706	0.3442	0.0716	0.5815	0.3609	0.4651	0.4706	0.5146	0.3719
Nat. fruity/floral	0.0598	0.7439	0.7778	**0.0112**	0.4959	0.4959	0.4706	0.3475	0.6836	0.2703	0.6252	0.3119	0.0554	0.7819	0.4091
Earthy soil	0.3278	**0.0136**	0.1162	0.6321	0.9603	0.2985	0.4706	0.561	0.4706	0.3367	0.5492	0.4013	0.5284	0.1496	0.4360
Rancid	0.09477	0.2697	0.2386	0.5123	0.3915	0.6038	0.4706	0.3878	0.4706	0.4822	0.5793	0.1436	0.5866	0.6142	0.4706
Art. fruity/floral	**4.4×10^{-5}**	0.4568	0.3301	0.5810	0.0916	**0.0457**	**0.0279**	0.2441	0.4706	0.0880	0.4715	0.4706	0.4876	0.7257	0.3934
Metallic	0.2645	0.5882	**0.0167**	0.4980	0.5781	0.8309	0.5770	0.3776	0.4706	0.4706	0.3236	0.3508	0.4074	0.5429	0.4706
Alcohol	0.3289	0.4027	0.1303	0.5559	0.7270	0.2080	0.5836	0.4118	0.4706	0.3147	0.3841	0.4706	0.4706	0.7604	0.4412
Cheesy smell	0.5312	0.5468	0.1270	0.6647	0.3881	0.7116	0.4706	0.3594	0.4706	0.7706	0.1850	0.5121	0.0275	0.7174	0.4914
Fishy smell	0.7693	0.1763	0.1578	0.5570	0.2083	0.7899	0.4706	0.3009	0.4706	**0.0289**	0.4288	0.1892	0.7001	0.5644	0.4497
Soapy smell/med	0.6524	0.5964	0.1828	0.5373	0.7019	0.6411	0.6237	0.8154	0.4706	**0.0145**	0.4853	0.5051	0.7512	0.4998	0.5669
Median	0.1299	0.1763	0.1828	0.1950	0.2170	0.2307	0.2312	0.2583	0.2662	0.2703	0.3556	0.3582	0.4706	0.5097	0.2448

Note: Significant values at $p \leq 0.05$ are indicated in bold.

Table 3. Panelist repeatability as assessed by the ANOVA residuals according to descriptors.

	A1	A10	A11	A12	A13	A14	A2	A3	A4	A5	A6	A7	A8	A9
Skin red	1.63	**2.27**	0.88	1.85	1.91	1.03	0.60	1.06	0.94	**2.74**	1.69	**2.06**	1.41	1.03
Skin green	1.50	**2.06**	1.26	1.47	1.96	1.37	0.31	1.30	0.73	1.75	1.23	**2.02**	1.03	**2.42**
Skin sheen	0.98	1.57	1.27	**3.20**	1.61	1.15	0.51	1.43	1.03	1.39	1.82	1.81	0.77	1.63
Flesh red	1.74	0.06	1.41	**2.89**	**3.10**	1.23	0.84	1.64	0.94	1.58	1.11	1.61	1.86	1.27
Flesh yellow	1.22	1.90	1.38	**2.77**	1.16	1.16	0.67	0.21	0.67	1.55	**2.41**	0.78	1.75	0.91
Flesh green	1.76	**2.05**	1.46	**2.76**	**2.64**	1.65	0.54	1.06	0.88	**2.88**	1.40	1.11	1.85	**2.08**
Briny	1.37	1.58	1.09	**2.32**	1.34	0.91	0.46	1.00	0.39	1.41	1.45	1.74	1.27	1.75
Mushroom	0.58	0.91	1.06	0.09	1.35	1.09	0.75	1.13	0.53	1.08	**2.19**	1.65	1.41	0.88
Earthy soil	0.78	1.66	**2.00**	0.09	0.98	0.29	0.54	0.68	0.86	1.21	1.86	0.59	<0.01	0.77
Oak barrel	0.41	1.93	0.65	1.33	1.03	0.36	0.89	0.53	0.59	1.63	1.88	1.06	1.93	0.79
Nutty	0.21	1.19	0.80	0.10	0.64	0.40	0.54	0.77	0.54	0.51	1.07	0.96	**2.06**	1.33
Artificial fruity/floral	0.18	0.93	***0.02***	0.12	0.72	0.36	0.59	0.15	0.78	1.14	1.19	1.19	0.79	0.96
Natural fruity/floral	1.13	1.74	0.07	1.03	1.06	0.87	0.87	0.79	0.80	1.61	1.95	1.50	**<0.01**	1.05
Vinegary	0.24	***<0.01***	0.28	1.55	1.20	0.38	0.53	0.39	0.53	**2.31**	**1.98**	0.59	0.41	0.61
Alcohol	0.20	0.13	0.07	0.14	0.61	0.67	0.55	0.26	0.60	**2.03**	1.73	1.09	1.04	1.00
Fishy smell/ocean	0.48	0.85	0.15	0.14	0.96	1.08	0.31	0.38	0.84	1.37	1.46	1.71	0.35	1.58
Cheese smell	0.44	1.00	0.09	0.20	0.86	0.09	0.29	0.09	0.47	0.76	1.39	0.55	***0.02***	0.15
Sourness	0.14	0.56	0.53	1.28	0.35	1.89	0.14	0.20	0.70	**1.98**	0.84	1.65	0.65	0.28
Bitterness	0.80	1.06	0.30	1.27	1.54	0.87	0.42	0.55	0.87	**1.99**	1.07	**2.73**	1.59	0.37
Saltiness	0.78	0.13	0.65	1.75	0.76	0.27	0.30	0.40	0.86	1.19	1.19	**2.35**	1.33	0.93
Ripeness	1.27	**2.81**	0.52	**2.12**	1.95	1.73	0.45	0.98	0.73	1.43	1.49	**2.07**	1.69	1.04
Buttery	0.71	**2.26**	0.68	1.41	1.71	0.91	0.53	1.18	0.77	1.09	1.87	1.57	1.26	**2.12**
Metallic	1.62	1.37	0.54	0.25	0.82	1.45	0.52	1.09	0.95	1.84	**2.00**	1.69	0.82	0.80
Rancid	0.36	0.73	***0.04***	***<0.01***	0.45	0.56	0.37	0.33	0.74	0.70	**2.01**	0.26	***<0.01***	***0.05***
Soapy smell/medicinal	0.30	1.37	0.46	0.42	1.21	0.05	0.30	0.85	0.81	0.91	**2.50**	1.14	0.66	0.71
Gassy smell	0.12	0.59	0.05	0.71	***<0.01***	**<0.01**	0.12	0.47	0.70	0.12	1.41	0.13	***0.022***	0.07
Firmness	1.29	1.36	0.64	1.12	**2.36**	1.68	0.38	1.14	0.82	1.85	1.83	1.10	1.03	1.01
Fibrousness	1.02	0.77	0.62	1.25	**2.02**	1.43	0.25	1.51	0.71	1.14	1.56	1.35	1.53	1.17
Moisture release	1.36	2.02	0.61	1.20	1.85	1.34	0.29	1.01	0.67	0.99	1.41	1.32	1.03	1.42
Mouth coating	0.85	1.73	0.60	0.65	1.30	1.32	0.52	1.20	0.64	0.72	1.15	1.00	0.84	0.92
Chewiness	0.75	1.46	0.64	1.10	1.94	1.11	0.36	1.10	1.26	1.36	1.73	1.04	0.84	0.92
Astringency	0.05	1.60	0.53	1.19	**1.99**	0.07	0.12	0.91	0.62	0.54	1.37	1.45	0.91	0.10
Residual	0.62	**2.14**	0.13	0.13	1.41	1.89	1.36	0.34	0.95	0.82	1.35	1.28	1.04	0.55

Notes: Significant higher values at $p \leq 0.05$ are indicated in bold while an agreement is showed as bold and italic.

Table 4. Panelist agreement with panel as assessed by the correlation coefficient.

	A6	A8	A14	A5	A1	A7	A13	A10	A9	A3	A2	A4	A12	A11	Median
Skin green	0.674	0.828	0.792	0.810	0.660	0.603	**0.932**	0.547	0.814	0.808	0.853	0.874	0.791	**0.860**	0.800
Skin sheen	0.762	0.787	0.840	**0.962**	**0.809**	**0.869**	0.677	0.820	0.050	0.220	0.092	0.883	0.762	0.323	0.744
Flesh red	**0.907**	**0.904**	0.499	0.841	0.667	**0.956**	0.190	***−0.322***	0.816	0.268	0.759	0.524	0.722	0.420	0.695
Firmness	0.335	0.678	0.600	0.296	0.754	0.758	0.789	0.696	0.500	0.216	0.720	**0.875**	0.335	0.803	0.687
Flesh green	**0.960**	0.601	0.152	0.198	0.752	0.550	0.727	0.615	0.505	0.751	0.344	0.764	0.097	0.572	0.637
Fibrousness	0.644	0.639	0.152	0.198	0.752	0.550	0.727	0.615	0.505	0.751	0.344	0.833	0.965	***−0.658***	0.577
Flesh yellow	0.309	0.673	0.293	0.529	***−0.635***	0.598	0.801	0.556	0.653	0.306	0.649	0.538	***−0.226***	0.504	0.558
Moist. release	0.781	0.425	0.579	0.588	0.667	0.605	0.355	***−0.078***	0.653	0.306	0.649	0.538	***−0.226***	0.504	0.558
Fishy smell	***−0.018***	0.777	**0.885**	0.434	0.336	0.366	0.723	0.216	0.650	**0.886**	***−0.532***	**0.886**	0.774	0.245	0.542
Nutty	***−0.206***	0.831	0.606	0.401	***−0.049***	***−0.445***	0.674	***−0.262***	0.640	0.706	0.665	0.566	***−0.157***	***−0.580***	0.484
Astringency	***−0.387***	**0.921**	0.488	0.769	0.470	0.850	***−0.412***	0.576	***−0.254***	0.497	0.730	0.373	***−0.249***	***−0.471***	0.479
Briny	***−0.250***	0.478	***−0.043***	0.470	0.620	0.560	0.386	***−0.477***	0.530	0.513	***−0.133***	0.183	0.609	0.735	0.474
Ripeness	0.420	0.093	0.674	0.454	0.760	***−0.309***	0.690	0.559	0.618	***−0.106***	0.832	***−0.057***	0.444	0.373	0.473
Buttery	0.036	0.065	0.691	***−0.266***	0.600	0.405	0.751	0.537	0.407	***−0.353***	0.607	0.309	0.480	0.582	0.444
Skin red	**0.862**	0.356	0.551	***−0.063***	0.676	0.257	0.444	0.154	0.585	0.387	0.517	0.228	0.776	0.223	0.416
Chewiness	0.135	0.269	−0.045	0.485	0.235	0.568	0.646	0.476	0.311	0.414	0.046	0.677	0.406	0.811	0.410
Vinegary	**0.989**	0.255	0.798	**0.853**	0.392	0.079	0.422	-	***−0.277***	***−0.723***	***−0.511***	***−0.696***	**0.982**	0.594	0.392
Oak barrel	0.696	0.401	0.556	0.511	0.250	0.117	***−0.058***	0.367	0.546	0.118	0.640	0.334	***−0.426***	***−0.209***	0.351
Bitterness	***−0.145***	0.754	0.433	***−0.312***	0.590	0.533	0.735	0.127	0.158	0.740	0.245	0.053	***−0.039***	0.450	0.339
Saltiness	0.682	**0.850**	0.061	0.397	0.440	0.607	***−0.033***	0.792	0.531	0.060	***−0.312***	0.244	0.314	0.170	0.279
Earthy soil	0.703	-	***−0.843***	0.013	0.462	***−0.133***	0.478	0.544	0.328	0.239	***−0.283***	***−0.067***	***−0.392***	0.540	0.329
Mouth coating	0.770	0.770	***−0.558***	0.172	***−0.189***	0.283	***−0.143***	***−0.511***	0.406	**0.870**	0.434	0.817	***−0.393***	***−0.255***	0.228
Natural fruity/floral	***−0.514***	-	0.773	0.765	0.546	0.210	***−0.700***	0.685	***−0.482***	**0.952**	0.464	***−0.013***	0.055	***−0.798***	0.210
Mushroom	***−0.064***	0.473	0.023	0.834	0.386	0.408	0.286	0.152	0.076	0.343	***−0.070***	0.054	0.003	0.229	0.190
Cheese smell	0.237	0.050	0.258	***−0.058***	***−0.037***	0.379	***−0.241***	0.830	***−0.403***	0.846	0.478	***−0.652***	0.090	0.358	0.164
Gassy smell	0.695	0.537	-	***−0.099***	0.155	***−0.132***	-	0.138	***−0.037***	0.104	***−0.464***	0.192	0.341	0.180	0.146
Alcohol	***−0.294***	***−0.119***	0.526	0.805	***−0.192***	0.684	0.735	***−0.306***	0.389	0.011	***−0.292***	***−0.423***	0.798	0.260	0.135
Soapy smell/medicinal	0.134	***−0.387***	***−0.256***	**0.865**	0.279	0.795	**0.928**	***−0.370***	0.354	**0.865**	***−0.099***	***−0.056***	0.075	0.070	0.105
Sourness	0.605	***−0.037***	0.532	0.540	0.320	***−0.213***	0.123	***−0.149***	0.061	0.241	***−0.056***	***−0.414***	0.667	***−0.287***	0.092
Rancid	**0.932**	-	***−0.407***	0.694	0.038	0.038	***−0.108***	0.423	***−0.397***	0.568	0.589	***−0.132***	-	0.077	0.058
Metallic	0.466	***−0.117***	***−0.169***	0.307	0.700	**0.915**	***−0.601***	0.543	***−0.311***	***−0.061***	***−0.153***	0.776	0.087	***−0.527***	0.028
Artificial fruity/floral	0.597	0.579	0.310	0.359	0.441	***−0.337***	***−0.058***	0.103	***−0.217***	***−0.219***	***−0.137***	0.517	***−0.054***	***−0.054***	0.024
Residual	0.680	0.842	***−0.859***	0.068	***−0.500***	***−0.204***	***−0.731***	0.522	***−0.727***	0.760	***−0.680***	**0.876**	***−0.384***	***−0.234***	**−0.217**
Median	0.597	0.558	0.493	0.470	0.441	0.408	0.404	0.395	0.385	0.387	0.344	0.334	0.324	0.245	0.400

Notes: Significant agreement is indicated in bold while opposed behavior is shown in bold and italic.

Overall, the descriptors that had the best agreement between panelists and panel, sorted by the median, were (in decreasing order of relationship) skin green, skin sheen, flesh red, firmness, flesh green, fibrousness, flesh yellow, and moisture release (Table 4). They were also among the descriptors with the most discriminant power. On the contrary, those with more discrepancies among the panelists were residual, artificial fruit/floral, metallic, rancid, sourness, or soapy smell/medical (Table 4), all of them with no discriminant influence.

These results show that the overall behaviour of the panelists was reasonable, although there was still margin for some improvement in their performance, particularly regarding those panelists with strongly opposed correlation to the mean of the panel. Alternatively, they could be candidates for further rejection.

Lanza and Amoruso [17] used line plot according to the attribute and deviation index (DI_t) to evaluate the agreement between panelists and whole panel. Their results are in line with those described above, since they also found some panelists who clearly deviated from the consensus. According to these authors, this type of results helps the panel leader to identify repeatability problems of specific assessors as compared to the whole panel and correct the deviation by the corresponding training.

3.4. *Multivariate Study of Panelists and Panel*

3.4.1. Clustering

A first multivariate approach of the similarity among panelists was achieved by hierarchical clustering analysis based on the scores given to the sample descriptors by each of them. The study was performed in XLSTAT, while using Wards' aggregation criterion [28]. Three groups of panelists were formed when comparing the panelists' behaviour (Figure 7A). The greatest dissimilarity was found between the group that was formed by A4 and A6 with respect to the other panelists. The dissimilarity within the groups of other panelists was sensibly lower, leading to three groups. Two of them were composed of four and seven panelists, while the third only included panelist A8, who had a peculiar behaviour. Therefore, in this case, the cluster analysis, which considers the overall panelist performance, showed that the panelists followed a somewhat similar trend when evaluating the black ripe olive samples, but not reveal their peculiarities. In line with this result, the hierarchical classification is more usually applied for the classification of products or studying the association among descriptors. Francois et al. [28] used this technique for assessing the astringency of different beers while Pense-Lheritier et al. [29] applied it to link the sensory changes induced by the addition of drugs to different beverages. Alasalvar et al. [6] found similarity among the flavor of natural and roasted Turkish hazelnut cultivars. Clustering was also used to segregate different consumers segments according to their overall liking scores [34].

3.4.2. Panelist Reproducibility

The multivariate study of the agreement among panelists and the whole panel [18], while using bootstrapping, was made in SensoMiner, by considering the results of a virtual panel that was obtained by taking successive samples (500 simulations) from the real data and applying Principal Component Analysis. Only two eigenvalues ≥ 1 were found and they accounted for ~42 and 26% of the variance, respectively. The analysis was made while using the function panelipse·session. The resampling technique has been described in detail elsewhere [31].

The closeness of the whole panel and panelists' answers was evaluated by projecting them onto the first two PCs. A PCA on the consensus allows for visualizing the strength of the consensus and the global discrimination of the products; besides, treatments identification shows the observed differences between the products [35]. In this work, the distance from each panelist to the situation of the corresponding sample assessed the agreement between the whole panel (squares symbols and different colours for the samples) and the panelists' acronyms (associated to samples by circle symbols using the same colours) (Figure 7B).

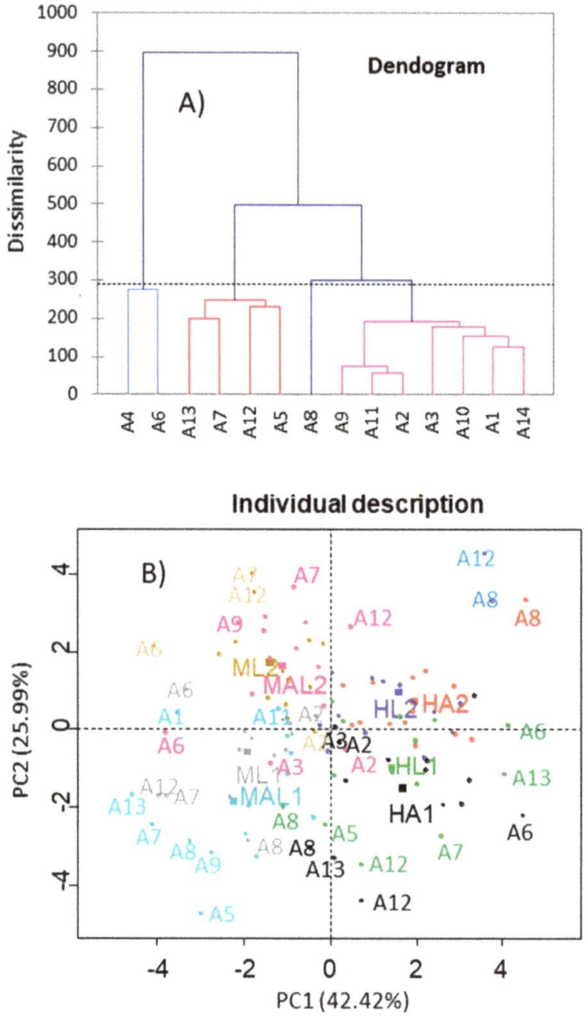

Figure 7. Panelist performance as assessed by multivariate analysis. (**A**) Clustering of panelists according to their performance. (**B**) Projection of panelists' loads (individual description) and samples' scores onto the first two Principal Components.

PC1 was highly efficient for segregating samples from Manzanilla (on the left) and Hojiblanca (on the right) and it could be associated to cultivar, while PC2 was able to distinguishing samples as a function of growing area and storage. In general, the projections of panelists for each sample were situated around that of the whole panel (sample associated to the same colour); although, there were some of them far for their respective samples. The discrepant panelists were (as identified by the corresponding acronyms) the same already mentioned in previous sections, mainly: A12, A8 for HL2; A8 for HA2; A13, A12, A8 and A6 for HA1; A12, A7, A9, A6, A3, and A2 for MAL2; A12, A7, A6, and A2 for ML2; A13, A11, A9, A8, A7, A5, and A1 for MAL1; and, A12, A8, A7, A6, and A2 for ML1. The panelist who scored the samples differently more times was A12, followed by A8, A7, and A6. Lower discrepancies were observed for A2, A9, A13, A3, and A5. However, they represent just a few cases of divergences, while most of the panelists' scores are jointly distributed around their corresponding samples. Additionally, panelists had greater ability (closeness to the sample average) to evaluate long stored Hojiblanca samples (HL2 and HA2) than any other sample. In conclusion, this plot has identified the panelists who will require particular training, but the performance of the others will also benefit from training. Our results are in agreement to those that were presented by Tomic et al. [21], who also found underperformance panelists and emphasized the need for a detailed study of their behavior while using the established statistical methods for the evaluation. Lanza and Amoruso [17] studied the performance of panelist against the whole panel using Eggsshell plots, concluding that there were also a few panelists that ranked some of the descriptors quite differently from the consensus, while there was a good agreement in others, like hardness.

3.4.3. Panel Repeatability

Study by Variables Projection on the Correlation Circle According to Sessions

The analysis was carried out using the virtual panel described above [31]. A first approach of the panel repeatability was observed by projecting the descriptors (only those more relevant, contribution >0.20) onto the first two PC according to sessions. Close situations of descriptors in the correlation circle for the different sessions indicate good repeatability. The panel was particularly repeatable among sessions for some descriptors, like skin green, astringency, flesh green, moisture release, fibrousness, flesh red, skin sheen, or flesh yellow. However, others had sensible distances from one session to another, like fishy smell/ocean, saltiness, or chewiness (Figure 8A). The interpretation of the relationships among variables is not straightforward due to these oscillations on the variables' projections. Nevertheless, it is possible to establish overall associations, mainly in those variables with high repeatability among sessions. For example, firmness, fibrousness, or chewiness are opposed to moisture release, ripeness, or flesh green. Additionally, those black ripe olives with high astringency could also present flesh yellow or skin green notes, but low vinegar or ripeness scores.

Galán Soldevilla et al. [14] associated bitter, sour, and wood with *Green*, *Cured*, and *Traditional Aloreña de Málaga* table olives, respectively. In black ripe olives, discrimination among the samples from different origins was mainly based on the 2nd and 3rd PCs, which were the components linked to aroma and flavour characteristics; however, the more linear behaviour of panelists was related to a textural dimension that was strongly connected to PC1 [9]. Kinesthetic sensations were also critical for the segregation between defected and un-defected samples by PCA [12].

Study by Sample Projections According to Sessions

The analysis was also carried out using the virtual panel described above. In this case, the median scores of the virtual panel perception of the samples (the same of the real panel) were projected onto the plane of the two first PCs according to sessions. Subsequently, 95% of the closest points of the generated cloud of points were used to draw their confidence ellipses (p-value = 0.05), which were built according to the procedure that was described by Husson et al. [31] (Figure 8B). The repeatability of the panel to the session can be assessed by the displacement of the sample centres. In general, the separation between the sample centres due to session was limited, indicating a good panel agreement between sessions, which is also corroborated by the overlapping of their confidence ellipses. Incidentally, the plot also indicates that the long stored fruits showed lower dispersion by sessions than the just processed fruits (one-month storage).

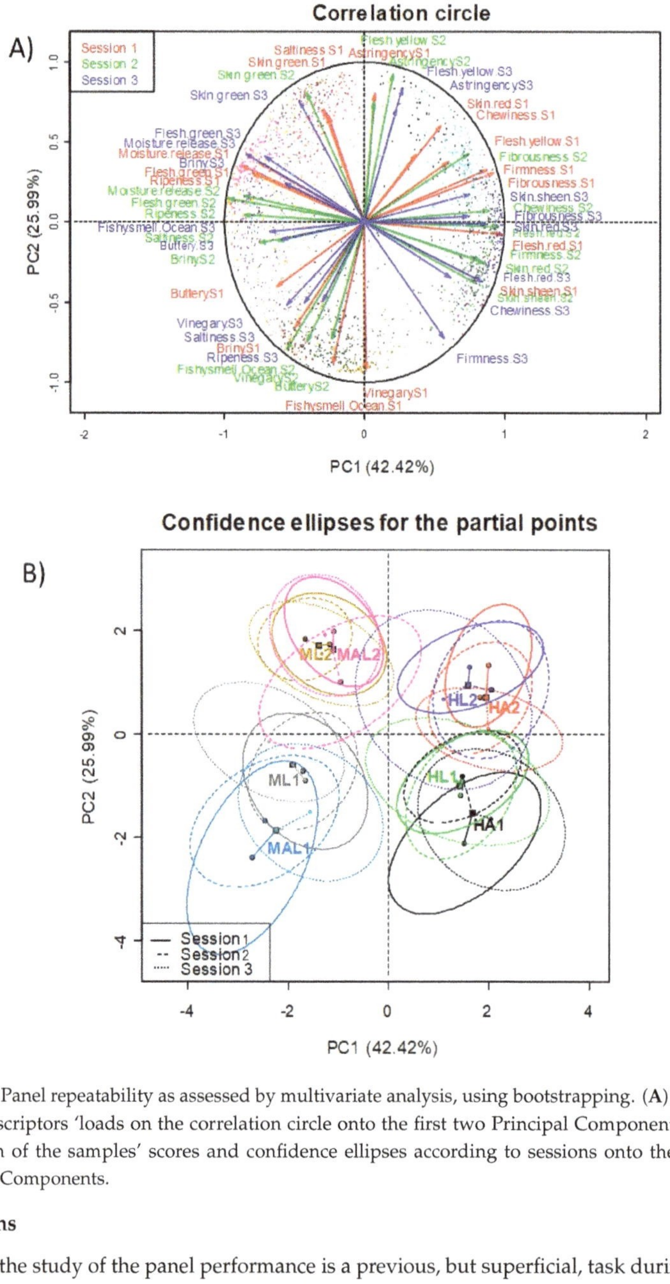

Figure 8. Panel repeatability as assessed by multivariate analysis, using bootstrapping. (**A**) Projection of the descriptors 'loads on the correlation circle onto the first two Principal Components, and (**B**) Projection of the samples' scores and confidence ellipses according to sessions onto the first two Principal Components.

4. Conclusions

Usually, the study of the panel performance is a previous, but superficial, task during the sensory evaluation of products. However, a detailed investigation of the panel and panelist performance is a convenient tool to uncover the details of their evaluation. In this work, such study allowed for the assessment of the panel performance as a whole, as well as detecting the panelist with the lowest discriminant power, those that have interpreted the scale in a different way than the panel and, therefore, require further training or even discovery that the stored black ripe olive products are more similarly perceived by the panelists over sessions. Besides, the study identified the descriptors of hard evaluation (skin green, vinegar, bitterness, or natural fruity/floral). Therefore, panelists would require particular training on them or, in case of not reaching the appropriate level of discrimination, be replaced by some other/s with higher sensitivity. In summary, the work has confirmed that such studies are an essential tool for the appropriate panel control and training, which should be a permanent concern of the panel leader.

Author Contributions: Conceptualization: A.L.-L. and A.G.-F.; Methodology: A.L.-L. and A.H.S.-G.; Software: A.G.-F.; Validation: A.L.-L. and A.G.-F.; Formal analysis: A.G.-F. and A.L.-L.; Investigation: A.C.-D., A.H.S.-G., A.M., A.L.-L.; Resources: A.L.-L., A.H.G.-S. and A.M.; Data curation: A.L.-L. and A.G.-F.; Writing—original draft preparation: A.L.-L. and A.G.-F.; Writing—review and editing: A.L.-L. and A.G.-F.; Visualization: A.L.-L. and A.G.-F.; Supervision: A.L.-L.; Project administration: A.M. and A.L.-L.; Funding acquisition: A.M. and A.L.-L.

Funding: This research was funded in part by the Ministry of Economy and Competitiveness from the Spanish government through Project AGL2014-54048-R, partially financed by the European Regional Development Fund (ERDF).

Acknowledgments: We thank Elena Nogales Hernández for her technical assistance.

Conflicts of Interest: The authors declare no conflict of interest.

References

1. International Olive Council (IOC). Online Reference Included in World Table Olives Figures: Production. 2018. Available online: http://www.internationaloliveoil.org/estaticos/view/132-world-table-olive-figures (accessed on 30 April 2019).
2. Garrido-Fernández, A.; Fernández-Díez, M.J.; Adams, R.M. *Table Olive Production and Processing*; Chapman & Hall: London, UK, 1997.
3. International Olive Council (IOC). *Trade Standards Applying to Table Olives. IOC/OT/NC No. 1*; International Olive Council: Madrid, Spain, 2004.
4. International Olive Council (IOC). *Method for the Sensory Analysis of Table Olives. COI/OT/MO No. 1/Rev.2 November 2011*; International Olive Council: Madrid, Spain, 2011; Available online: http://www.internationaloliveoil.org/estaticos/view/70-metodos-de-evaluacion (accessed on 30 April 2018).
5. Department of Agriculture. *United States Standards for Grades of Ripe Olives*; Agricultural Marketing Order, Department of Agriculture: Colombia, WA, USA, 1983.
6. Alasalvar, C.; Pelvan, E.; Bahar, B.; Korel, F.; Ölmez, H. Flavour of natural and roasted Turkish hazelnut varieties (*Corylus avellana* L.) by descriptive sensory analysis, electronic nose and chemometrics. *Int. J. Food Sci. Technol.* **2012**, *47*, 122–131. [CrossRef]
7. Dabbou, S.; Issaoui, M.; Brahmi, F.; Nakbi, A.; Chelab, H.; Mechri, R.; Hammani, M. Changes in volatile compounds during processing of Tunisian-style table olives. *J. Am. Oil Chem. Soc.* **2012**, *89*, 347–354. [CrossRef]
8. Heyman, H.; Hopfer, H.; Bershaw, D. An exploration of the perception of minerality in white wines by projective mapping and descriptive analysis. *J. Sens. Stud.* **2014**, *29*, 1–13. [CrossRef]
9. Lee, S.M.; Kitsawad, K.; Sigal, A.; Flynn, D.; Guinard, J.X. Sensory properties and consumer acceptance of imported and domestic sliced black ripe olives. *J. Food Sci.* **2012**, *77*, 439–448. [CrossRef] [PubMed]
10. López-López, A.; Sánchez-Gómez, A.H.; Montaño, A.; Cortés-Delgado, A.; Garrido-Fernández, A. Sensory characterisation of black ripe table olives from Spanish Manzanilla and Hojiblanca cultivars. *Food Res. Int.* **2019**, *116*, 114–125. [CrossRef] [PubMed]
11. López-López, A.; Sánchez-Gómez, A.H.; Montaño, A.; Cortés-Delgado, A.; Garrido-Fernández, A. Sensory profile of Green Spanish-style table olives according to cultivar and origin. *Food Res. Int.* **2018**, *108*, 347–356. [CrossRef] [PubMed]
12. Lanza, B.; Amoruso, F. Sensory analysis of natural table olives: Relationships between appearance of defect and gustatory-kinaesthetic sensation changes. *LWT-Food Sci. Technol.* **2016**, *68*, 365–372. [CrossRef]
13. Yilmaz, E.; Aydeniz, B. Sensory evaluation and consumer perception of some commercial green table olives. *Br. Food J.* **2012**, *114*, 1085–1094. [CrossRef]
14. Galán Soldevilla, H.; Ruiz Pérez-Cacho, P.; Hernández Campuzano, J.A. Determination of the sensory profiles of Aloreña table olives. *Grasas Aceites* **2013**, *64*, 442–452. [CrossRef]

15. Marsilio, V.; Campestre, C.; Lanza, B.; De Angelis, M.; Russi, F. Sensory analysis of green table olives fermented in different saline solutions. *Acta Hortic.* **2002**, *586*, 617–620. [CrossRef]
16. Lombardi, S.J.; Macciola, V.; Iorizzo, M.; De Leonardis, A. Effect of different storage conditions on the shelf life of natural green table olives. *Ital. J. Food Sci.* **2018**, *30*, 414–427.
17. Lanza, B.; Amoruso, F. Panel performance, discrimination power of descriptors, and sensory characterization of table olive samples. *J. Sens. Stud.* **2019**, e12542. [CrossRef]
18. Rossi, F. Assessing sensory panelist performance using repeatability and reproducibility measures. *Food Qual. Pref.* **2001**, *12*, 467–479. [CrossRef]
19. Kermit, M.; Lengard Almli, V. Assessing the performance of a sensory panel-panellist monitoring and tracking. *J. Chemom.* **2005**, *19*, 154–161. [CrossRef]
20. Tomic, O.; Nilsen, A.; Martens, M.; Naes, T. Visualization of sensory profiling data for performance monitoring. *LWT-Food Sci. Technol.* **2007**, *40*, 262–269. [CrossRef]
21. Tomic, O.; Forde, C.; Delahunty, C.; Naes, T. Performance indices in descriptive sensory analysis-A complimentary screening tool for assessor and panel performance. *Food Qual. Pref.* **2013**, *28*, 122–133. [CrossRef]
22. Sipos, L.; Ladányi, M.; Gere, A.; Kókai, Z.; Kovács, S. Panel performance monitoring by Poincaré plot: A case study on flavoured bottled waters. *Food Res. Int.* **2017**, *99*, 198–205. [CrossRef] [PubMed]
23. López-López, A.; Rodríguez-Gómez, F.; Cortés-Delgado, A.; Montaño, A.; Garrido-Fernández, A. Influence of ripe table olive processing on oils characteristics and composition as determined by chemometrics. *J. Agric. Food Chem.* **2009**, *57*, 8973–8981. [CrossRef] [PubMed]
24. International Olive Council (IOC). *Sensory Analysis of Olive Oil Standard Glass for Oil Tasting. COI/T20/Doc No. 5*; International Olive Council: Madrid, Spain, 1987.
25. Husson, F.; Lê, S. SensoMineR: Sensory Data Analysis with R. R Package Version 1.07. 2007. Available online: http://agrocampus-rennes.fr/math/SensoMinR (accessed on 08 November 2019).
26. R Development Core Team. *R: A Language and Environment for Statistical Computing*; The R Foundation for Statistical Computing: Vienna, Austria, 2011.
27. XLSTAT. *Data Analysis and Statistical Solution for Microsoft Excel*; Addinsoft: Paris, France, 2017.
28. François, N.; Guyot-Declerck, C.; Hug, B.; Callemien, D.; Govaerst, B.; Collin, S. Beer astringency assessed by time-intensity and quantitative descriptive analysis: Influence of pH and accelerated aging. *Food Qual. Pref.* **2006**, *17*, 445–452. [CrossRef]
29. Pense-Lheritier, A.-M.; Vallet, T.; Aubert, A.; Courne, M.-A.; Lavarde, M. Descriptive analysis of a complex product space: Drug-beverage mixtures. *J. Sens. Stud.* **2016**, *31*, 101–113. [CrossRef]
30. Nyambaka, H.; Ryley, J. Multivariate analysis of the sensory change in the dehydrated cowpea leaves. *Talanta* **2004**, *64*, 23–29. [CrossRef] [PubMed]
31. Husson, F.; Lê, S.; Pagés, J. Confidence ellipse for the sensory profiles obtained by Principal Components Analysis. *Food Qual. Pref.* **2005**, *16*, 245–250. [CrossRef]
32. Stone, H.; Sidel, J.; Oliver, S.; Woolsey, A.; Singleton, R.C. Sensory evaluation by quantitative descriptive analysis. *Food Technol.* **1974**, *28*, 24–34.
33. Powers, J.J. Current practices and application of descriptive methods. In *Sensory Analysis of Foods*; Piggott, J.R., Ed.; Elsevier Applied Science: London, UK, 1988.
34. Kim, M.K.; Lee, Y.-J.; Kwat, H.S.; Kang, M.W. Identification of sensory attributes that drive consumer liking of commercial orange juice products in Korea. *J. Food Sci.* **2013**, *78*, 1451–1458. [CrossRef] [PubMed]
35. Rodrigue, N.; Guillet, M.; Fortin, J.; Martin, J.F. Comparing information obtaining for ranking and descriptive tests of four sweet corn products. *Food Qual. Pref.* **2000**, *11*, 47–54. [CrossRef]

foods

MDPI

Article

Table Olives Fermented in Iodized Sea Salt Brines: Nutraceutical/Sensory Properties and Microbial Biodiversity

Barbara Lanza [1,*], Sara Di Marco [1], Nicola Simone [1], Carlo Di Marco [1] and Francesco Gabriele [2]

1 Council for Agricultural Research and Economics (CREA), Research Centre for Engineering and Agro-Food Processing (CREA-IT), Via Nazionale 38, 65012 Cepagatti (PE), Italy; sarettadimarco87@gmail.com (S.D.M.); nicola.simone@crea.gov.it (N.S.); carlo.dimarco@crea.gov.it (C.D.M.)

2 Azienda Agricola Francesco Gabriele, Via Praino Agostino 1, 87076 Villapiana (CS), Italy; info@francesco-gabriele.com

* Correspondence: barbara.lanza@crea.gov.it

Received: 25 January 2020; Accepted: 4 March 2020; Published: 6 March 2020

Abstract: This research aimed to study the influence of different brining processes with iodized and noniodized salt on mineral content, microbial biodiversity, sensory evaluation and color change of natural fermented table olives. Fresh olives of *Olea europaea* Carolea and Leucocarpa cvs. were immersed in different brines prepared with two different types of salt: the PGI "Sale marino di Trapani", a typical sea salt, well known for its taste and specific microelement content, and the same salt enriched with 0.006% of KIO_3. PGI sea salt significantly enriches the olive flesh in macroelements as Na, K and Mg, and microelements such as Fe, Mn, Cu and Zn. Instead, Ca decreases, P remains constant, while iodine is present in trace amounts. In the olives fermented in iodized-PGI sea salt brine, the iodine content reached values of 109 μg/100 g (Carolea cv.) and 38 μg/100 g (Leucocarpa cv.). The relationships between the two varieties and the mineral composition were explained by principal component analysis (PCA) and cluster analysis (CA). Furthermore, analyzing the fermenting brines, iodine significantly reduces the microbial load, represented only by yeasts, both in Carolea cv. and in Leucocarpa cv. Candida is the most representative *genus*. The sensory and color properties weren't significantly influenced by iodized brining. Only Carolea cv. showed significative difference for b* parameter and, consequently, for C value. Knowledge of the effects of iodized and noniodized brining on table olives will be useful for developing new functional foods, positively influencing the composition of food products.

Keywords: table olives; minerals; sea salt; PGI; iodized salt; functional food

1. Introduction

Table olives are a typical food product in the "Mediterranean diet", edible as finger food directly or as an ingredient for more complex dishes. In Italy, during the last five years, the average consumption of table olives was approximately 115,000 tons/year with a per capita consumption of 1.9 kg/year. Italian production covers only 50.9% of consumer demand; the remaining part is imported from Spain, Greece and Tunisia. Italy is rich in typical table olive products [1], obtained by traditional methods, and many of those have obtained or are likely to obtain to the recognition of the PDO (Protected Designation of Origin) or PGI (Protected Geographical Indication) trademarks. Currently, four Italian PDO are recognized: "Nocellara del Belice" (Reg. EC 134/1998), "La Bella della Daunia" (Reg. EC 1904/2000), "Oliva Ascolana del Piceno" (Reg. EU 1855/2005) and "Oliva di Gaeta" (Reg. EU 2016/2252). The processing of table olives also has long been part of Mediterranean traditional food and food industry. Table olives can be produced using different processes that vary according to many parameters; the most applied processes in Italy are the "Sevillan-style" and "Castelvetrano-style" for green treated olives and the "Greek-style" for green, turning color and black olives [2]. Referring to the new tendencies in "functional foods" that are intended to produce foods that contain value-added compounds as vitamins, microelements and other healthy substances (i.e., radical scavenging molecules), it appears to be quite clear that table olives already have most of the characteristics required to properly join the group due to their composition, i.e., high bio-phenol content with antioxidant and radical scavenging activity, vitamins, MUFA, PUFA, minerals and other nutraceutical compounds. Table olives contain simple and complex phenolic compounds (at least 30 different phenolic compounds) in amounts ranging between 100 and 350 mg/100 g of e.p. (edible portion). This quantity is the same of 1 kg of extra virgin olive oil. The polyphenol content and composition depend on several factors such as cultivar, stage of ripening, location and processing [1,3]. The ratios MUFA/SFA (3.3–6.8), PUFA/SFA (0.2–0.8), cis-MUFA + cis-PUFA/SFA + TFA (3.5–7.3), oleic acid/palmitic acid (3.6–8.0) and $\omega 6/\omega$ 3 (6.7–23.5) are used to assess the nutritional quality of the lipid fraction in foods having a regulatory influence on certain thrombogenic and fibrinolytic markers during the postprandial state in healthy subjects [1].

Iodine is an important micronutrient element and is required for the synthesis of T4 and T3 thyroid hormones. An iodine-deficient diet causes a wide spectrum of illnesses, including goiter and mental retardation. Adolescents and adults need iodine in amounts of 150 µg/day (Table 1). The oral intake also includes iodine from water and beverages; however, food provides by far the most to the total iodine absorption. The iodine contents in the principal categories of foods are summarized in Table 2. The iodine content in sea fishes and seafood is high, reflecting the content in the water they inhabit [4].

Table 1. Intakes and allowances recommended by SINU [5], EFSA [6] and EU [7].

Mineral	SINU-LARN [1]	EFSA-DRVs [2]	EU-RDA [3]
NaCl (g/day)			6
Sodium (g/day)	1.2–1.5 (AI [4])	1.5 (AI)	
Potassium (mg/day)	3900 (AI)	3500 (AI)	2000
Calcium (mg/day)	1000–1200 (PRI [5])	950 (PRI)	800
Magnesium (mg/day)	240 (PRI)	350 (AI)	375
Phosphorus (mg/day)	700 (PRI)	550 (AI)	700
Iron (mg/day)	10 (PRI)	11–16 (PRI)	14
Manganese (mg/day)	2.3–2.7 (AI)	3 (AI)	2
Copper (mg/day)	0.9 (PRI)	1.3–1.6 (AI)	1
Iodine (µg/day)	150 (AI)	150 (AI)	150
Zinc (mg/day)	9–12 (PRI)	7.5–16.3 (PRI)	10

[1] LARN: Italian reference intake levels for nutrients; [2] DRVs: Dietary Reference Values for nutrients; [3] RDA: Recommended Daily Allowance; [4] AI: Adequate Intake; [5] PRI: Population Reference Intakes.

Table 2. Iodine content in foods by FAO/WHO [8].

Food	FAO/WHO (µg/g)
Fish (marine)	163–3180
Fish (fresh water)	17–40
Shellfish	308–1300
Eggs	93
Milk	35–56
Meat	27–97
Cereal grains	22–72
Legumes	23–36
Vegetables	12–201
Fruits	10–29

Iodine content in vegetable foods is lower compared to those of animal origin due to a low iodine concentration in soil. As consequence, 80% of the vegans suffer from iodine deficiency [9].

Iodized salt has been used in food processing to prevent iodine deficiency disorders. In Italy, current legislation requires the iodization of salt for direct human consumption or as an ingredient in the preparation and storage of food products (Law n.55 of 21 March 2005). Moreover, table olives are mostly fermented, so they contain a good quantity of yeasts and bacteria involved in the fermentation processes [10–12] that could have a probiotic activity on human organisms [13–15]. Little information is available on the effects of iodine in the fermentation process and associated microbiota [16].

In this work, we focused our research on a typical production, consisting of natural processing fermented table olives obtained by two different cultivars, i.e., cv. Leucocarpa (white olives) and cv. Carolea, using two different types of salt for brine preparation, i.e., the PGI "Sale marino di Trapani", a typical sea salt, well known for its taste and specific content in microelements, and the same salt enriched with 0.006% of KIO_3. The olive samples were analyzed by looking for differences in chemical composition with regard to macro–micro element enrichment, in particular iodine content, to develop a new functional food which is well characterized from the sensory point of view. We also investigated the microbiological composition of the different fermentation brines, looking for differences in the microbial pools involved in the fermentation process.

2. Material and Methods

2.1. Samples and Treatments

Fresh olives of *Olea europaea* "Carolea" and "Leucocarpa" cvs. were harvested in December at full ripening (Carolea fruits were purple-black while Leucocarpa fruits were ivory-white). Two samples for each cultivar were submerged in two different 8% brines prepared with (a) PGI "Sale marino di Trapani" sea salt; (b) the same sea salt enriched with 0.006% of KIO_3 (corresponding to 3.7 mg of iodine/100 g of salt). The Protected Geographical Indication (PGI) "Sale marino di Trapani" (Reg. EU 1175/2012) is a sea salt obtained through the fractional precipitation of the compounds contained in seawater by evaporation, within the salt pans of Trapani (Sicily, Italy), without additives, bleaches, preservatives or anticaking agents. It is very rich in mineral macro and microelements (Table 3).

Table 3. Composition of PGI "Sale Marino di Trapani" sea salt.

Composition	Units of Measurement	PGI "Sale Marino di Trapani" Sea Salt	Limits (Reg. EU 1175/2012)
Insoluble residue	%	0.07	<0.2
Residual moisture	%	<0.1	<8
Sodium chloride (NaCl)	g/kg	99.6	>97.0
Magnesium (Mg)	g/kg	0.05	<0.70
Potassium (K)	g/kg	0.07	<0.30
Calcium (Ca)	g/kg	0.094	<0.40
Iron (Fe)	mg/kg	6	<20
Copper (Cu)	mg/kg	<0.5	<1
Phosphorus (P)	mg/kg	<0.5	nd [1]
Zinc (Zn)	mg/kg	<0.5	<1
Manganese (Mn)	mg/kg	<0.01	nd [1]
Iodine (I)	mg/100 g	0.1	>0.07

[1] Data not available.

At the end of fermentation (after 8 months), representative samples were analyzed for the mineral content, microbial biodiversity, sensory attributes and color change.

2.2. Mineral Composition

Representative samples were analyzed in order to quantify the minerals content, according to the procedures described by Lopez, Garcia and Garrido [17]. Iodine levels were determined according to a method described by Amr and Jabay [18]. The analyses were carried out in duplicate and the results expressed as mg/kg of olive fresh pulp.

2.3. Microbiological Monitoring

To study the microbial diversity (total microflora, yeasts, molds, and lactic acid bacteria), serial dilutions with distilled water were prepared from each brine and plated on agar media. Total microflora was grown on Plate Count Agar (PCA; Oxoid, Basingstoke, UK) incubating the plates at 30 °C for 72 h; yeasts on Malt Extract Agar (MEA; Oxoid) at 30 °C for 48 h; lactic acid bacteria on Man, De Rogosa and Sharpe (MRS; Oxoid) at 30 °C for 48 h in anaerobic atmosphere. Culture responses were expressed as colonies forming units (CFU) per ml of brine.

At least five yeast colonies from each brine were isolated and subcultured in MEA. Biochemical identification was carried out by API 20 C AUX (bioMerieux SA, Marcy-l'Etoile, France) and, in case of uncertain classification, by RapID Yeast Plus System (Remel, Lenexa, KS, USA).

The API 20 C AUX system consists of 20 cupules containing dehydrated substrates which enabled us to perform 19 assimilation tests. After 72 h of incubation, in case of positivity, inoculum suspensions generate turbidity changes. The API on-line database (api*web*^{TM}) was used for species identification and an associated probability was assigned to each culture.

RapID Yeast Plus system uses a qualitative micromethod with 18 conventional and chromogenic substrates. Based on chromogenic changes, a microcode, which was entered into the Remel database (ERIC^{TM}) for species identification with an associated probability for each culture.

2.4. Sensory Evaluation of Table Olives

The sensory characteristics of table olives were evaluated by tasters of the CREA-IT Panel, according to the COI/OT/MO No 1/Rev. 2 [19]. The evaluated attributes were negative sensations (abnormal fermentation, cooking effect, rancid, musty, or other defects), gustatory sensations (salty, bitter, sour) and kinesthetic sensations (hardness, fibrousness, and crunchiness). The table olive profile sheet uses a 10 cm intensity scale ranging from 1 (no perception) to 11 (extreme).

2.5. Determination of Color

The surface color of the fruits was measured using a Color-view spectrophotometer (Konica Minolta Optics, 2970 Ishikawa-machi, Hachioji, Tokyo, Japan; Model CM-2600D). Color was expressed in terms of CIE (Commission Internationale de l'Eclairage) L* (lightness), a* (redness/greenness), b* (yellowness/blueness) and their derivative Chroma ($C = \sqrt{a^{*2} + b^{*2}}$). The analysis of color was made on 20 uniformly sized olive fruits.

2.6. Statistical Analyses

All data significance was evaluated by one-way ANOVA using the F-test ($P \leq 0.05$).

Mineral data are processed by principal component (PCA) and cluster (CA) analyses, carried out in the Past PAleontological STatistics software (Version 2.12, Øyvind Hammer, Natural History Museum, University of Oslo). For data preprocessing, the variables were rescaled from 0 to 1.

In order to elaborate the sensory data, a method was applied to calculate the median (Me), the robust standard deviation (DSr), the robust coefficient of variation percentage (CVr%), and the confidence intervals of the median at 95% ($C.I._{upper}$ and $C.I._{lower}$) contained in Annex 1 [19], taking into account those attributes with a robust coefficient of variation of 20% or less.

3. Results and Discussion

In the fresh fruits of Carolea cv. (Figure 1), the main composition (expressed in mg/kg) was: K (408), Ca (130), P (73), Mg (12), and Na (8). Fe, Mn, Cu, Zn, and I are <1. In the fresh fruits of Leucocarpa cv. (Figure 2), the main composition (expressed in mg/kg) is: K (382), Ca (114), P (53), Na

(9), Mg (9), and Fe (2). Mn, Cu, Zn, and I are <1. Variation in mineral levels of fresh fruits depends of olive variety, ripening, and growing conditions (soil, water, fertilizers). With the addition of sea salts, minerals infuse into the olive flesh. The results of analysis indicated that the Na, K, Mg, Fe, Mn, Cu, and Zn contents increased during brining.

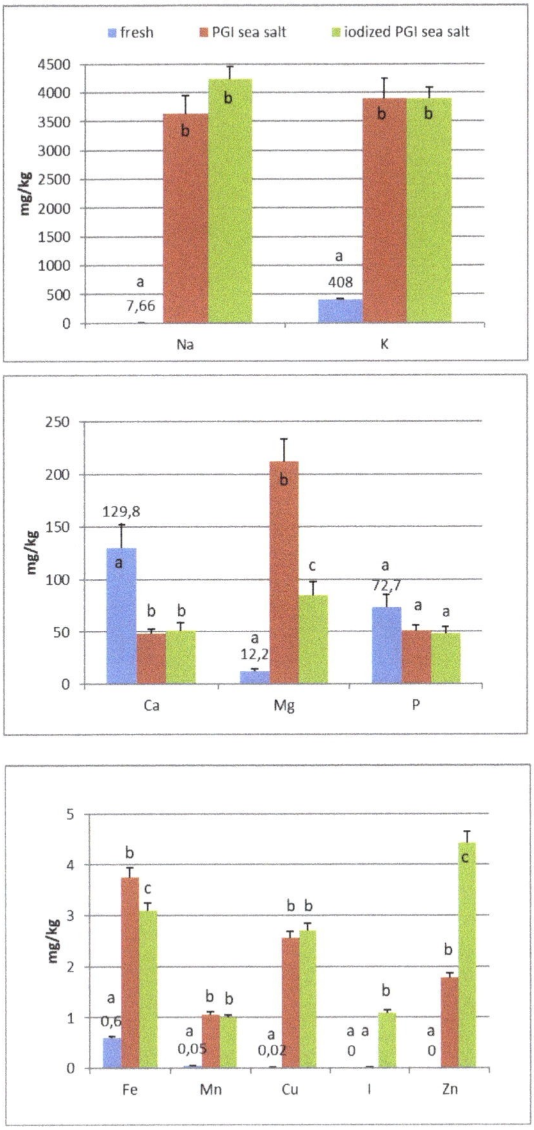

Figure 1. Mineral composition of fresh and processed Carolea olives. Data are expressed in mg/kg. Bars represent mean values of two replicates ± SD. Significant differences are indicated by different letters ($P < 0.05$).

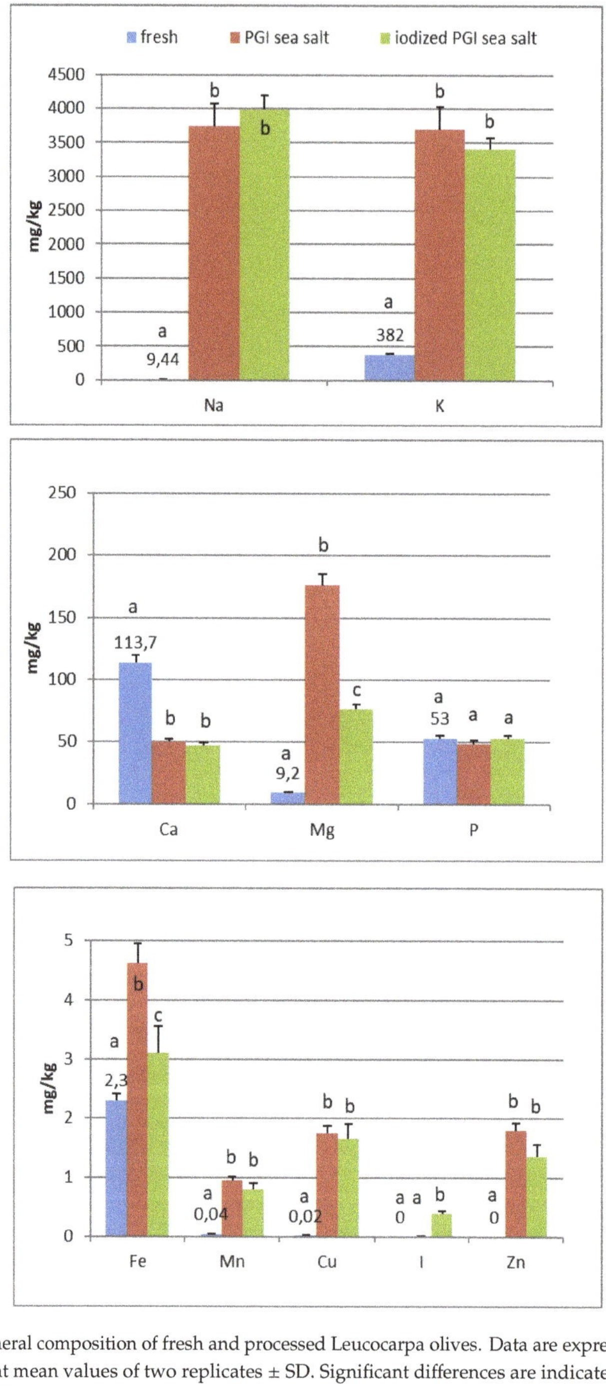

Figure 2. Mineral composition of fresh and processed Leucocarpa olives. Data are expressed in mg/kg. Bars represent mean values of two replicates ± SD. Significant differences are indicated by different letters ($P < 0.05$).

At the end of fermentation (Figures 1 and 2), PGI sea salt brine principally enriched the olive flesh of macroelements Na, K, and Mg, reaching values of 3624, 3897, and 212 mg/kg (Carolea cv.) and 3733, 3696, and 176 mg/kg (Leucocarpa cv.) respectively.

The sodium content (<450 mg/100 g of edible portion) was shown to be below the allowances recommended by Italian Society of Human Nutrition [5], European Food Safety Authority [6], and the European Parliament and the Council of the European Union [8] (Table 1). The consumption of table olives would not be recommended only in cases of hypertension, and, in any case, there are production technologies to reduce salt content (low-sodium olives) [20–22]. As reported by other authors, the potassium concentration is higher in directly brined olives (571 to 1176 mg/kg) [17], which are not subjected to lye treatments. Values reported by De Castro Ramos et al. [23] for green olives ranged from 640 to 1090 mg/kg. Unal and Nergiz [24] found 3760 mg/kg for natural black olives. Biricik and Basoglu [25] reported a concentration of 4123–7401 mg/kg for green olive brine. The K content of different Turkish table olives varied between 2814 and 3386 mg/kg [26]. Despite daily intakes for K being high (2000–3500 mg) (Table 2), our fermented olives may be considered a main source for the daily allowance. From literature, magnesium concentration ranged from 51 to 197 mg/kg [17]. Its wide interval of concentration reflects that its presence may be greatly affected by processing. De Castro Ramos et al. [23] found 60–400 mg/kg in green olives and 47–360 mg/kg in Biricik and Basoglu [25]. The Mg content of different Turkish table olives varied between 83 and 156 mg/kg [26]. Sahan et al. [27] found values for black olives that range between 36 and 125 mg/kg.

PGI sea salt also significantly enriches the olive flesh of the microelements Fe (3.8 and 4.6 mg/kg for Carolea and Leucocarpa respectively), Mn (1.06 and 0.95), Cu (2.56 and 1.76), and Zn (1.78 and 1.80). Instead, Ca significantly decrease (about 50 mg/kg for cultivars), P content remains practically constant, while I content was not detectable (Figures 1 and 2). The decrease in calcium can be explained by the subtraction of calcium ions from the olive flesh by Na+ and the formation of calcium salts (mainly CaCl2) in brine [28].

Iron concentrations in natural table olives are relatively low. Values reported by other authors were 3.49–7.70 [17], 6.4–10.9 mg/kg (green olives) [24], and 3.23–15.10 mg/kg for Turkish cultivars [26,27]. Manganese contents were always low in the same order as those given by other authors, i.e., 0.24–1.10 mg/kg [17], 1.40–2.72 [26], and 0.70–2.90 mg/kg for black table olives [29]. Copper concentrations in directly brined olives ranged between 3.99 and 10.93 mg/kg [17], 0.53 and 7.19 mg/kg for Turkish olives [26,27], and 7.00 and 30.00 mg/kg for black olives [29]. Zinc concentrations were like iron and copper in the same order as those given by other authors: 2.18–4.10 mg/kg [26], 4.25–14.30 [27], 1.55–3.20 [17], and 1.00–6.80 for black table olives [29]. Differences among cultivars within elaboration types were found. Analyzing the data present in the bibliography, phosphorus had a concentration that ranged from 57 to 144 mg/kg. Its highest average concentration was found in directly brined olives [17]. The P content varied between 116 and 250 mg/kg in different Turkish table olives [27]. The content of Ca in table olives ranged from 337 to 850 mg/kg [17], 422 to 850 mg/kg in green Turkish olives [26], 460 to 860 mg/kg in the green Spanish cultivar [23], 270 to 450 mg/kg for Kalamata, or 110 to 230 mg/kg for natural black olives after fermentation [24]. We haven't found data on iodine content in either fresh and processed olives.

In the olives fermented in iodized PGI sea salt, the iodine content reached values of 109 μg/100 g (Carolea cv.) and 38 μg/100 g (Leucocarpa cv.). The iodine retention in olive flesh contributes to meeting the recommended daily level for I (Table 1). Similar values were found in pickled vegetables (carrots, cucumbers, turnips, and cauliflowers). In this case, the iodine content ranged from 1.6 to 1.8 mg/kg [18]. The addition of KIO_3 to the salt used for brining affects the redistribution of macro and microelements and, in particular, of magnesium, iron and zinc. The magnesium and iron contents significantly decreased (Figures 1 and 2); this decrease in the olive flesh can be explained by the formation of $Fe(IO_3)_3$ and $Mg(IO_3)_2$ tetrahydrate or decahydrate. The concentrations of the various elements in the flesh depend on a favorable Ksp (solubility product constant) of the different salts that can be formed. In contrast, the zinc content significantly increased in Carolea cv. (Figure 1), but not in

Leucocarpa cv. (Figure 2). However, the attained values are in the same order as those given by other authors [24,28]. The simultaneous presence of several counterions determined the final concentrations of the different elements in the flesh, depending on the relative concentrations of IO_3^-, K^+, Mg^{++}, Fe^{+++}, and Zn^{++}.

The relationships between the two varieties and the three different processing steps were shown by principal component analysis (PCA) and cluster analysis (CA).

PCA *bi*-plot (Figure 3) shows a good separation in relation to the processing steps for the variables (minerals). The variance was determined to be 93.40% by summing PC1 and PC2, reaching a value of 97.68% with PC3. In Figure 4 the influence of the variables on the construction of PC1 and PC2 is represented. With PC1, the higher variance contribution (about 0.40) was due to Na, K, Mn, Cu, and Ca. The latter is negatively correlated with other descriptors. With PC2, the major modules belonged to I (>0.70), Mg (−0.47), and Zi (0.39). I and Zn contribute to segregate olives processed with iodized brines. Magnesium is negatively correlated to the PC2.

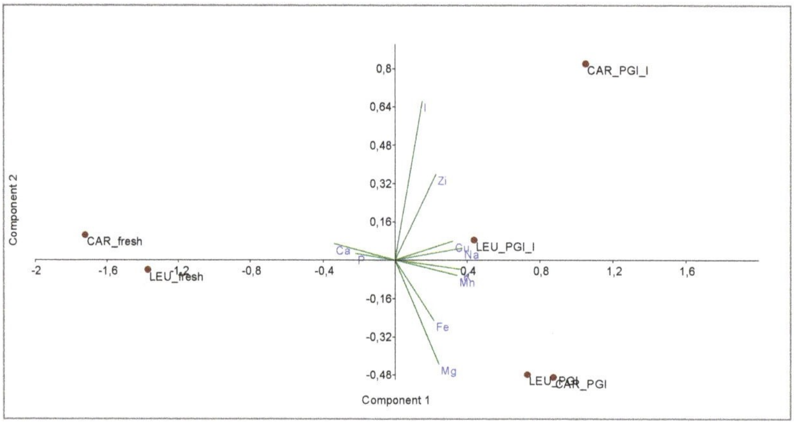

Figure 3. *Bi*-plot obtained by PCA of data-set based on mineral composition. CAR_fresh: Carolea fresh, LEU_fresh: Leucocarpa fresh; CAR_PGI: Carolea fermented in PGI sea salt; LEU_PGI: Leucocarpa fermented in PGI sea salt; CAR_PGI_I: Carolea fermented in PGI iodized sea salt; LEU_PGI_I: Leucocarpa fermented in PGI iodized sea salt.

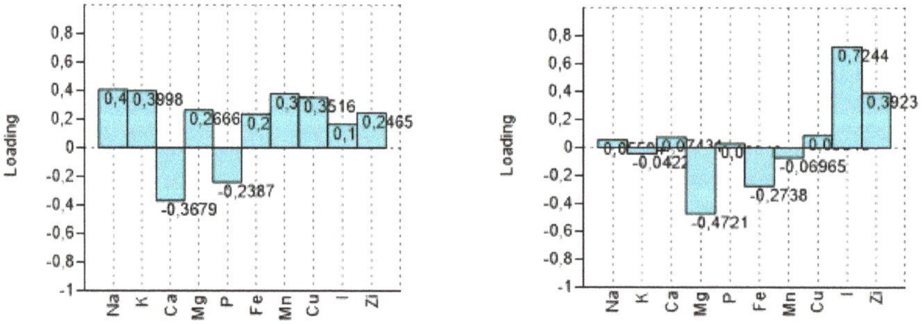

(**a**) (**b**)

Figure 4. Loadings of variables (minerals) on the first two components (PC1 and PC2). (**a**) loading plot on PC1 and (**b**) loading plot on PC2.

The new data matrix of PCs was subjected to Cluster Analysis. Using the Ward's method of clustering, the samples were grouped in three clusters (calculated at 1.0 of similarity distance) related to the processing steps. The obtained dendrogram is shown in Figure 5. The samples fermented in the iodized brines (CAR_PGI_I and LEU_PGI_I) were close to those fermented in simple sea salt, but were grouped into a separate cluster; they were very different from the fresh samples. This indicates considerable similarities between treatments but not between cultivars, i.e., the cultivar is a nondiscriminant attribute.

From the literature, there is no evidence that the use of iodized salt in processed foods production, including olives, could cause adverse changes in color and kinesthetic properties [18,30,31].

From our color data, Leucocarpa olives processed with PGI sea salt or with iodized PGI sea salt showed significative differences for all color parameters established by CIE (Table 4). In contrast, Carolea cv. showed significant differences only for b* and, consequently, for C values between PGI sea salt and iodized PGI sea salt, respectively (Table 4).

Concerning sensory aspects (Figure 6), none of samples presented any defects. This is very important because the occurrence of negative sensations has a negative impact on the gustative and kinesthetic attributes [32]. Olives prepared with iodized sea salt were harder and more bitter than those prepared with noniodized sea salt, but these differences were not significant (Table 5). The bitterness level of Leucocarpa cv. is high, depending on the variety, but some groups of consumers prefer natural olives with high bitterness values. The low level of hardness highlighted by a sensory evaluation could be related to olive softening. From the data in our possession, there is no evidence that this level is problematic for commercialization.

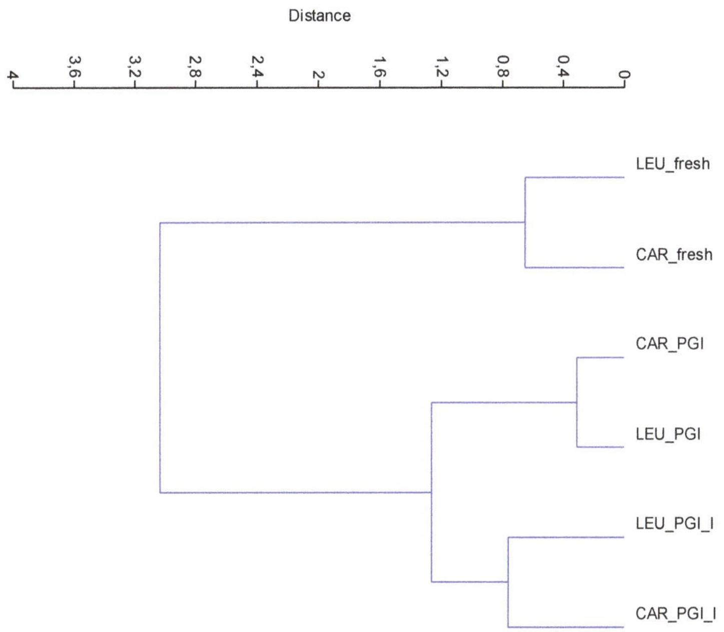

Figure 5. CA dendrogram of data-set based on mineral composition. Dendrogram is obtained using the Ward's algorithm and the Euclidean distance similarity. CAR_fresh: Carolea fresh, LEU_fresh: Leucocarpa fresh; CAR_PGI: Carolea fermented in PGI sea salt; LEU_PGI: Leucocarpa fermented in PGI sea salt; CAR_PGI_I: Carolea fermented in iodized PGI sea salt; LEU_PGI_I: Leucocarpa fermented in iodized PGI sea salt.

Table 4. Change in color values of processed olives. NS = not significant; * = significant. Significant differences are indicated by different letters ($P < 0.05$).

Parameters	Carolea cv. (PGI)	Carolea cv. (Iodized-PGI)	ANOVA	Leucocarpa cv. (PGI)	Leucocarpa cv. (Iodized-PGI)	ANOVA
L	26.79 ± 16.40 ^{a}	25.89 ± 13.87 ^{a}	NS	65.09 ± 3.25 ^{a}	62.45 ± 4.97 ^{b}	*
a	16.07 ± 6.33 ^{a}	15.98 ± 5.08 ^{a}	NS	11.23 ± 1.75 ^{a}	11.56 ± 2.01 ^{b}	*
b	24.39 ± 13.67 ^{a}	20.05 ± 11.04 ^{b}	*	39.89 ± 3.00 ^{a}	37.91± 3.93 ^{b}	*
C	31.39 ± 9.55 ^{a}	27.28 ± 7.66 ^{b}	*	41.48 ± 3.00 ^{a}	39.70 ± 3.80 ^{b}	*

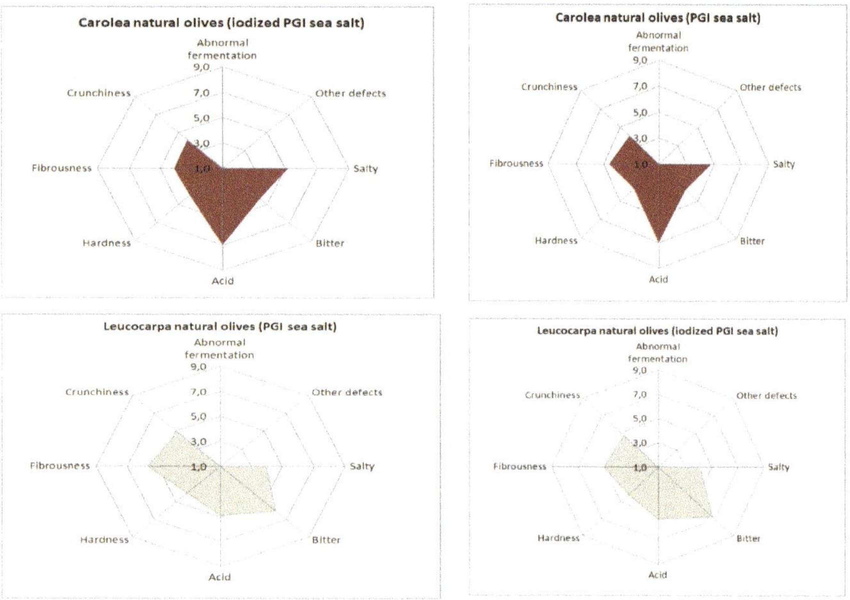

Figure 6. Sensory profiles of processed table olives.

Table 5. Evaluation of sensory attributes. NS = not significant.

	Median	DSr	CVr %	CI Upper	CI Lower	Median	DSr	CVr %	CI Upper	CI Lower	
	Carolea natural olives (PGI sea salt)					Carolea natural olives (iodized PGI sea salt)					ANOVA
Salty	4.80	0.84	17.56	6.45	3.15	5.15	0.63	12.24	6.39	3.91	NS
Bitter	3.75	0.65	17.46	5.03	2.47	4.35	0.76	17.50	5.84	2.86	NS
Acid	7.00	0.55	7.83	8.07	5.93	7.00	0.44	6.31	7.87	6.13	NS
Hardness	3.60	0.71	19.78	5.00	2.20	3.95	0.77	19.48	5.46	2.44	NS
Fibrousness	4.60	0.39	8.54	5.37	3.83	4.10	0.34	8.38	4.77	3.43	NS
Crunchiness	4.05	0.58	14.35	5.19	2.91	4.15	0.70	16.76	5.51	2.79	NS
	Leucocarpa natural olives (PGI sea salt)					Leucocarpa natural olives (iodized PGI sea salt)					
Salty	4.00	0.29	7.37	4.58	3.42	4.15	0.40	9.66	4.94	3.36	NS
Bitter	6.05	0.36	5.95	6.76	5.34	6.75	0.47	7.03	7.68	5.82	NS
Acid	4.90	0.65	13.36	6.18	3.62	5.30	0.52	9.73	6.31	4.29	NS
Hardness	4.00	0.77	19.23	5.51	2.49	4.20	0.66	15.78	5.50	2.90	NS
Fibrousness	5.70	0.75	13.21	7.18	4.22	5.10	0.38	7.54	5.85	4.35	NS
Crunchiness	4.85	0.86	17.72	6.53	3.17	4.70	0.45	9.58	5.58	3.82	NS

A microbiological analysis of the fermented brines showed that lactobacilli and aerobic bacteria were absent; the total microflora consisted exclusively of yeasts. Iodized brines significantly reduce the microbial load both in Carolea cv. (8.7×10^4 vs. 3.5×10^5 CFU/mL) and in Leucocarpa cv. (1.4×10^2 vs. 3.1×10^2 CFU/mL) (Table 6).

Table 6. Microbial monitoring. Data are expressed in CFU/mL. * = significant. Significant differences are indicated by different letters ($P < 0.05$).

Samples	Carolea cv. (PGI)	Carolea cv. (Iodized-PGI)	ANOVA	Leucocarpa cv. (PGI)	Leucocarpa cv. (Iodized-PGI)	ANOVA
Total aerobic bacteria	-	-	-	-	-	-
LAB	-	-	-	-	-	-
Yeasts	$3.5 \times 10^{5\,a}$	$8.7 \times 10^{4\,b}$	*	$3.1 \times 10^{2\,a}$	$1.4 \times 10^{2\,b}$	*

At the end of fermentation, *Candida* was the most representative genus, followed only by the genus *Cryptococcus*. As indicated in Table 7, the genus *Candida* was present in all four samples. In the Carolea cultivar only, *Candida krusei* was present (Table 7). The Leucocarpa cultivar showed a greater diversity in yeasts (*Candida famata*, *C. boidinii*, *C. intermedia*, *C. krusei*, and *Cryptococcus albidus*) (Table 7), indicating that these microorganisms are more related to the cultivar than to the environment. These yeasts have pectinolytic and cellulolytic enzymes and contribute to the degradation of the pectin that forms the middle lamella, which leads to cell separation and acts on cellulose, hemicellulose, and polysaccharides, giving texture to the pulp [33]. The low Ca content in processed olives (Figures 1 and 2) could be related to the loss of Ca^{2+} bridging between residues of galacturonic acid of adjacent pectic chains, which, in turn, is related to the softening of the olive fruit [34].

Table 7. Tentative identification of the isolated yeasts. Identification percentages are shown between parentheses. [1] API 20 C AUX; [2] RapID Yeast Plus System.

Samples	Microorganisms
Carolea natural olives (PGI sea salt)	
1	*Candida krusei* [99.1] [1]
2	*C. krusei* [99.1] [1]
3	*C. krusei* [99.1] [1]
4	*C. krusei* [99.1] [1]
5	*C. krusei* [99.1] [1]
Carolea natural olives (iodized PGI sea salt)	
1	*C. krusei* [99.1] [1]
2	*C. krusei* [99.1] [1]
3	*C. krusei* [99.1] [1]
4	*C. krusei* [99.1] [1]
5	*C. krusei* [99.1] [1]
Leucocarpa natural olives (PGI sea salt)	
1	*Cryptococcus albidus* [95.0] [2]
2	*Candida boidinii* [99.0] [1]
3	*Candida famata* [99.0] [2]
4	*C. boidinii* [99.0] [1]
5	*Candida intermedia* [99.0] [2]
Leucocarpa natural olives (iodized PGI sea salt)	
1	*C. boidinii* [99.0] [1]
2	*C. intermedia* [99.0] [2]
3	*C. krusei* [99.1] [1]
4	*C. boidinii* [99.0] [1]
5	*C. intermedia* [99.0] [2]

C. boidinii contributes to the decline of olive tissue structural integrity [35] through its pectinolytic enzymes that act on pectic substances that form the middle lamella, and on cellulose, hemicellulose, and polysaccharides, that form the cell walls [36].

Each species of *Candida* requires a carbon source from different carbohydrates to provide energy for cell growth. The need of different carbohydrates sources becomes the basic identification for the assimilation method by every *Candida* species [37].

C. krusei identified by API 20 C AUX, deriving from Carolea brine samples, showed a positive reaction to glucose, *p*-Nitrophenyl phosphate, proline β-naphthylamide, and histidine β-naphthylamide. Sample 3 of Leucocarpa (iodized sea salt) brine had the same profile. *Cryptococcus albidus* and *C. famata* from Leucocarpa brine, identified only by RapID, grew on glucose, threalose, *p*-nitrophenyl-β,D-glucoside, *p*-nitrophenyl phosphate. *Cr. albidus* shows a positive reaction to *p*-nitrophenyl-β,D-fucoside (not for *C. famata*), and *C. famata* uses proline β-naphthylamide (not for *Cr. albidus*). Sample 5 (PGI sea salt) and samples 2 and 5 (iodized PGI sea salt) of Leucocarpa were identified as *C. intermedia* by RapID. These yeasts grow on glucose, glycerol, threalose, D-cellobiose, L-arabinose, adonitol, D-sorbitol, *N*-acetyl-Glucosamine, *p*-nitrophenyl-β,D-glucoside, *p*-nitrophenyl phosphate, proline β-naphthylamide, and histidine β-naphthylamide. Four cultures from Leucocarpa brine were identified by API as *C. boidinii*. These yeasts species showed a positive reaction to glucose, glycerol, D-xylose, adonitol, xylitol, D-sorbitol, and *ϱ*-nitrophenyl phosphate. Furthermore, sample 4 of Leucocarpa (PGI sea salt) used histidine β-naphthylamide; samples 1 and 4 of Leucocarpa (iodized sea salt) used proline β-naphthylamide and histidine β-naphthylamide.

4. Conclusions

This research aimed to study the influence of different brining processes on mineral content, microbial biodiversity, sensory evaluation, and color change of natural fermented table olives. Fresh olives of *Olea europaea* Carolea and Leucocarpa cvs. were submerged in two different 8% brines prepared with iodized and noniodized PGI sea salts. Noniodized sea salt brines principally enriched the olive flesh with macroelements such as Na, K, and Mg, and microelements as Fe, Mn, Cu, and Zn. In contrast, a decrease in Ca was observed, while the P content remained practically constant, and I was present in trace amounts. In the olives fermented in iodized PGI sea salt, the iodine content reached the values of 109 μg/100 g (Carolea cv.) and 38 μg/100 g (Leucocarpa cv.).

The addition of KIO_3 to the salt used for brining affects the redistribution of macro and microelements and, in particular, of magnesium, iron, and zinc. The concentrations of the various elements in the flesh depend on the favorable Ksp (solubility product constant) of the different salts that can form.

The PGI sea salt enriches the olive fruit flesh with numerous microelements compared to simple brine prepared with only NaCl (data not shown). Many Italian companies utilize PGI sea salt only, instead of simple NaCl. The aim of our study was to determine whether the addition of KIO_3 transfers iodine to the olive flesh.

Analyzing the fermenting brines, iodine significantly reduces the microbial load, represented only by yeasts, both in Carolea cv. and in Leucocarpa cv. *Candida* is the most representative genus. The sensory and color properties were not significantly influenced by iodized brining. Only Carolea cv. showed significant differences for the b* parameter and, consequently, for the C value.

This explorative research involved a small number of olive samples, but it compels us to undertake further investigations concerning the side effects of different salts on the microorganisms involved in the fermentation process, and the development of new functional foods that merge tradition and innovation. The results obtained represent the first data on the enrichment of iodine in table olives.

Author Contributions: B.L. planned the experimental trial; B.L., N.S. and C.D.M. harvested olive samples; F.G. carried out the lab-scale fermentation of olives in his farm; B.L. and C.D.M. performed the color analysis; S.D.M. performed the microbiological analyses; B.L., S.D.M., N.S. and F.G. performed sensory and mineral analyses; B.L. and S.D.M. performed statistical analyses; B.L., S.D.M. and N.S. wrote the manuscript. All authors have read and agreed to the published version of the manuscript.

Funding: This research was supported by the projects INFOLIVA (D.M. 12479/7110/2018) and DEAOLIVA (D.M. 93882) funded by the Italian Ministry of Agricultural, Food and Forestry Policies (MiPAAF).

Conflicts of Interest: No potential conflict of interest was reported by the authors.

References

1. Lanza, B. Nutritional and Sensory Quality of Table Olives. In *Olive Germplasm—The Olive Cultivation, Table Olive and Olive Oil Industry in Italy*; Muzzalupo, I., Ed.; IntechOpen: Rijeka, Croatia, 2012; pp. 343–372. [CrossRef]
2. CODEX STAN 66-1981 (rev. 2013). Standard for Table Olives. Available online: http://www.fao.org/fao-who-codexalimentarius/sh-proxy/en/?lnk=1&url=https%253A%252F%252Fworkspace.fao.org%252Fsites%252Fcodex%252FStandards%252FCXS%2B66-1981%252FCXS_066e.pdf (accessed on 25 January 2020).
3. Blekas, G.; Vassilakis, C.; Harizanis, C.; Tsimidou, M.; Boskou, D.G. Biophenols in table olives. *J. Agric. Food Chem.* **2002**, *50*, 3688–3692. [CrossRef] [PubMed]
4. Haldimann, M.; Alt, A.; Blanc, A.; Blondeau, K. Iodine content of food groups. *J. Food Compos. Anal.* **2005**, *18*, 461–471. [CrossRef]
5. SINU. *Intake Levels of Reference of Nutrients and Energy*; IV revision; Italian Society of Human Nutrition: Milan, Italy, 2014.
6. EFSA. Dietary reference values for nutrients. Summary report. *Eur. Food Saf. Auth.* **2017**, *14*, e15121E. [CrossRef]
7. Regulation (EU) No 1169/2011 of the European parliament and of the council of 25 October 2011 on the provision of food information to consumers, amending Regulations (EC) No 1924/2006 and (EC) No 1925/2006 of the European Parliament and of the Council, and repealing Commission Directive 87/250/EEC, Council Directive 90/496/EEC, Commission Directive 1999/10/EC, Directive 2000/13/EC of the European Parliament and of the Council, Commission Directives 2002/67/EC and 2008/5/EC and Commission Regulation (EC) No 608/2004. *Off. J. Eur. Union* **2011**, *L 304*, 18–63.
8. FAO/WHO. Chapter 12. Iodine. In *Expert Consultation on Human Vitamin and Mineral Requirements*; FAO: Rome, Italy, 2001.
9. Krajčovičová-Kudláčková, M.; Bučková, K.; Klimeš, I.; Šeboková, E. Iodine Deficiency in Vegetarians and Vegans. *Ann. Nutr. Metab.* **2003**, *47*, 183–185. [CrossRef]
10. Arroyo-Lopez, F.N.; Querol, A.; Bautista-Gallego, J.; Garrido-Fernandez, A. Role of yeasts in table olive production. *Int. J. Food Microbiol.* **2008**, *128*, 42–49. [CrossRef]
11. Bautista-Gallego, J.; Rodriguez-Gomez, F.; Barrio, E.; Querol, A.; Garrido-Fernandez, A.; Arroyo-Lopez, F.N. Exploring the yeast biodiversity of green table olive industrial fermentations for technological application. *Int. J. Food Microbiol.* **2011**, *14*, 89–96. [CrossRef]
12. Zago, M.; Lanza, B.; Rossetti, L.; Muzzalupo, I.; Carminati, D.; Giraffa, G. Selection of *Lactobacillus plantarum* strains to use as starters in fermented table olives: Oleuropeinase activity and phage sensitivity. *Food Microbiol.* **2013**, *34*, 81–87. [CrossRef]
13. Bonatsou, S.; Karamouza, M.; Zoumpopoulou, G.; Mavrogonatou, E.; Kletsas, D.; Papadimitriou, K.; Tsakalidou, E.; Nychas, G.E.; Panagou, E.Z. Evaluating the probiotic potential and technological characteristics of yeasts implicated in cv. Kalamata natural black olive fermentation. *Int. J. Food Microbiol.* **2018**, *271*, 48–59. [CrossRef]
14. Argyri, A.A.; Zoumpopoulou, G.; Karatzas, K.G.; Tsakalidou, E.; Nychas, G.E.; Panagou, E.Z.; Tassou, C.C. Selection of potential probiotic lactic acid bacteria from fermented olives by in vitro test. *Food Microbiol.* **2013**, *33*, 189–196. [CrossRef]
15. Botta, C.; Langerholc, T.; Cencic, A.; Cocolin, L. In Vitro Selection and characterization of new probiotic candidates from table olive microbiota. *PLoS ONE* **2014**, *9*, e94457. [CrossRef] [PubMed]
16. Muller, A.; Rosch, N.; Cho, G.-S.; Meinhardt, A.-K.; Kabish, J.; Habermann, D.; Bohnlein, C.; Brinks, E.; Greiner, R.; Franz, C.M. Influence of iodized table salt on fermentation characteristics and bacterial diversity during sauerkraut fermentation. *Food Microbiol.* **2018**, *76*, 473–480. [CrossRef] [PubMed]
17. Lopez, A.; Garcia, P.; Garrido, A. Multivariate characterization of table olives according to their mineral nutrient composition. *Food Chem.* **2008**, *106*, 369–378. [CrossRef]
18. Amr, A.S.; Jabay, O.A.J. Effect of salt iodization on the quality of pickled vegetables. *Food Agric. Environ.* **2004**, *2*, 151–156.
19. IOC. *Method for the Sensory Analysis of Table Olives COI/OT/MO/Doc. No 1/Rev. 2*; International Olive Oil Council: Madrid, Spain, 2011.

20. Panagou, E.Z.; Hondrodimou, O.; Mallouchos, A.; Nychas, G.J.E. A study on the implications of NaCl reduction in the fermentation profileof Conservolea natural black olives. *Food Microbiol.* **2011**, *28*, 1301–1307. [CrossRef] [PubMed]
21. Mateus, T.; Santo, D.; Saúde, C.; Pires-Cabral, P.; Quintas, C. The effect of NaCl reduction in the microbiological quality of crackedgreen table olives of the Maçanilha Algarvia cultivar. *Int. J. Food Microbiol.* **2016**, *218*, 57–65. [CrossRef] [PubMed]
22. Zinno, P.; Guantario, B.; Perozzi, G.; Pastore, G.; Devirgiliis, C. Impact of NaCl reduction on lactic acid bacteria during fermentation ofNocellara del Belice table olives. *Food Microbiol.* **2017**, *63*, 239–247. [CrossRef]
23. De Castro Ramos, R.; Nosti Vega, M.; Vazquez Ladron, R. Composicion y valor nutritivo de algunas variedades espanolas de aceitunas de mesa. I. Aceitunas verdes aderezadas al estilo sevillano. *Grasas Aceites* **1979**, *30*, 83–91.
24. Unal, K.; Nergiz, C. The effect of table olive preparing methods and storage on the composition and nutritive value of table olives. *Grasas Aceites* **2003**, *54*, 71–76. [CrossRef]
25. Biricik, G.F.; Basoglu, F. Determination of mineral contents in some olives (Samanli, Domat, Manzanilla, Ascolana) varieties. *GIDA* **2006**, *31*, 67–75.
26. Uylaşer, V.; Yıldız, G. Fatty acid profile and mineral content of commercial table olives from Turkey. *Not. Bot. Horti Agrobot. Cluj Napoc.* **2013**, *41*, 518–523. [CrossRef]
27. Sahan, Y.; Basoglu, F.; Gucer, S. ICP-MS analysis of a series of metals (namely: Mg, Cr, Co, Ni, Fe, Cu, Zn, Sn, Cd and Pb) in black and green olive samples from Bursa, Turkey. *Food Chem.* **2007**, *105*, 395–399. [CrossRef]
28. Sanchez-Rodriguez, L.; Corell, M.; Hernandez, F.; Sendra, E.; Moriana, A.; Carbonell-Barrachina, A.A. Effect of Spanish-style processing on the quality attributes of *HydroSOStainable* green olives. *J. Sci. Food Agric.* **2019**, *99*, 1804–1811. [CrossRef] [PubMed]
29. Di Giacomo, F.; Marsilio, V. Qualità e contenuto di oligoelementi nelle olive da tavola. In Proceedings of the Atti XV Congresso di Merceologia, Roma, Italy, 24–26 September 1992; pp. 501–511.
30. Blankenship, J.L.; Garrett, G.S.; Ahmad Khan, N.; Maria De-Regil, L.; Spohrer, R.; Gorstein, J. Effect of iodized salt on organoleptic properties of processed foods: A systematic review. *J. Food Sci. Technol.* **2018**, *55*, 3341–3352. [CrossRef] [PubMed]
31. West, C.E.; Merx, R.J. *Effect of Iodized Salt on the Colour and Taste of Food*; UNICEF: New York, NY, USA, 1995.
32. Lanza, B.; Amoruso, F. Sensory analysis of natural table olives: Relationship between appearance of defect and gustatory-kinaesthetic sensation changes. *LWT Food Sci. Technol.* **2016**, *68*, 365–372. [CrossRef]
33. Hommel, R.K. Candida. In *Encyclopedia of Food Microbiology*, 2nd ed.; Batt, C.A., Tortorello, M.L., Eds.; Elsevier: London, UK, 2014; pp. 367–373. [CrossRef]
34. De Castro, A.; Garcia, P.; Romero, C.; Brenes, M.; Garrido, A. Industrial implementation of black ripe olive storage under acid conditions. *J. Food Eng.* **2007**, *80*, 1206–1212. [CrossRef]
35. Golomb, B.L.; Morales, V.; Jung, A.; Yau, B.; Boundy-Mills, K.L.; Marco, M.L. Effects of pectinolytic yeast on the microbial composition and spoilage of olive fermentations. *Food Microbiol.* **2013**, *33*, 97–106. [CrossRef]
36. Lanza, B. Abnormal fermentations in table-olive processing: Microbial origin and sensory evaluation. *Front. Microbiol.* **2013**, *4*, 91. [CrossRef]
37. Hermansyah, H.; Adhiyanti, N.; Julinar, J.; Rahadiyanto, K.Y.; Susilawati, S. Identification of *Candida* species by assimilation and Multiplex-PCR methods. *J. Chem. Technol. Metall.* **2017**, *52*, 1070–1078.

foods

MDPI

Article

In Vitro Bioaccessibility of Ripe Table Olive Mineral Nutrients

Antonio López-López *, José María Moreno-Baquero and Antonio Garrido-Fernández

Food Biotechnology Department, Instituto de la Grasa (CSIC), Campus Universitario Pablo de Olavide, Edificio 46, Ctra. Utrera km 1, 41013 Sevilla, Spain; jose.moreno.baquero@gmail.com (J.M.M.-B.); garfer@cica.es (A.G.-F.)
* Correspondence: all@cica.es

Received: 7 January 2020; Accepted: 27 February 2020; Published: 3 March 2020

Abstract: For the first time, the bioaccessibility of the mineral nutrients in ripe table olives and their contributions to the recommended daily intake (RDI), according to digestion methods (Miller's vs. Crews' protocols), digestion type (standard vs. modified, standard plus a post-digest re-extraction), and mineralisation system (wet vs. ashing) were studied. Overall, when the standard application was used, Miller's protocol resulted in higher bioaccessibilities of Na, K, Ca, Mg, and Fe than the Crews' method. The modified protocols improved most of these values, but the Crews' results only approximated the Miller's levels in the case of Na and K. The bioaccessibility of P was hardly affected by the factors studied, except that the modified Miller's protocol led to higher levels when ashing. No significant effect of the mineralisation system was found. The modified Miller's protocol, regardless of the mineralisation system, led to the overall highest bioaccessibility values in ripe olives, which were: Na (96%), K (95%), Ca (20%), Mg (73%), Fe (45%), and P (60%). Their potential contributions to the RDI, based on these bioaccessibilities and 100 g olive flesh service size, were then 29, 0.5, 4, 3, 33, and 1% respectively. The investigation has led to the development of a method for assessing the bioaccessibility of the mineral nutrients not only in ripe but also in the remaining table olive presentations and opens a new research line of great interest for producing healthier products.

Keywords: sodium; potassium; calcium; magnesium; iron; phosphorus; darkened by oxidation olives; Miller's protocol; Crews' protocol; post-digest re-extraction

1. Introduction

The concentrations of mineral elements can be declared in the nutritional labelling of foods [1,2]. The Food and Drug Administration (FDA) and European Union (EU) standards also include recommended daily intakes for minerals. A detailed description of the individual requirements of these nutrients can likewise be found in the Dietary References Intakes Tables and Application issued by the Health and Medicine Division of the National Academies of Sciences, Engineering and Medicine [3].

Table olives are well known all over the world. The consolidated balance issued by the International Olive Council established a global production of 3.28×10^6 t for the 2018/2019 season [4]. As with many other vegetables, the fruit storage/fermentation process takes place in brine with a NaCl concentration in an equilibrium ≥50 g/L [5]. As all the solutions used for processing are aqueous, marked leaching of minerals from the flesh into the brine (except for Na, which moves in the opposite direction) usually occurs [5]. As a result, the Na level increases while the contents of the other elements in the final products, despite these losses, remain moderately high. The concentrations reported in the literature depend on processing conditions, cultivars, and preparation styles and range between the following values: Na, 571–17,221 mg/kg; K, 81–1176 mg/kg; Ca, 337–850 mg/kg; Mg, 13–133 mg/kg; Fe, 4–132 mg/kg; and P, 57–118 mg/kg. [6]. The most significant differences were found among green

(Spanish-style), directly brined (natural olives) and ripe olives (darkened by oxidation). However, the concentrations of minerals only provide information on their potential contributions to the diet. Assessment of their effective intake by consumers requires an estimation of their bioaccessibilities, defined as the proportions of the elements converted into soluble forms in the gastrointestinal tract [7,8].

Several methods have been proposed to assess the mineral bioaccessibility in foods. Miller's protocol uses low amounts of food and reduced volumes of enzymatic solutions [7] and has been slightly modified by Mesias, Seiquer, and Navarro for studying the calcium bioavailability of diets rich in Maillard reaction products [8]. In contrast, the protocol developed by Crews, Burrell, and McWeeny [9,10] is characterised by the use of relatively high amounts of samples and the addition of substantial volumes of solutions (125 mL of intestinal juice) to mimic the liquids incorporated into the food during its passage through the gastrointestinal tract. Apart from these differences, both methodologies are based on a sequential enzymolysis [7,9,10].

At the moment, no information is available on the relative behaviour of both methods when applied to table olives. The high proportion of fat in these fruits [11] may require high proportions of bile salts. In addition, the abundant presence of Na might interfere in the solubilisation of the other elements or require a more intense extraction to reduce its presence in the final solid residue as much as possible. Therefore, the development of a method for studying the bioaccessibility of selected mineral nutrients adapted to the high fat and Na contents of table olives is an essential first step in any study of their actual contributions to RDI values and nutritional valorisation.

This work aimed to investigate the bioaccessibilities of mineral nutrients in table olives according to digestion methods (Miller's vs. Crews' protocols), digestion type (standard vs. modified, standard plus a post-digest re-extraction), and mineralisation systems (wet vs. ashing). This study may help the selection/adaptation of a protocol compatible with the high fat and Na content of these products and lead to results that approach the real bioaccessibilities of their minerals. Its development could facilitate further studies on other presentations and promote the nutritional value of table olives. Furthermore, as far as we know, this is the first time that an investigation on the bioaccessibility of mineral nutrients in fermented vegetables is carried out. Therefore, the work represents pioneer information in this field.

2. Materials and Methods

2.1. Samples and Experimental Design

Samples were of the Cacereña cultivar, picked at the green maturation stage (Caceres, Extremadura, Spain). Fruits of size 201/290 were selected and processed as ripe olives, according to the standard procedure, which consisted of three lye treatments, which progressively penetrated the flesh, followed by immersion in tap water to remove the excess alkali, and aeration. After oxidation, the olives were submerged overnight in a 0.1% ferrous lactate solution, packed in a 2.0% NaCl and 0.1% acetic acid cover brine, and sterilised at 121 °C for 45 min to reach an $Fo^{10}_{121\ °C}$ (cumulative sterility value) of 15 [5]. The experiment consisted of a complete factorial design at two levels, with the variables being: gastrointestinal digestion protocol (Miller vs. Crews), digestion type (standard vs. modified, that is standard plus an additional post-digest re-extraction, using distilled-deionised water (onwards water), and mineralisation system (wet vs. ashing). Due to the impossibility of running the complete design simultaneously, each combination of variables (treatment) was carried out independently, with its raw material from the same oxidation process batch. In this way, the experiment consisted of 2^3 different treatments, illustrated in Figure 1 for Miller's protocol. As an example, the first treatment consisted of subjecting the olive sample to the Miller's protocol, following the standard method, using the wet mineralisation for both the supernatant solution and the solid residue (Figure 1). All treatments were carried out in triplicate, using, for each, 100 g of homogenised olive flesh as raw material and one blank, prepared with only the reagents and run in parallel to the sample. The blank was used to evaluate the contribution of enzymes and other chemicals to the final mineral content in the digestion fractions.

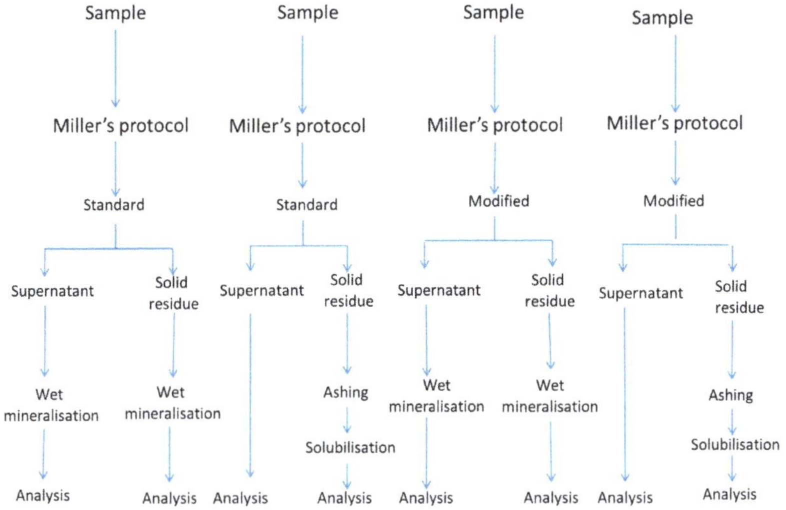

Figure 1. In vitro gastrointestinal digestion of ripe olives. The effect of digestion protocol (Miller vs. Crews), digestion type (standard vs. modified), and mineralisation system (wet vs. ashing) on the mineral bioaccessibility. Schema of the experimental design for Miller's protocol. A similar one for Crews' protocol can be obtained just by substituting Miller's by Crews'.

The introduction of the post-digest re-extraction with water was due to the high levels of Na and K in table olives, which could hardly be solubilised in the volume of liquid used in the standard protocols. This modification may contribute to improving the solubilisation of these minerals and evaluate the strength of the complexes formed by some of them with the flesh components; furthermore, its application is in agreement with the intense nutrient exchanges between phases that take place during the gastrointestinal passage of foods and the re-extraction steps used in other works [12].

2.2. Cleaning of the Material

All glassware used for the determination of the minerals was immersed in 10% (*w*/*w*) nitric acid overnight and then rinsed several times with water.

2.3. In Vitro Digestion of Olives

2.3.1. Miller's Protocol

This method was based on Miller, Schicker, Rasmussen, and Campen [7]. A flowchart of its application is shown in Figure 2 (standard). Briefly, 2 g of homogenised olive pulp (an aliquot from the 100 g ripe olive sample) was suspended in 18 mL of water. For the gastric digestion, its pH was adjusted to 2.0 with 6 N HCl, and the mixture was added to 625 µL of simulated gastric juice (prepared by dissolving 80 mg of pepsin in 5 mL of 0.1 N HCl). The suspension was then placed in a shaking water bath incubator at 37 °C and 110 rpm for 2 h. For the intestinal digestion, the pH of the digest was raised to 6.0 with 1 M $NaHCO_3$ and 5 mL of simulated intestinal juice (prepared by dissolving 10 mg of pancreatin and 62.5 mg of bile salts in 25 mL of 0.1 M $NaHCO_3$) was added. The pH was then adjusted to 7.5 with 1 M $NaHCO_3$, and the suspension incubated at 37 °C and 110 rpm for 2 h. After the gastrointestinal digestion, the digestive enzymes were inactivated in an oven at 100 °C for 4 min. The sample was then cooled in an ice bath and centrifuged at 15,550× *g* and 4 °C (5804R centrifuge, Eppendorf, Hamburg, Germany) for 40 min. The supernatant and the solid residue were separated and weighed, and the mineral concentration in each fraction was analysed.

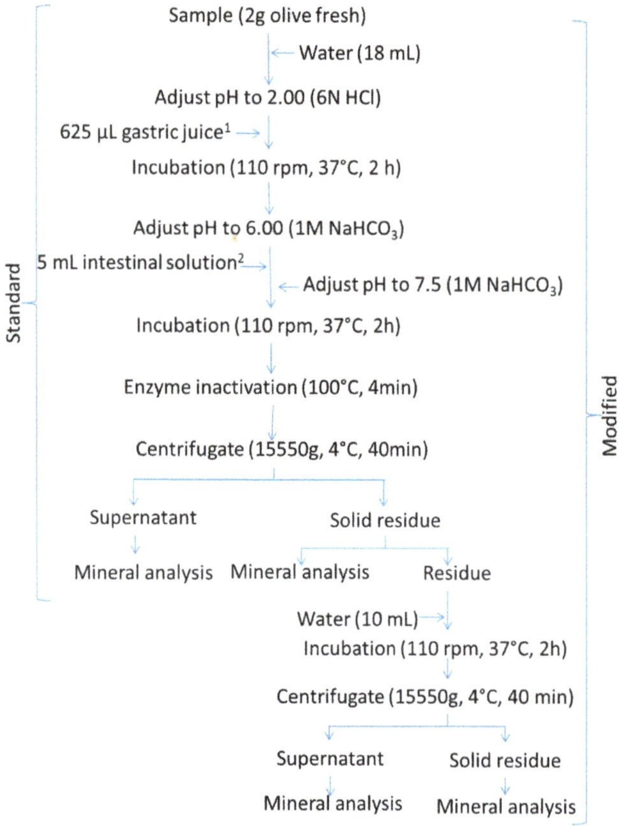

Figure 2. In vitro gastrointestinal digestion of ripe olives. Schema of the Miller's standard and modified (standard plus a post-digest re-extraction) protocols. [1] gastric solution consisted of 0.8 g pepsin dissolved in 5 mL 0.1 N HCl. [2] Intestinal solution consisted of 0.1 g pancreatin and 0.625 g bile salts dissolved in 25 mL 0.1 M $NaHCO_3$. Water stands for deionized-distilled water.

2.3.2. Crews' Protocol

The procedure described in Crews, Burrell, and McWeeny [9,10] was followed (Figure 3, standard). Briefly, 25 g of homogenised olive pulp (an aliquot of the 100 g ripe olive sample) was weighed and suspended in 50 mL of simulated gastric juice, prepared by dissolving 10 mg/mL of pepsin in saline hydrochloric acid (0.15 M sodium chloride; 0.02 M hydrochloric acid) at pH 1.8. The suspension was incubated at 37 °C and 150 rpm for 2 h and 6 N HCl was added as necessary to maintain pH $\leq$ 3.5. After incubation, the suspension pH was adjusted to 7.4 with a saturated $NaHCO_3$ solution and was added to 50 mL of simulated intestinal juice, prepared by mixing equal volumes of (a) 30 mg/mL pancreatin plus 10 mg/mL of amylase and (b) 1.5 g/L of bile salts in 0.15 M NaCl. The mixture was again incubated at 37 °C and 150 rpm for 2 h and centrifuged at 30,000× *g* and 4 °C for 60 min (Sorvall RC6 plus centrifuge, Thermo Scientific, Langenselbold, Germany). The weights and mineral concentrations in the supernatant and the solid residue were calculated as described in Miller's protocol.

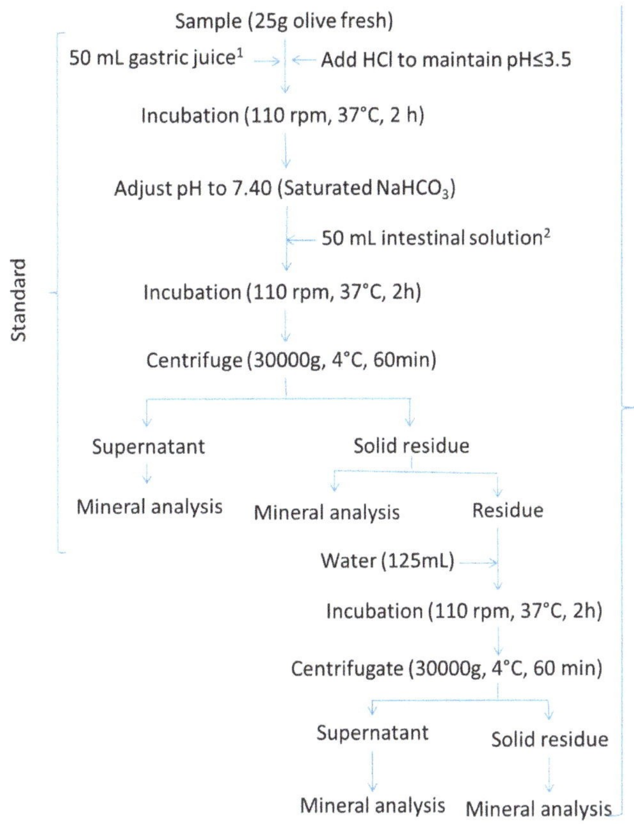

Figure 3. In vitro gastrointestinal digestion of ripe olives. Schema of the Crews' standard and modified (standard plus a post-digest re-extraction) protocols. [1] gastric solution consisted of 1% pepsin in saline HCl (0.15 M NaCl; 0.02 M HCl) at pH 1.8. [2] Intestinal solution consisted of a mixture of equal volumes of (a) 3% pancreatin, 1% amylase, and (b) bile salts in saline solutions (0.15 M NaCl). Water stands for deionized-distilled water.

2.3.3. Modified Protocols

The modification consisted of adding 10 mL, or 125 mL, of water to the digested residues from the standard Miller's or Crews' protocols, respectively, incubating the suspension again in a shaking water bath at 37 °C and 110 rpm for 2 h, and centrifuging at 15,000× *g* (Miller's protocol) or 30,000× *g* (Crews'protocol) and 4 °C for 60 min (Sorvall RC6 plus centrifuge). The supernatant was combined with that from the standard protocol to form the supernatant of the modified technique. The mineral content in these supernatants and their respective re-extracted solid residues were determined. The resulting methodology will be onwards referred to as the modified protocol.

2.4. Mineralisation

The analysis of most fractions requires previous mineralisation. Due to the diversity of samples studied (olive paste, supernatant solutions, and post-digestion solid residues), evaluation of the effect of the mineralisation system was considered of interest. Two options were assayed.

2.4.1. Wet Mineralisation

For this process, 20–25 mL of the supernatants were concentrated to 15 mL in a flask, and then added to 5 mL of 65% HNO_3. The container was then heated in a shaking sand bath at 180–220 °C until the liquid was clear or pale straw-coloured and orange fumes ceased. Then, 5 mL of a mixture of HNO_3 (65%)–$HClO_4$ (60%) (1:4) was added, and the solution was heated at 180–220 °C until discolouration and white fumes evolved. The samples were cooled, transferred into a 25 mL volumetric flask and made up to volume with water.

For the homogenised olive flesh (raw material) and the solid residues of the digestions, 2.5 g of the paste were weighed into a 50 mL Erlenmeyer flask and added to 5 mL of HNO_3 (65%). Then, the suspensions were subjected to the same steps described above for liquids.

2.4.2. Ashing

This method was applied only to solids as the solutions were directly submitted to the analysis. In short, 2.5 g of sample (homogenised olive flesh or solid residues from the digestions) was weighed in a quartz capsule and placed in a muffle oven whose temperature was quickly brought to 100 °C, followed by a slow increase up to 550 °C. After incineration for 8–10 h, the ashes, greyish-white in colour, were moistened and dissolved (slightly warming the capsule) in three portions of 2 mL 6 N HCl and filtered through a filter paper into a 25 mL volumetric flask, using a suction hood. After washing the filter three times with 3 mL of water, the solution was made up to volume with water.

2.5. Mineral Analysis

Na, K, Ca, Mg, and Fe were analysed by atomic absorption spectrophotometry, using an air-acetylene flame and the analytical conditions were recommended by the equipment manufacturer [13]. The value for each triplicate was the average of three determinations.

To prevent interferences and ionisation of the air-acetylene flame, the aliquots for analysis and the calibration standards were added to lantane (0.5%, *w*/*v*), when analysing Ca and Mg, or potassium (0.1%, *w*/*v*) and sodium (0.1%, *w*/*v*), in the case of Na and K, respectively.

Phosphorus was analysed following the official method of the AOAC n° 970-39 Phosphorus in Fruits and Fruit Products (spectrophotometric molybdovanadate method) [14]. This method is based on the absorbance at 400 nm of the yellow phospho-molybdovanadate complex formed in the presence of V^{5+} and Mo^{6+}. The value for each triplicate was also the average of three determinations.

Calibration curves were obtained daily from successive dilutions of the stock solutions. Interpolation was always made after subtracting the signal of the blank from those of the samples. Furthermore, samples of standard solutions were also periodically included in the determinations.

2.6. Apparatus and Reagents

The equipment included a GBC model 932 AA (GBC, Braeside, VIC, Australia) atomic absorption spectrometer equipped with three hollow multi-element cathode lamps, (Na and K) (Photron, Narre Warren, VIC, Australia), (Cu, Fe, and Mn) (GBC, Braeside, VIC, Australia) and (Ca, Mg, Cu, and Zn) (Photron, Narre Warren, VIC, Australia); a Cary UV/Visible spectrophotometer model 1E (Varian Australia, Mulgrave, Victoria); a shaking water bath incubator (WY-200 COD. 5312091, COMECTA, S.A., Barcelona, Spain); and a shaking sand bath incubator (Combiplac-Sand 6000709; J.P. Selecta, Barcelona, Spain).

All reagents were of analytical grade. The enzymes and bile salts were purchased from Sigma-Aldrich (pepsin from porcine gastric mucosa Cat N° P7000; pancreatin from porcine pancreas Cat N° P1750; α-amylase from porcine pancreas Cat N° A3176; and bile salts Cat N° B8756).

2.7. Mineral Recovery and Bioaccessibility Estimation

The calculous were carried out independently for each treatment (the combination of factors). Considering the concentrations in the raw material, the supernatant solutions (including that from the modified protocols), the solid residue, and the blank, the amount of each mineral nutrient in them (RM, S, SR, and B, respectively) were estimated. The amount of mineral in each of these fractions was calculated taking into account the weight of each of them, which allows a correct mass balance. Bioaccessibility and recovery (expressed as percentages) were estimated using the following formulae:

$$\text{Bioaccessibility } (\%) = \left(\frac{(\text{S} - \text{B})}{\text{RM}}\right) \times 100 \quad (1)$$

$$\text{Recovery } (\%) = \left(\frac{(\text{S} - \text{B} + \text{SR})}{\text{RM}}\right) \times 100 \quad (2)$$

An approach to the contribution of the ripe olives to the RDI of the minerals studied (Na, K, Ca, Mg, Fe and P), based on their bioaccessible amount in 100 g olive flesh serving size, were also deduced.

2.8. Statistical Analysis

The effect of the different factors (digestion protocols, digestion type, and mineralisation system) on bioaccessibility was studied by General Linear Model (GLM). The effects were considered significant at $p \leq 0.05$ when the corresponding confidence limits (CL) of their averages did not overlap. The study was carried out using Statistic v. 8.0 (StatSoft, Inc., Tulsa, OK, USA) [15].

3. Results and Discussion

Through processing, ripe olives are usually in contact with solutions containing Na (brine and lye), Ca (mainly during storage and final packaging), and Fe (for fixing the colour) [5], which increase their contents in the flesh. The concentrations of these minerals in the different samples used as raw material were high. The Na ranged from 7085 to 7181 mg/kg (Tables 1 and 2), lower than the levels reported for any other table olive presentation [6]. The Ca content, 1666–1711 mg/kg (Tables 1 and 2), was higher than in green plain Spanish-style or directly brined olives [6]. However, the most notable difference was found for Fe because its content was particularly high (102 to 105 mg/kg) with respect to any other non-oxidized olive product (3.49–7.70 mg/kg) [6].

Regarding other mineral nutrients not intentionally incorporated during processing, their contents suffer a progressive diminution during the elaboration [5]. However, the ripe olives of this experiment still retained substantial levels of Mg (129 to 138 mg/kg), K (104 to 109 mg/kg), and P (91 to 100 mg/kg) (Tables 1 and 2).

The differences among the mineral contents in diverse raw materials (ripe olive samples) were relatively close since they came from the same batch; but, even in such circumstances, the use of a specific raw material for each treatment (a combination of factors) was considered convenient to eliminate this source of variability on bioaccessibility.

Most of the enzymes and bile salts used for the digestion also contained nutrient elements that contributed to the mineral levels in the final fractions (see contents in the blanks) (Tables 1 and 2). The levels were particularly high in Na, whose concentrations ranged from 1002 to 1309 mg/kg in Miller's protocol but was markedly higher (5008–6139 mg/kg) in Crews' method. Furthermore, the levels of P (25.9–28.1 mg/kg, Miller; 37.7–38.8 mg/kg, Crews) and K (18.4–20.3 mg/kg, Miller; 31.3–33.4 mg/kg, Crews) were also relevant, but not the presence of the other nutrients, which were low (Tables 1 and 2). In any case, the concentrations in the supernatants were always corrected by subtracting the corresponding blanks.

Table 1. In vitro gastrointestinal digestion of ripe olives, using the Miller's protocol. Effect of the digestion type (standard vs. modified) and mineralisation system (wet vs. ashing) on the mineral concentrations in the supernatants and solid residues. The contents in the raw material (samples) and blanks are also provided since they are required for the estimations of mineral recoveries and bioaccessibilities. Concentrations in mg/kg.

		Standard				Modified			
Element	**Mineralisation**	**Raw Material**	**Supernatant Solution**	**Solid Olive Residue**	**Blank**	**Raw Material**	**Supernatant Solution**	**Solid Olive Residue**	**Blank**
Na	Wet	7132 (20)	1861 (8)	2089 (110)	1309 (8)	7104 (30)	1184 (1)	292 (14)	1002 (5)
	Ashing	7181 (5)	1752 (5)	2738 (89)	1215 (6)	7085 (11)	1257 (11)	272 (9)	1111 (6)
K	Wet	109.4 (0.7)	28.0 (0.1)	41.0 (1.7)	20.3 (0.4)	104.8 (1.0)	20.2 (<0.1)	5.2 (0.4)	18.4 (<0.1)
	Ashing	107.4 (1.0)	26.9 (<0.1)	55.1 (3.1)	19.4 (0.2)	107.0 (1.5)	21.2 (<0.1)	4.8 (0.1)	19.6 (<0.1)
Ca	Wet	1689.0 (8.0)	29.7 (0.6)	1275.7 (66.4)	1.7 (0.0)	1666.2 (3.6)	20.7 (0.1)	1449.9 (63.8)	1.4 (<0.1)
	Ashing	1700.6 (6.0)	29.5 (0.2)	1751.5 (104.4)	1.9 (<0.1)	1698.4 (6.0)	21.0 (0.1)	1713.4 (4.1)	1.7 (<0.1)
Mg	Wet	133.4 (0.4)	12.4 (<0.1)	38.3 (2.0)	3.9 (<0.1)	138.4 (0.2)	9.01 (<0.1)	45.0 (1.6)	3.9 (<0.1)
	Ashing	129.7 (0.9)	12.1 (<0.1)	48.9 (2.7)	3.8 (<0.1)	128.7 (0.8)	8.4 (<0.1)	50.1 (0.8)	3.7 (<0.1)
Fe	Wet	102.6 (0.8)	4.8 (<0.1)	55.7 (2.5)	0.9 (<0.1)	101.8 (0.9)	3.0 (<0.1)	66 (3.7)	0.4 (<0.1)
	Ashing	104.5 (1.1)	4.5 (<0.1)	77.6 (5.4)	0.6 (<0.1)	104.6 (2.4)	3.4 (<0.1)	77.0 (0.6)	0.7 (<0.1)
P	Wet	94.3 (0.4)	33.2 (0.1)	34.7 (1.0)	25.9 (<0.1)	91.3 (1.4)	25.1 (<0.1)	38.2 (1.3)	28.1 (<0.1)
	Ashing	94.8 (0.6)	NA	48.6 (2.7)	NA	99.8 (0.8)	NA	48.4 (0.5)	NA

Average of three independent experiments; standard error in parentheses; NA, not available.

Table 2. In vitro gastrointestinal digestion of ripe olives, using the Crews' protocol. Effect of the digestion type (standard vs. modified) and mineralisation system (wet vs. ashing) on the mineral concentrations in the supernatants and solid residues. The contents in the raw material (samples) and blanks are also provided since they are required for the estimations of mineral recoveries and bioaccessibilities. Concentrations in mg/kg.

		Standard				Modified			
Element	**Mineralisation**	**Raw Material**	**Supernatant Solution**	**Solid Olive Residue**	**Blank**	**Raw Material**	**Supernatant Solution**	**Solid Olive Residue**	**Blank**
Na	Wet	7120 (7)	6139 (63)	7940 (220)	6139 (63)	7121 (25)	3653 (11)	981 (20)	5008 (11)
	Ashing	7116 (5)	6076 (85)	7148 (27)	5100 (37)	7157 (12)	3539 (21)	950 (30)	5098 (13)
K	Wet	104.0 (0.8)	44.7 (0.4)	112.1 (2.0)	33.4 (0.2)	107.6 (0.3)	28.6 (0.1)	19.0 (1.2)	31.3 (0.1)
	Ashing	105.5 (1.6)	42.8 (0.6)	103.2 (2.0)	32.1 (<0.1)	105.0 (0.7)	27.2 (0.1)	16.2 (0.2)	31.3 (0.2)
Ca	Wet	1710.5 (8.3)	1.3 (<0.1)	2949.3 (56.8)	0.6 (<0.1)	1693.9 (4.6)	3.4 (<0.1)	2686.0 (4.6)	0.6 (<0.1)
	Ashing	1702.6 (7.3)	1.2 (<0.1)	2585.0 (26.7)	0.6 (<0.1)	1709.9 (8.3)	3.3 (<0.1)	2580.7 (120.1)	0.5 (<0.01)
Mg	Wet	132.0 (2.1)	15.0 (<0.1)	163.3 (5.3)	6.1 (<0.1)	132.6 (1.0)	11.6 (<0.1)	82.4 (3.3)	5.3 (<0.1)
	Ashing	134.5 (0.7)	14.3 (0.2)	143.0 (1.0)	5.6 (<0.1)	133.5 (1.5)	11.8 (<0.1)	79.3 (2.9)	6.1 (<0.1)
Fe	Wet	103.3 (2.1)	6.0 (<0.1)	131.0 (4.5)	0.35 (<0.1)	103.5 (0.7)	3.7 (<0.1)	125.1 (11.9)	0.7 (<0.1)
	Ashing	104.6 (1.6)	6.3 (0.1)	119.6 (2.3)	0.8 (<0.1)	103.7 (1.0)	3.7 (<0.1)	114.1 (4.7)	0.6 (<0.1)
P	Wet	96.2 (0.8)	53.6 (0.5)	69.5 (0.7)	37.7 (0.2)	96.1 (0.4)	29.1 (0.1)	59.9 (0.9)	38.8 (0.1)
	Ashing	94.7 (1.1)	NA	57.2 (0.7)	NA	95.4 (1.6)	NA	59.0 (1.9)	NA

Average of three independent experiments; standard error in parentheses; NA, not available.

3.1. Mineral Linkage to Olive Flesh Components and Post-Digestion Extraction

The response of the mineral nutrients in foods to digestion is strongly related to their aqueous solubility and linkage to the structural components of the products. According to the literature [16], the bioaccessibility of Na is usually considered complete but, due to the high salt concentration in table olives, its response could not be straightforward but requires investigation.

In the standard protocols, the concentration of Na in the solid residue was higher than in the supernatant solution (average 2413 vs. 1806 mg/kg for Miller; average 7544 vs. 6108 mg/kg for Crews), regardless of the mineralisation system (Tables 1 and 2). However, applying modified protocols, the Na concentrations in both fractions reversed, being markedly higher in the supernatant fractions than in the solid residues (average 1221 vs. 282 mg/kg, Miller; average 3596 vs. 966 mg/kg, Crews) (Tables 1 and 2). Hence, the application of one post-digest re-extraction to the standard protocols constitutes a closer approach to reality [9,10]. Furthermore, this also means that Na is weakly linked to the ripe olive flesh components.

The distributions of K in the two fractions of the digestion (standard or modified) followed a similar trend to Na, regardless of treatments (Tables 1 and 2). This behaviour may also indicate that K could be, in practice, completely bioaccessible when subjected to the conditions prevailing in the gastrointestinal tract [16] and that it is weakly retained in the ripe olive flesh.

On the contrary, the concentrations of Ca in the supernatant solutions after digestion were very low with respect to the solid fraction, regardless of the technique applied (Tables 1 and 2) Furthermore, when the modified protocol was applied, the release was not improved despite the still high Ca levels in solid residues (Tables 1 and 2). In consequence, no equilibrium between solid residues and supernatant solutions could be expected even in the case of a large number of post-digest re-extractions. Such behaviour means a strong Ca linkage to the olive flesh components, resistant to the digestive enzymes. These data are in agreement with the literature reports on Ca absorption by olives [17]. Furthermore, this flesh capacity for Ca bounding is used for improving texture during green olive fermentation and ripe olive storage [5].

Magnesium is not added during processing; on the contrary, leakage through elaboration is common [5]. After standard digestion, its concentrations in the supernatant solutions (Tables 1 and 2) were markedly lower than the levels in the solid residues, regardless of digestion technique, following, in this case, a similar trend to Na, K or even Ca. However, applying post-digest re-extraction, the contents in the solid residues remained similar (Miller) or slightly decreased (Crews). Therefore, its behaviour in the modified protocol was completely different from that followed by Na and K but approached that of Ca, with a higher solubilisation. Therefore, Mg was retained in the olive flesh more than Na and K but less than Ca, meaning that at least part of it can also be bound to the ripe olive flesh.

As inferred from the low concentrations of iron found in the standard and modified protocol supernatants and its high contents in the solid residues, strong retention of iron by the ripe olive flesh is evident (Tables 1 and 2). This behaviour is somewhat similar to that of Ca, although may have a different origin since such absorption has been related to the formation of complexes between this element and polymers from hydroxytyrosol and caffeic acid, which are produced during the ripe olive darkening (oxidation) process [5].

Due to the relatively high content of P (not added during elaboration) in the raw material, it is evident that there is a strong link between this element and the olive flesh, which has resisted the successive alkali treatment and tap washings applied throughout processing [5]. The enzymes used for the digestion produced a marked solubilisation of P (Tables 1 and 2), although still left a sensible proportion of it in the solid residue, which was not solubilised by the post-digest re-extraction. Therefore, P was relatively resistant to the digestion attack but weaker than that observed for Ca or Fe, stronger than Na and K, and slightly weaker than Mg.

Hence, the application of a post-digestion extraction was useful not only for a more exhaustive removal of some minerals (Na and K) from the solid residue but also for assessing the different linkage degrees of the studied minerals and the olive flesh components. The results from the modified protocol re-affirm the hypothesis that the resistance to solubilisation of Ca, Fe (mainly) or Mg and P (in lower proportions) was not just a matter of equilibrium between phases but was also related to their bounding strength to the olive flesh. Crews, Burrell, and McWeeny [9,10] also observed similar behaviour for some of these nutrients in the cereal food group [12].

3.2. *Effect of Diverse Factors on the Mineral Bioaccessibility*

The weights of the raw materials, blank solutions and the different digestion fractions required for evaluating the mineral recovery are shown in Table 3. The markedly higher weights of olive samples and digested fractions in the Crews' technique are apparent. When applying the modified protocols, the weights of supernatants are the sum of those from the standard supernatant solutions of enzymes plus those from the post-digest re-extraction. From the data in Tables 1–3, the bioaccessibility of the minerals were estimated (Equation (1)). Overall, the mineralisation system had scarce or no effect on the bioaccessibility results. Regarding digestion methods, the bioaccessibilities of Na and K (Figure 4a,b) were markedly higher in Miller's than in Crews' standard protocols, but the application of the modified protocol considerably increased them to values greater than 90% (mainly in Crews' methodology), with only slight differences between protocols in the case of Na and a significant difference in favour of Miller's for K. Hence, the notable increase in the recuperation of these elements with the post-digest re-extraction means that they could be completely bioaccessible from ripe olives (progressive dilution and absorption from the human gut), as confirmed by their weak interaction with the flesh components and as already suggested in the literature for other foods [9,10].

In contrast, the bioaccessibility of Ca was always low (Figure 4c), although it was higher when applying Miller's protocol. However, the use of the modified protocols hardly led to any further improvement. As a result, the potential contribution of ripe olives, and possibly other presentations, to Ca in the diet could be minimal, despite the relatively high proportion of this element usually present in the product. However, this problem is not exclusive of olives because low bioaccessibility levels of Ca have also been observed in other foods, such as school meals (0.75%), with the lowest bioaccessibility found in vegetables [18] or milk, where calcium was partially soluble and ranged from 48% to 62% [19].

The highest bioaccessibility of Mg (Figure 5a) was observed using the standard Miller's protocol (above 70%) without any effect of post-digest re-extraction. The modified Crews' protocol increased the bioaccessibility versus the standard but without reaching Miller's levels. Therefore, Crews' conditions were scarcely efficient for studying Mg bioaccessibility. According to the literature, Mg bioaccessibility may be influenced by the compound used in the diet; the ingestion of citrate was more efficient that oxide [20]. Mg absorption in healthy women is reported to be incremental (11–14%) when mineral water was consumed alone or in combination with meals [21]. Furthermore, the mineralisation level (sulfate, bicarbonate, or calcium) in mineral waters did not influence the Mg bioavailability [22].

Table 3. In vitro gastrointestinal digestion of ripe olives. Weights of the raw materials (samples), the different fractions obtained after digestion, and the blank solutions, according to digestion protocols (Miller vs. Crews), digestion type (standard vs. modified), and mineralisation system (wet vs. ashing). The information allows estimation of the mineral recovery and bioaccessibility. Data are expressed in g.

Technique	Type of Digestion	Mineralisation	Sample Weight	Supernatant Solution	Solid Residue	Blank Solution
Miller	Standard	Wet	2.011 (0.004)	24.020 (0.096)	2.163 (0.094)	26.406 (–)
		Ashing	2.030 (0.012)	24.070 (0.086)	1.599 (0.121)	26.345 (–)
	Modified	Wet	2.022 (0.002)	34.005 (0.088)	1.849 (0.086)	26.454 (–)
		Ashing	2.041 (0.009)	34.218 (0.109)	1.622 (0.021)	26.372 (–)
Crews	Standard	Wet	25.060 (0.025)	123.160 (1.194)	14.533 (0.405)	136.530 (–)
		Ashing	25.017 (0.038)	125.802 (1.931)	16.272 (0.265)	138.090 (–)
	Modified	Wet	25.170 (0.067)	236.286 (0.850)	15.753 (0.469)	139.150 (–)
		Ashing	25.407 (0.194)	244.145 (1.659)	16.424 (0.594)	136.510 (–)

Average of three independent experiments; standard error in parentheses. Each digestion had its blank.

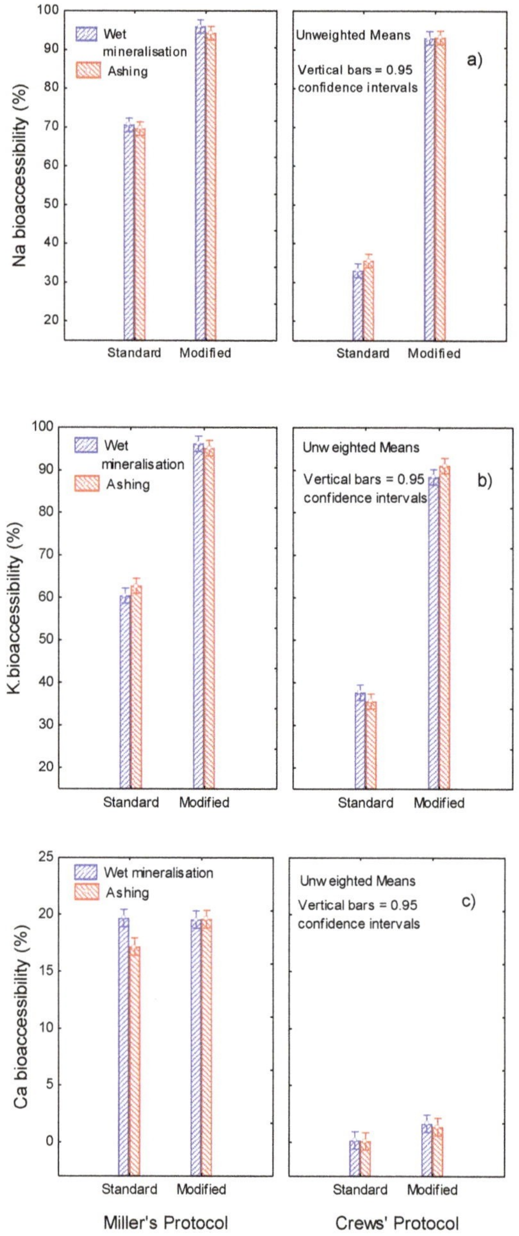

Figure 4. In vitro gastrointestinal digestion of ripe olives. Effect of digestion protocol (Miller vs. Crews), and digestion type (standard vs. modified) and mineralisation system (wet vs. ashing) on bioaccessibility (%) of (**a**) Na, (**b**) K and (**c**) Ca.

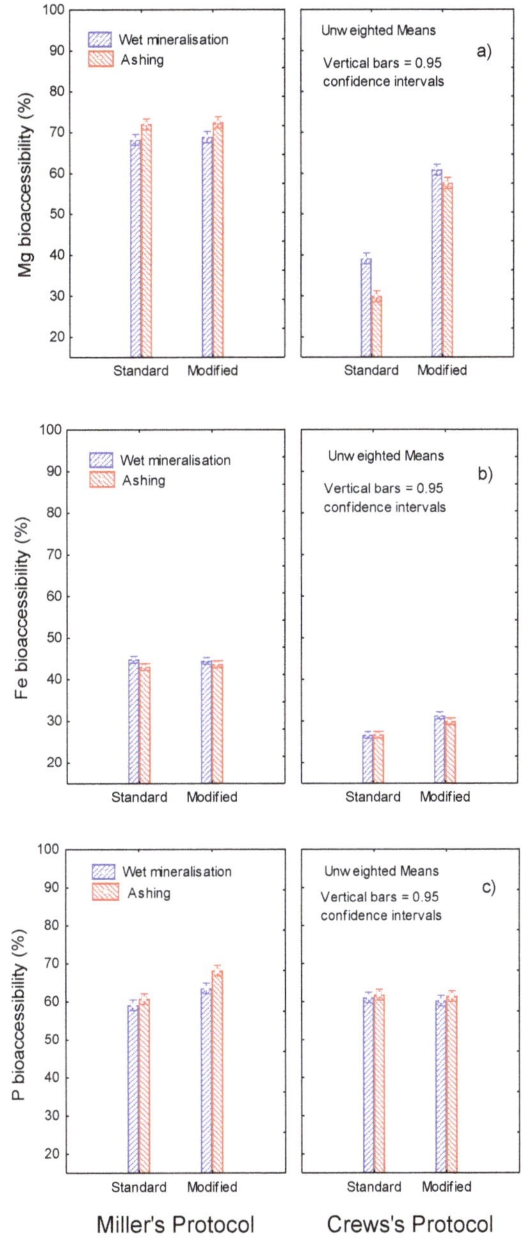

Figure 5. In vitro gastrointestinal digestion of ripe olives. Effect of digestion protocol (Miller vs. Crews), and digestion type (standard vs. modified) and mineralisation system (wet vs. ashing) on bioaccessibility (%) of (**a**) Mg, (**b**) Fe, and (**c**) P.

The highest bioaccessibility of Fe (approx. 45%) was observed when applying Miller's protocols (standard or modified) (Figure 5b). The re-extraction had a limited effect on the Crews'protocol, changing from approx. 27% (standard) to just above 30% (modified). As in the case of Ca and Mg,

Miller's protocol was more efficient (approx. 45% bioaccessibility) and reached an intermediate level between them (approx. 20 and 70%, respectively). In previous studies by Crews, Burrell, and McWeeny [12], high percentages of iron solubility were reported; about 25% (cereals) and 75% (vegetables). The percentage solubility of iron in whole-meal bread was approx. 35%, while in crab it was sensibly lower (5%) [9]. In selected foods (meat as well as bread, milk, and beverage substitutes), the iron available was in general low (below 10%), reaching only slightly higher proportions in orange juice (about 25%) and cheese plus orange juice (about 17%) [7]. Therefore, iron bioaccessibility in ripe olives could be higher than in more common foods. Its low bioaccessibility has been related to the presence of oxalic acid and egg proteins, while meat had a favourable effect [18].

Phosphorus bioaccessibilities (Figure 5c) were relatively high (about 60%) and improved (5–10%) when applying the post-digest re-extraction only in the case of Miller's protocol. A study of the in vitro P digestible in meat and milk products showed better absorbability in foods of animal origin than, for example, legumes [23]. However, legumes may be a relatively poor source of P, while in products containing phosphates additives, the digestible P was easily available [24]. In general, P from plants is not well absorbed because this element is stored in the form of phytic acid or phytate, which may interfere with its absorption [25]. The relative high bioaccessibility of P in table olives may be related to the low/absence presence of phytate.

3.3. Mineral Recovery during Digestion

The data in Tables 1–3 allow for a complete estimation of the mineral recovery, regardless of the supernatant type and solid residue. The comparison of their sum with the weights of the minerals initially present in the samples is straightforward (Equation (2)). The results show good overall recovery for all the mineral nutrients analysed (Table 4).

3.4. Contribution of Ripe Olives to Daily Recommended Mineral Intake

The ripe olive minerals' bioaccessibilities can be used for estimating the potential contribution of the product to their RDI [2] (Table 5). Their values are affected by the three factors involved in the experiment, similar to their bioaccessibilities (they are just linear combinations of these), and do not require further comments. Overall, Miller's protocol led to higher contributions. To emphasise that the ripe olives have an outstanding contribution to the RDI of Fe in the diet, which reaches about 34%, far above the 15% limit required to be considered as a significant source of this mineral and be declared in the label. Furthermore, its impact is higher than that of Na (approx. 28%).

Table 4. In vitro gastrointestinal digestion of ripe olives. Overall mineral recovery (expressed as %), according to digestion protocols (Miller vs. Crews), digestion type (standard vs. modified), and mineralisation system (wet vs. ashing).

Technique	Type of Digestion	Mineralisation	Na	K	Ca	Mg	Fe	P
Miller	Standard	Wet	102.20 (0.26)	101.32 (0.36)	100.61 (0.48)	103.30 (0.34)	102.45 (0.28)	99.45 (0.49)
		Ashing	99.84 (0.52)	102.43 (0.33)	99.90 (0.42)	102.32 (0.24)	102.17 (0.24)	100.00 (NA) *
	Modified	Wet	99.59 (0.58)	99.65 (0.28)	99.02 (0.38)	102.11 (0.23)	104.03 (0.24)	98.33 (0.42)
		Ashing	97.90 (0.31)	99.70 (0.52)	99.86 (0.55)	103.41 (0.17)	103.74 (0.14)	100.00 (NA) *
Crews	Standard	Wet	97.98 (0.35)	98.95 (0.62)	100.06 (0.49)	101.99 (0.48)	100.16 (0.56)	102.17 (0.40)
		Ashing	98.95 (0.74)	99.95 (0.68)	98.91 (0.59)	99.54 (0.43)	101.84 (0.79)	100.00 (NA) *
	Modified	Wet	101.30 (0.49)	99.02 (0.26)	100.65 (0.35)	98.27 (0.28)	100.03 (0.54)	99.80 (0.50)
		Ashing	100.94 (0.49)	99.08 (0.38)	98.93 (0.42)	98.91 (0.38)	101.87 (0.28)	100.00 (NA) *

n = 3 for each treatment; * bioaccessibility estimated by difference.

Table 5. Contribution (expressed as %) of ripe olives to the RDI of the nutrient mineral studied, according to digestion protocols (Miller vs. Crews), digestion type (standard vs. modified), and mineralisation system (wet vs. ashing). Data are based on their bioaccessibilities and 100 g olive flesh.

Technique	Type of Digestion	Mineralisation	Na	K	Ca	Mg	Fe	P
Miller	Standard	Wet	21.04 (0.08)	0.336 (0.001)	4.16 (0.06)	1.09 (0.01)	32.40 (0.14)	0.808 (0.003)
		Ashing	20.91 (0.07)	0.334 (<0.001)	4.08 (0.02)	2.52 (<0.01)	32.88 (0.07)	0.811 (0.005) *
	Modified	Wet	28.37 (0.15)	0.499 (0.001)	4.12 (0.02)	2.68 (<0.01)	32.76 (0.07)	0.785 (0.004)
		Ashing	28.00 (0.08)	0.512 (0.002)	4.12 (0.02)	2.48 (<0.01)	33.58 (0.08)	0.871 (0.002) *
Crews	Standard	Wet	9.91 (0.10)	0.190 (<0.001)	0.04 (<0.01)	1.07 (<0.01)	19.74 (0.06)	0.828 (0.002)
		Ashing	9.96 (0.14)	0.192 (<0.001)	0.04 (<0.01)	1.09 (<0.01)	19.62 (0.05)	0.822 (0.004) *
	Modified	Wet	27.52 (0.13)	0.477 (0.001)	0.36 (<0.01)	2.12 (<0.01)	22.46 (0.08)	0.835 (0.003)
		Ashing	27.55 (0.13)	0.468 (0.001)	0.36 (<0.01)	2.16 (0.01)	22.98 (0.12)	0.820 (0.006) *

n = 3 for each treatment; * based on bioaccessibility estimated by difference.

4. Conclusions

The digestion protocols had significant effects on the bioaccessibility estimation of ripe olive mineral nutrients. Overall, Miller's protocol led to higher values than Crews' protocol. The application of a post-digest re-extraction improved (vs. standard digestion) the potential bioaccessibility of the most soluble minerals. The application of this modification was useful to evaluate the strength of the linkage between some elements and olive flesh components. Monovalent minerals (Na and K) were hardly bound and were completely bioaccessible. In contrast, the noticeable presence of divalent (and P) elements in the final solid residue indicated that at least some of them can still be strongly linked to olive flesh even after digestion. Among these cations, Ca was the most vigorously retained and, as a result, showed the lowest bioaccessibility (a maximum of approx. 20%); Mg was weakly bound and showed a high bioaccessibility level (>70%). P reached intermediate values of 60–70%. Fe was moderately retained and showed a bioaccessibility of about 45%. Based on these data, the contribution of 100 g ripe olive flesh to RDI of Fe can be estimated as approx. 34%, which allowed consideration of the product as a source of this element while maintaining a moderate Na level (about 28%) and a negligible impact of the other elements.

The modified Miller's protocol, which includes a post-digest re-extraction, uses less sample, produces a lower volume of supernatant solutions (and solid residues), and, overall, leads to higher bioaccessibility values; therefore, it is proposed for further studies on the bioaccessibility of mineral nutrients in table olives in general.

Author Contributions: Conceptualization: A.L.-L. and A.G.-F.; Methodology: A.L.-L.; Software: A.G.-F.; Validation: A.L.-L. and A.G.-F.; Formal analysis: A.G.-F. and A.L.-L.; Investigation: J.M.M.-B., A.L.-L.; Resources: A.L.-L. and A.G.-F.; Data curation: A.L.-L. and A.G.-F.; Writing—original draft preparation: A.L.-L. and A.G.-F.; Writing—review and editing: A.L.-L. and A.G.-F.; Visualization: A.L.-L. and A.G.-F.; Supervision: A.L.-L.; Project administration: A.G.-F. and A.L.-L.; Funding acquisition: A.G.-F. and A.L.-L. All authors have read and agreed to the published version of the manuscript.

Funding: This work was funded in part by the Ministry of Economy and Competitiveness from the Spanish government through projects AGL2009-07436/ALI and AGL2010-15494/ALI, partially financed by European regional development funds (ERDF) and Junta de Andalucía through financial assistance to group AGR-125.

Acknowledgments: J.M.M.-B. thanks CSIC for their JAE fellowship. We thank Elena Nogales Hernández for her technical assistance.

Conflicts of Interest: The authors declare no conflict of interest.

References

1. Food and Drug Administration, Department of Health and Human Services. Part 101.9, Nutrition Labeling of Food. In *Title 21, Foods and Drugs. Electronic Code of Federal Regulations*; US. Government Printing Office: Washington DC, USA, 2019. Available online: https://www.ecfr.gov/cgi-bin/text-idx?SID=c9d1fccd36e3cef779a7ecf6019b2e04&mc=true&node=se21.2.101_19&rgn=div8 (accessed on 12 December 2019).
2. European Parliament and Council of the European Union. Regulation (EU) No 1169/2011 of the European Parliament and of the Council of 25 October 2011 on the provision of food information to consumers. *Off. J. Eur. Union* **2011**, *54*, 18–61.
3. The National Academies of Sciences, Engineering and Medicine, Health and Medicine Division. Dietary Reference Intakes Tables and Application. 2019. Available online: https://http://nationalacademies.org/hmd/Activities/Nutrition/SummaryDRIs/DRI-Tables.aspx (accessed on 19 December 2019).
4. International Olive Council (IOC). World Table Olives Figures: Production. 2020. Available online: https://www.internationaloliveoil.org/what-we-do/economic-affairs-promotion-unit/ (accessed on 16 February 2020).
5. Garrido-Fernández, A.; Fernández-Díez, M.J.; Adams, R.M. *Table Olive Production and Processing*; Chapman & Hall: London, UK, 1997.
6. López, A.; García, P.; Garrido, A. Multivariate characterization of table olives according to their mineral nutrient composition. *Food Chem.* **2008**, *106*, 369–378. [CrossRef]
7. Miller, D.D.; Schicker, B.R.; Rasmussen, R.R.; Campen, D.V. An in vitro method for estimation of iron availability from meals. *Am. J. Clin. Nutr.* **1981**, *34*, 2248–2256. [CrossRef] [PubMed]

8. Mesias, M.; Seiquer, I.; Navarro, P. Influence of diets rich in Maillard reaction products on calcium bioavailability. Assays in male adolescents and in Caco-2 cells. *J. Agric. Food Chem.* **2009**, *57*, 9532–9538. [CrossRef] [PubMed]
9. Crews, H.M.; Burrell, J.A.; McWeeny, D.J. Trace element solubility from food following enzymolysis. *Z. Lebensm. UntersForsch* **1985**, *180*, 221–226. [CrossRef] [PubMed]
10. Crews, H.M.; Burrell, J.A.; McWeeny, D.J. Comparison of trace element solubility from food items treated separately and in combination. *Z. Lebensm. UntersForsch* **1985**, *180*, 405–410. [CrossRef] [PubMed]
11. López, A.; Montaño, A.; García, P.; Garrido, A. Fatty acid profile of table olives and its multivariate characterization using unsupervised (PCA) and supervised (DA) chemometrics. *J. Agric. Food Chem.* **2006**, *54*, 6747–6753. [CrossRef] [PubMed]
12. Crews, H.M.; Burrell, J.A.; McWeeny, D.J. Preliminary enzymolysis studies on trace element extractability from food. *J. Sci. Food Agric.* **1983**, *34*, 997–1004. [CrossRef]
13. Athanasopoulos, N. *GBC 932/933 Atomic Absorption Spectrophotometers*; Operation Manual: Braeside, VIC, Australia, 1994.
14. AOAC International. Phosphorus in fruits and fruit products no. 970.39. In *Official Methods of Analysis of AOAC International*, 18th ed.; Horwitz, W., Latimer, G.W., Eds.; AOAC International: Gaithersburg, MD, USA, 1993.
15. StatSoft. *Statistica for Windows (Computer Program Manual)*; StatSoft: Tulsa, Oklahoma, 2007.
16. Miller, D.D. Minerals. In *Food Chemistry*; Fennema, O.R., Ed.; Marcel Dekker: New York, NY, USA, 1996; Chapter 9; pp. 617–649.
17. Jiménez, A.; Heredia, A.; Guillén, R.; Fernández-Bolaños, J. Correlation between soaking conditions, cation content of cell wall, and olive firmness during "Spanish green olive" processing. *J. Agric. Food Chem.* **1997**, *45*, 1653–1658. [CrossRef]
18. Cámara, F.; Amaro, M.A.; Barberá, R.; Clemente, G. Bioaccessibility of minerals in school meals: Comparison between dialysis and solubility methods. *Food Chem.* **2005**, *92*, 481–489. [CrossRef]
19. Seiquer, I.; Delgado Andrade, C.; Haro, A.; Navarro, M.P. Assessing the effects of severe heat treatment of milk on calcium bioavailability; in vitro and in vivo studies. *J. Dairy Sci.* **2010**, *93*, 5635–5643. [CrossRef] [PubMed]
20. Kappeler, D.; Heimbeck, I.; Herpich, C.; Naue, N.; Höfler, J.; Timmer, W.; Michalke, B. Higher bioavailability of magnesium citrate as compared to magnesium oxide shown by evaluation of urinary excretion and serum levels after single-dose administration in a randomized cross-over study. *BMC Nutr.* **2017**, *3*, 7. [CrossRef]
21. Sabatier, M.; Arnaud, M.J.; Kastenmayer, P.; Rytz, T.Z.; Barclay, D.V. Meal effect on mineral bioavailability from mineral water in healthy women. *Am. J. Clin. Nutr.* **2002**, *75*, 65–71. [CrossRef] [PubMed]
22. Schenider, I.; Greupner, T.; Hahn, A. Magnesium bioavailability from mineral waters with different mineralization levels in comparison to bread and supplement. *Food Nutr. Res.* **2017**, *61*, 1384686. [CrossRef] [PubMed]
23. Karp, H.; Ekholm, P.; Kemi, V.; Hirvonen, T.; Lamberg-Allardt, C. Differences among total and in vitro digestible phosphorus content in meat and milk products. *J. Ren. Nutr.* **2012**, *22*, 344–349. [CrossRef] [PubMed]
24. Karp, H.; Ekholm, P.; Kemi, V.; Itkonen, S.; Hirvonen, T.; Nárkki, S.; Lamberg-Allardt, C. Differences among total and in vitro digestible phosphorus content in plant foods and beverages. *J. Ren. Nutr.* **2012**, *22*, 416–422. [CrossRef] [PubMed]
25. Schlemmer, U.; Frolich, W.; Prieto, R.M.; Grasses, F. Phytate in foods and significance for humans: Food sources, intake, processing, bioavailability, protective role and analysis. *Mol. Nutr. Food Res.* **2009**, *53*, S330–S375. [CrossRef] [PubMed]

foods

MDPI

Article

Influence of Alkaline Treatment on Structural Modifications of Chlorophyll Pigments in NaOH—Treated Table Olives Preserved without Fermentation

Marta Berlanga-Del Pozo, Lourdes Gallardo-Guerrero and Beatriz Gandul-Rojas *

Chemistry and Biochemistry of Pigments, Food Phytochemistry, Instituto de la Grasa (CSIC), Campus Universitario Pablo de Olavide, Edificio 46, Ctra. Utrera km 1, 41013 Sevilla, Spain; mberlanga@ig.csic.es (M.B.-D.P.); lgallardo@ig.csic.es (L.G.-G.)

* Correspondence: gandul@ig.csic.es

Received: 28 April 2020; Accepted: 18 May 2020; Published: 1 June 2020

Abstract: Alkaline treatment is a key stage in the production of green table olives and its main aim is rapid debittering of the fruit. Its action is complex, with structural changes in both the skin and the pulp, and loss of bioactive components in addition to the bitter glycoside oleuropein. One of the components seriously affected are chlorophylls, which are located mainly in the skin of the fresh fruit. Chlorophyll pigments are responsible for the highly-valued green color typical of table olive specialties not preserved by fermentation. Subsequently, the effect on chlorophylls of nine processes, differentiated by NaOH concentration and/or treatment time, after one year of fruit preservation under refrigeration conditions, was investigated. A direct relationship was found between the intensity of the alkali treatment and the degree of chlorophyll degradation, with losses of more than 60% being recorded when NaOH concentration of 4% or greater were used. Oxidation with opening of the isocyclic ring was the main structural change, followed by pheophytinization and degradation to colorless products. To a lesser extent, decarbomethoxylation and dephytylation reactions were detected. An increase in NaOH from 2% to 5% reduced the treatment time from 7 to 4 h, but fostered greater formation of allomerized derivatives, and caused a significant decrease in the chlorophyll content of the olives. However, NaOH concentrations between 6% and 10% did not lead to further time reductions, which remained at 3 h, nor to a significant increase in oxidized compounds, though the proportion of isochlorin e_4-type derivatives was modified. Chlorophyll compounds of series *b* were more prone to oxidation and degradation reactions to colorless products than those of series *a*. However, the latter showed a higher degree of pheophytinization, and, exclusively, decarbomethoxylation and dephytylation reactions.

Keywords: chlorophyll; pigments; allomerization; table olive; alkaline treatment; phytyl-chlorin; phytyl-rhodin

1. Introduction

Table olives may be considered one of the most nutritious and least caloric snacks, thanks to their balanced fat composition, in which monounsaturated oleic acid predominates and includes essential fatty acids, and to their fiber, vitamin and mineral content [1–3]. In addition, table olives contain phytochemicals such as polyphenols [4], chlorophylls, carotenoids [5], and triterpenic acids [6], which give them functional value. In fact, the table olive is recognized as an essential component of the Mediterranean diet, having been explicitly included in the second level of its nutritional pyramid [7] as an aperitif or culinary ingredient, with a recommended daily consumption of one or two portions (15–30 g). According to the International Olive Council [8], worldwide table olive consumption for the 2018/2019 season will be around 2,667,000 t, which means an increase of around 21% in the last 10 years.

Unlike most fruits, olives must be processed to be edible, since the fresh fruit has an extremely bitter taste, due to its high phenolic compound content, mainly oleuropein. These bitter components can be totally or partially eliminated by various procedures including both hydrolysis—chemical and/or enzymatic—and brine diffusion mechanisms [9]. Of all the procedures, alkaline treatment is the most widely used [10], since it is the method applied to olives processed in the Spanish or Sevillian style (green table olives) and in the Californian style (black table olives), which are two of the main commercial table olive preparations worldwide. It is also the de-bittering method used in the preparations called green ripe olives, which are widely consumed in the United States, as well as in other table olive processing specialties—the Castelvetrano, Picholine and Campo Real styles [11].

Alkaline treatment is a key stage in the preparation of table olives, and can be more or less intense depending on factors such as the variety and state of maturity of the olive, the temperature and quality of the water [9], as well as the subsequent preservation system. In the processing of Spanish-style green table olives, the fruits are treated with a diluted solution of NaOH in water, with a concentration between 2% and 5%, until this solution (lye) penetrates two thirds or three quarters of the pulp towards the stone, which usually takes 4–11 h [12]. Subsequently, after several washings with water, the olives are put in brine (NaCl solution), where they undergo an acid-lactic fermentation. In the preparations of olives in the Castelvetrano [13,14], Picholine [9], Campo Real [15] and green ripe styles [16,17], the fruits remain in alkaline conditions for longer, and the lye can penetrate to the stone. After alkaline treatment and washing, the fruits are kept in refrigerated conditions and/or subjected to heat treatment, thus avoiding any fermentation process.

During the preparation of the olives, the fruits undergo changes in their composition that are mediated by the different stages involved in the processing system. Thus, the way in which the alkaline treatment is applied, the number and time of subsequent washings of the fruits, the presence or otherwise of a fermentation stage, etc., will have an impact on the end product. The treatment with NaOH causes structural modifications in both the epicarp and the mesocarp of the olive, which will depend on the concentration and temperature of the alkaline solution, and which influence the physical-chemical composition of the fruit [18]. The lye solution that penetrates the pulp hydrolyses the oleuropein and ligstroside, producing non-bitter hydrolyzed phenols such as hydroxytyrosol and tyrosol. In addition, this alkaline solution changes the composition of the polysaccharides in the cell wall structure, reducing the firmness of the fruit [19]. The higher the concentration of the lye, and the longer the treatment, the greater the loss of firmness. Chemical damage to the olive's skin, and to its cell structure, allows for more rapid diffusion of the remaining phenolic compounds, and of the sugars, into the brine during the subsequent rinsing and fermentation stages.

Different treatment options with alkaline solutions have been studied to improve the organoleptic quality of the end product and the characteristics of the washing water. Transporting mechanically-harvested olives in low-grade lye, rather than resting prior to processing, avoids peeling and superficial dark spots (damage). On the other hand, the addition of calcium and/or sodium salts to the alkaline lye, and cooling it to 8 °C, gives rise to treatments that improve the texture and prevent the breakage of the olive skin [12,20], while the replacement of NaOH by KOH improves the potential use of the washing waters for agronomic purposes [21]. An inappropriate alkaline treatment—low NaOH concentration and/or insufficient alkaline penetration—leads to the presence of antimicrobial compounds in the brine, which inhibits the growth of *Lactobacillus pentosus*, negatively affecting the fermentation processes [22].

Alkaline treatment and subsequent washing lead to a high loss of volatile compounds [23] and bioactive substances in the olives, such as phenolic compounds and triterpenic acids, which are diffused from the fruit into the wastewater [6,24]. In this regard, García et al. [25] recently verified that black ripe olives produced in the USA have lower contents in phenolic compounds and triterpenic acids than those produced in Spain, and they attribute these differences to the alkaline treatment used in the former, which involves a higher number of alkali/washing cycles.

Chlorophylls (*a* and *b*) and carotenoids are the pigments responsible for the color of the olives in their green ripening state, and constitute another group of phytochemicals in olives that are affected by their processing: they undergo certain structural transformations that can have an impact on both the color and the functional value of the end product. All green table olive preparations are made with fruits of the same chlorophyll and carotenoid composition; however, the transformation these compounds undergo is different for each processing system, depending on whether or not they are treated with alkali and/or fermented. Thus, in each case, an end product with its own composition of pigments and a characteristic color will be obtained [26,27].

The transformations of chlorophylls and carotenoids during the processing of Spanish-style green table olives, which includes alkaline treatment and fermentation, have been widely studied [28–31]. However, for specialties of table olives treated with alkali and preserved without fermentation, only olives processed in the Castelvetrano style have been studied [27,32]. One of the main and most highly-valued characteristics of these specialties of table olives is a typical bright green color. For some years now, suspicions of the color adulteration of table olives by regreening practices—by addition of E141ii coloring additive or Cu^{2+} salts—have arisen. In this sense, several studies have been carried out aimed to characterize the chlorophyll pigment profile of commercial bright green table olives [33–36].

In general, it is known that alkaline treatment causes the partial degradation of chlorophylls *a* and *b* into more hydrosoluble derivatives, but with a green color similar to that of their respective precursors. Depending on the conditions in which this treatment is carried out—essentially the volume of fruits treated and the presence of oxygen—dephytylation reactions of the chlorophylls by the activation of endogenous chlorophyllase may be caused, and/or oxidation reactions affecting the isocyclic ring of the chlorophyll structure (allomerization reactions), originating mainly phytyl-chlorin or phytyl-rhodin derivatives, depending on whether they are from series *a* or *b*, respectively [5,27]. The carotenoid pigment fraction, however, is not affected, since they are alkali-stable compounds [37]. There are many ways in which alkaline treatment can be applied to olives, even reusing a lye solution to treat several batches. This is a very widespread practice aimed at reducing the polluting environmental impact of NaOH solutions [10,38]. In this regard, Gallardo-Guerrero et al. [39] investigated, for a fixed treatment time, the influence of three factors: concentration of NaOH, use of recycled alkaline solution and fruit size. It was shown that the use of recycled solution significantly intensified the oxidizing capacity of such treatment, to a greater extent than the concentration of alkali, and fruit size had a certain effect, with the greatest transformation in the smaller ones.

The aim of this study was to advance knowledge of the complex action that the alkaline treatment of olives has on the structural modifications of chlorophylls, bioactive components [40,41] and pigments responsible for the characteristic bright green color of table olives not preserved by fermentation. Specifically, the effect of nine combinations of two important parameters of the alkali treatment (NaOH concentration and treatment time) on green table olives processed in the Campo Real style—with penetration of the alkaline solution as far as the stone—and preserved for one year under refrigeration conditions, was investigated. A direct relationship between the intensity of the alkali treatment and the degree of degradation of the chlorophylls was shown.

2. Materials and Methods

All procedures were performed under dimmed green light to avoid any photo-oxidation of chlorophylls.

2.1. Raw Material and Preparation of Samples

The study was carried out on olives of the Verdial variety (*Olea europaea* L.) collected from olive trees in an orchard located in Paterna del Campo (Huelva, Spain). Around 5 kg of fruits were picked at the end of August in the intense green ripening stage. Green table olives were processed at laboratory scale, according to the Campo Real-style [15]. The experimental design consisted in the treatment of the fruits with an alkaline solution that included 3% NaCl, and NaOH in a concentration equal to or greater than 2%. Glass containers (370 mL of capacity) were filled approximately with 200 g of fruits and 150 mL of alkaline brine. Nine samples were prepared, increasing in the NaCl solution the percentage of NaOH from 2% to 10% (Table 1). Treatment time was adjusted for each sample in order to achieve a penetration of the alkaline solution into the fruits until reaching the stone. Longitudinal cuts were carried out on the fruits every 30 min for controlling the alkaline penetration. It was visible to the naked eye by the turn of the flesh greenish color to dark brown. The applied times varied from 3 h, in those samples processed with alkaline solutions of NaOH concentration ≥6% (Table 1, samples E, F, G, H and I), to 7 h in the sample treated with the lowest concentration of NaOH (sample A), which was considered the standard treatment.

Table 1. Sample codes and conditions (NaOH concentration and time) of the alkaline treatments applied to each sample.

Sample Code	A	B	C	D	E	F	G	H	I
[NaOH] (%*w*/*v*)	2	3	4	5	6	7	8	9	10
Time (h)	7	6	5	4	3	3	3	3	3

After alkaline treatment, olive fruits were washed twice with tap water. The first one was dynamic under running water, and the second by immersion for 20 h. Finally, the fruits were placed in 6% NaCl brine and kept for 12 months in a refrigerated chamber at 4 °C to avoid fermentation.

2.2. Chemicals and Standards

Ammonium acetate was supplied by Fluka (Zwijndrecht, The Netherlands). Solvents used for chromatography were HPLC grade (Prolabo, VWR International Eurolab, Barcelona, Spain). Analysis grade solvents were supplied by Scharlau (Microdur, Sevilla, Spain). The deionized water was obtained from a Milli-Q® 50 system (Millipore Corporation, Milford, MA, USA). For all purposes, analytical grade (American Chemical Society) reagents were used (Merck, Madrid, Spain). Standards of chlorophylls *a* and *b* were supplied by Sigma Chemical Co. (St. Louis, MO, USA), and standards of pheophytin *a* and pyropheophytin *a* were provided by Wako chemicals Gmbh (Neuss, Germany). Standards of pheophorbide *a*, pyropheophorbide *a*, and chlorine e_6 and rhodin g_7 sodium salts, were purchased from Frontier Scientific Europe Ltd. (Carnforth, Lancashire, UK). The C-13 epimers (chlorophylls *a'* and *b'*) were prepared by treatment of the respective chlorophyll with chloroform, and 13^2-OH-chlorophylls (*a* or *b*) were obtained by selenium dioxide oxidation of the corresponding chlorophyll at reflux-heating for 4 h in pyridine solution under argon [42]. Pheophytin *b* was prepared from a solution of chlorophyll *b* in ethyl ether by acidification with 13% HCl (v/v), and shaking the mixture for 5 min [43]. Standards purity, evaluated by HPLC, was ≥95% in all cases with the exception of rhodin g_7 sodium salt (~90%).

2.3. Pigment Extraction

Previously to the pigment analysis, the fruits were washed several times by immersion in water until the pH of the wash water was neutral. Pigments were extracted with N,N-dimethylformamide (DMF) according to the method of Mínguez-Mosquera and Garrido-Fernández [44], slightly modified as described in detail in Gandul-Rojas et al. [34]. The technique is based on the selective separation of pigments and lipids between DMF and hexane, respectively, which allows obtaining a fat–free pigment extract. From a homogenized triturate prepared with 5 destoned fruits (ca. 30 g), two samples of 2 g each were weighed to carry out the pigment extraction in duplicate. In this methodology, the fat-free pigments dissolved in the DMF phase are subsequently transferred to hexane/diethyl ether (1:1, v/v) mixture by adding a cold solution (4 °C) of 10% NaCl. In this way, the pigment extract may be concentrated to dryness without exceeding 30 °C. The dry residue was dissolved in 1 mL of acetone for its subsequent analysis by HPLC. Similarly, solvent of the hexane phase was evaporated and the remaining residue eluted in a known volume of hexane. In this way, casual losses of pheophytin *a* in the hexane phase can be quantified by direct absorbance measurement at 670 nm using the molar absorption coefficient *EmM* = 53.4.

2.4. Pigment Analysis by HPLC

Separation, identification and quantification of pigments were carried out by HPLC (HP 1100 Hewlett-Packard, Palo Alto, CA; fitted with an HP 1100 automatic injector and diode array detector). A stainless-steel column (20 × 0.46 cm i.d.), packed with a multifunctional endcapped deactivated octadecylsilyl (C18) Mediterranea™ Sea18, 3 μm particle size (Teknokroma, Barcelona, Spain) was used. The column was protected by precolumn (1 × 0.4 cm i.d.) packed with the same material. Solutions of pigment extract were centrifuged at 13,000× *g* prior to injection into the chromatograph. Pigment separation was performed using an elution gradient (flow rate 1.250 mL·min^{-1}) with the mobile phases (A) 0.5 M ammonium acetate in water/methanol (1/4, v/v) and (B) methanol/acetone (1/1, v/v). The gradient scheme is a modification of that of Mínguez-Mosquera et al. [45], as previously described by Gandul-Rojas and Gallardo-Guerrero [32]. The on-line UV-Vis spectra were recorded from 350 to 800 nm with the photodiode-array detector. Data were collected and processed with a LC HP ChemStation (Rev.A.05.04). Pigments were identified by co-chromatography with the corresponding standard and from the spectral characteristics as described in detail in previous publications [27,33,46].

Spectrophotometric detection of pigments was performed by absorbance at different wavelengths. For each pigment, the wavelength closest to its absorption maximum in the red region was chosen: 626 nm for the Mg complex of 15^2-Me-phytyl-rhodin g_7 ester; 640 nm for the Mg complex of both 15^2-Me-phytyl-chlorin e_6 and 15^2-Me-phytyl-isochlorin e_4 esters; 650 nm for chlorophylls *b* and *b'*, 13^2-OH-chlorophyll *b* and the Mg-free chlorophyll derivatives of the series *b*; and 666 nm for chlorophylls *a* and *a'*, 13^2-OH-chlorophyll *a* and the Mg-free chlorophyll derivatives of the series *a*.

Pigments were quantified using external standard calibration curves (amount versus integrated peak area) prepared with the pigment standards listed in Section 2.2. Calibration curves for chlorophylls *a* and *b* were used for their respective epimers and 13^2-OH-derivatives. Calibration curve obtained for pheophytin *a* was used for pheophytin *a'*. For pigments with chlorin- and rhodin-type structures, calibration curves obtained for chlorine e_6 and rhodin g_7 sodium salts were used, respectively. The calibration equations were obtained by least-squares linear regression analysis over a concentration range according to the levels of these pigments in green table olives. Injections in duplicate were made for five volumes of each standard solution (range of concentrations between 2 and 2500 ng; $R^2 < 0.9983$). Limit of detection (LOD) and limit of quantification (LOQ), defined as a signal-to-noise ratio of 3.3 and 10, respectively, were LOD 0.30–1.19 ng and LOQ 0.90–3.6 ng.

2.5. Statistical Analysis

Analyses in this study were performed in duplicated. Statistica software for Windows (version 6, StatSoft, Inc., Tulsa, OK, USA, 2001) was used for data processing. Data were expressed as mean values ± standard deviation (SD). The data were analyzed for differences between means using one-way analysis of variance (ANOVA). Post hoc comparisons were carried out according to Duncan's multiple range-test and the differences were considered significant when $p < 0.05$.

3. Results and Discussion

The alkaline treatments, which the olives were subjected to, caused—to a greater or lesser extent—the transformation of the chlorophylls *a* and *b* present in the fresh fruit. Figure 1 shows the HPLC chromatograms resulting from the separation of pigments from the olives, before (fresh fruit) and after having been subjected to alkaline treatment. All the alkali-treated samples showed a similar qualitative pigment profile, and the chromatogram corresponding to sample A (standard alkaline treatment) was selected by way of representation. Table 2 shows the chromatographic and spectroscopic characteristics of the different pigments identified, and Figure 2 shows the structure of each.

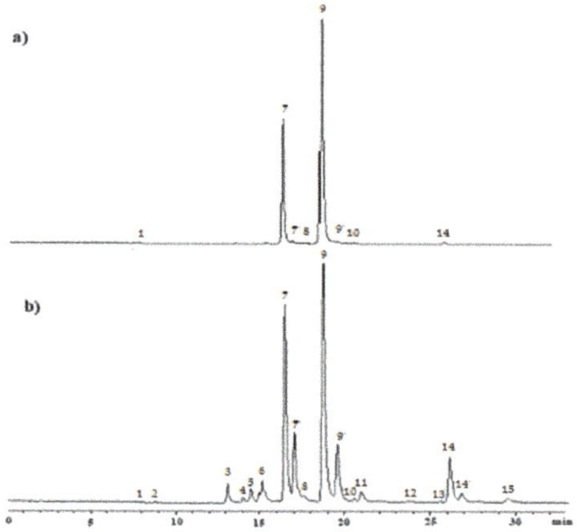

Figure 1. HPLC chromatograms at 640 nm of pigment extracts from: (**a**) fresh fruits; (**b**) fruits after alkaline treatment (sample A). Peaks: (1) pheophorbide *a*; (2) pyropheophorbide *a*; (3) Mg-15^2-Me-phytyl-rhodin g_7 ester; (4) 15^2-Me-phytyl-rhodin g_7 ester; (5) Mg-15^2-Me-phytyl-chlorin e_6 ester; (6) 15^2-Me-phytyl-chlorin e_6 ester; (7) chlorophyll *b*; (7′) chlorophyll *b*′; (8) 13^2-OH-chlorophyll *b*; (9) chlorophyll *a*; (9′) chlorophyll *a*′; (10) 13^2-OH-chlorophyll *a*; (11) Mg-15^2-Me-phytyl-isochlorin e_4 ester; (12) pheophytin *b*; (13) 15^2-Me-phytyl-isochlorin e_4 ester; (14) pheophytin *a*; (14′) pheophytin *a*′; (15) pyropheophytin *a*.

Table 2. Chromatographic and spectroscopic characteristics in the HPLC eluent of chlorophyll pigments.

Pigments	Peak No	t_r [1]	k_c' [2]	Spectroscopic Characteristics Absorption Maxima (nm)	
				Soret	Q [3]
Series *a*					
Pheophorbide *a*	1	7.6	2.7	410	666
Pyropheophorbide *a*	2	8.9	3.3	410	666
Mg-15^2-Me-phytyl-chlorin e_6 ester	5	14.9	6.2	416	638
15^2-Me-phytyl-chlorin e_6 ester	6	15.1	6.3	400	662
Chlorophyll *a*	9	18.8	8.1	432	666
Chlorophyll *a*′	9′	19.6	8.5	432	666
13^2-OH-chlorophyll *a*	10	20.2	8.8	434	664
Mg-15^2-Me-phytyl-isochlorin e_4 ester	11	21.0	9.2	416	638
15^2-Me-phytyl-isochlorin e_4 ester	13	26.1	11.7	400	662
Pheophytin *a*	14	26.3	11.8	410	666
Pheophytin *a*′	14′	26.8	12.0	410	666
Pyropheophytin *a*	15	29.6	13.4	410	666
Series *b*					
Mg-15^2-Me-phytyl-rhodin g_7 ester	3	13.1	5.4	450	626
15^2-Me-phytyl-rhodin g_7 ester	4	14.0	5.8	426	650
Chlorophyll *b*	7	16.4	7.0	466	650
Chlorophyll *b*′	7′	16.9	7.2	466	650
13^2-OH-chlorophyll *b*	8	17.6	7.5	466	646
Pheophytin *b*	12	24.0	10.7	436	654

[1] t_r: Retention time (min); [2] k_c': Retention factor = (tr − tm)/tm where tm is the retention time of an unretained component; [3] Q: maximum in the red region of the spectrum.

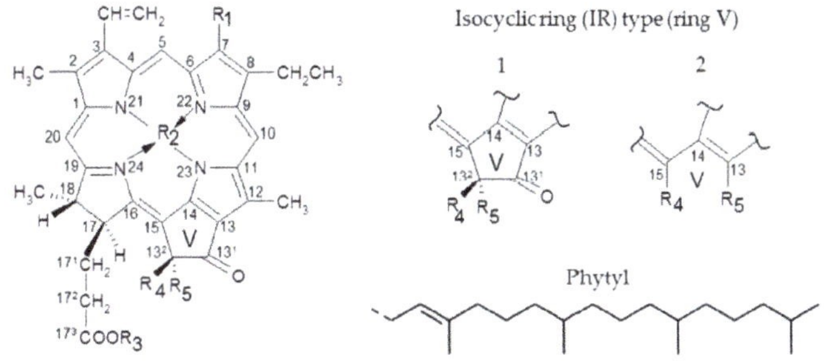

Pigment	IR	R_1	R_2	R_3	R_4	R_5
Series *a*						
Chlorophyll *a*	1	CH_3	Mg	Phytyl	H	$COOCH_3$
Chlorophyll *a'*	1	CH_3	Mg	Phytyl	$COOCH_3$	H
13^2-OH-chlorophyll *a*	1	CH_3	Mg	Phytyl	OH	$COOCH_3$
Pheophytin *a*	1	CH_3	2H	Phytyl	H	$COOCH_3$
Pheophytin *a'*	1	CH_3	2H	Phytyl	$COOCH_3$	H
Pyropheophytin *a*	1	CH_3	2H	Phytyl	H	H
Mg-15^2-Me-phytyl-chlorin e_6 ester	2	CH_3	Mg	Phytyl	CH_2COOCH_3	COOH
Mg-15^2-Me-phytyl-isochlorin e_4 ester	2	CH_3	Mg	Phytyl	CH_2COOCH_3	H
15^2-Me-phytyl-chlorin e_6 ester	2	CH_3	2H	Phytyl	CH_2COOCH_3	COOH
15^2-Me-phytyl-isochlorin e_4 ester	2	CH_3	2H	Phytyl	CH_2COOCH_3	H
Pheophorbide *a*	1	CH_3	2H	H	H	$COOCH_3$
Pyropheophorbide *a*	1	CH_3	2H	H	H	H
Series *b*						
Chlorophyll *b*	1	COH	Mg	Phytyl	H	$COOCH_3$
Chlorophyll *b'*	1	COH	Mg	Phytyl	$COOCH_3$	H
13^2-OH-chlorophyll *b*	1	COH	Mg	Phytyl	OH	$COOCH_3$
Pheophytin *b*	1	COH	2H	Phytyl	H	$COOCH_3$
Mg-15^2-Me-phytyl-rhodin g_7 ester	2	COH	Mg	Phytyl	CH_2COOCH_3	COOH
15^2-Me-phytyl-rhodin g_7 ester	2	COH	2H	Phytyl	CH_2COOCH_3	COOH

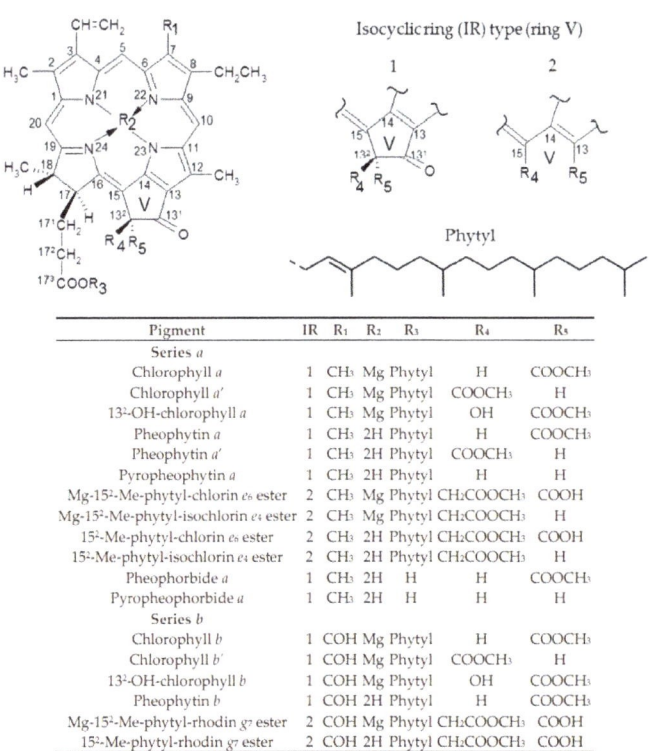

Figure 2. Structures of chlorophyll pigments present in green table olives.

Once olive fruits were alkali-treated, a greater presence of chlorophyll epimers (*a'* and *b'*) regarding their respective precursor was evidenced. Moreover, as expected from previous studies [27,32], chlorophylls *a* and *b* underwent allomerization reactions after the alkaline treatment of the olives, which produced, as well as the hydroxylated derivatives—13^2-OH-chlorophyll *a* and 13^2-OH-chlorophyll *b*—oxidized chlorophyll derivatives of the chlorin- and rhodin-types, which are characterized by their open isocyclic ring (V) of the chlorophyll structure (IR 2 in Figure 2). In series *a*, the compounds Mg-15^2-Me-phytyl-chlorin e_6 ester and Mg-15^2-Me-phytyl-isochlorin e_4 ester were detected, whose structures differ only in the C-13 substitute, which in the former is a carboxyl group, while in the latter it is a hydrogen. With respect to series *b*, Mg-15^2-Me-phytyl-rhodin g_7 ester was detected. In addition to these allomerization reactions, evidence of pheophytinization reaction was found, by substitution of the Mg ion by 2 H in the porphyrin ring, which gave rise to the formation of the magnesium–free derivatives of the previous compounds, i.e., the esters 15^2-Me-phytyl-chlorin e_6 and 15^2-Me-phytyl-isochlorin e_4 in series *a*, and 15^2-Me-phytyl-rhodin g_7 in series *b*. This modification also affected the original chlorophylls *a* and *b*, with the formation of pheophytins *a* and *b*, respectively. Although the pheophytinization reaction typically occurs under acidic conditions, it is also fostered by temperature increase [47]. Therefore, the heat generated during the alkaline treatment of the fruits, due to the alkaline hydrolysis reactions produced by the hydroxyl ions diffusing inside the olive [48], must have fostered the pheophytinization reaction. This same factor even caused decarbomethoxylation due to loss of the -$COOCH_3$ group at C-13^2, since pyroderivatives (pyropheophytin *a* and pyropheophorbide *a*) were also produced. The dephytylated derivatives pheophorbide *a* and the previously mentioned pyropheophorbide *a* were also detected, though at much lower proportion. In plant foods, pheophytins, pheophorbides and pyroderivatives are the chlorophyll compounds that originate widely during thermal processing or fermentation of fruits, vegetables and green algae [47] while chlorophyll derivatives with phytyl-chlorin or phytyl-rhodin structures are specifically associated with the processing of alkali-treated table olives [26,27,31]. In another field, this type of derivatives has also been identified during the senescence of microalgaes incubated under oxic conditions [49], and more recently as secondary metabolites of digestion in mice from a diet rich in chlorophylls [50]. Its presence in the liver indicates the existence of alternative metabolic pathways that modify the structure of the chlorophyll macrocycle increasing its polarity and, as indicated by the authors [50], to facilitate probably a greater metabolism and/or excretion.

As mentioned above, the profile of chlorophyll derivatives identified in the olives was practically the same in all the treatments tested. However, differences were found both in the total chlorophyll pigment content of the fruits and in the proportion of each of the derivatives after the alkaline treatment, with some of the precursor pigments even disappearing in the more intense procedures (Table 3). The outstanding presence of oxidized chlorophyll derivatives showed that the alkaline treatment was responsible for their formation in the fruit processing, as has been shown in previous studies carried out in mildly alkaline conditions [46]. Sample A, which had been subjected to the mildest alkaline treatment conditions, and normally used in the preparation of olives in the Campo Real style, did not show significant variation with respect to the total amount of chlorophyll pigments present in the fresh fruit ($p < 0.05$). However, for the rest of the samples, as the olives were subjected to a more intense alkaline treatment, in general, lower pigment content was found, which showed a transformation of the chlorophylls into colorless products (Figure 3). Thus, the increase in NaOH concentration from 2% to 3% caused about 16% destruction of chlorophyll pigments in the fruits (sample B), despite the fact that the treatment time required was reduced from 7 to 6 h. This effect was much more pronounced in sample C, which had been treated with 4% NaOH. The process was shortened to 5 h but a significant decrease of 68% in pigment content was noted, compared to sample B. However, the effects produced on the total chlorophyll content of the olives by the treatments with 5% and 6% NaOH, with time reduction at 4 and 3 h, respectively (samples D and E), were not significantly different from those of sample C. The alkaline treatments from sample F, in which the NaOH concentration was increased from 7% to 10% (samples F–I), did not lead to further reductions in the time needed for the total penetration of the alkaline lye, which was maintained at 3 h. In these samples, the quantity of pigments continued to decrease more gradually and, in some cases, not significantly.

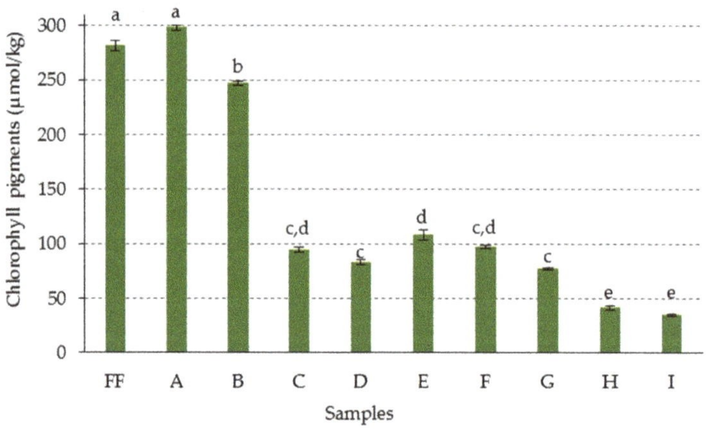

Figure 3. Total content (µmol/kg destoned fruit) of chlorophyll pigments in green olives. Abbreviations: FF fresh fruit; A–I: fruits with different alkaline treatments (see Table 1 for description of samples). Data represent mean values ± SD (n = 2). Different letters above the error bars indicate significant differences according to the Duncan's multiple-range test ($p < 0.05$).

Table 3. Chlorophyll pigment composition (μmol/kg destoned fruit) of green olives with different alkaline treatments [1,2]. See Table 1 for description of samples.

Pigment	Fresh Fruit	Samples								
		A	B	C	D	E	F	G	H	I
Chlorophyll *a*	223.70 ± 13.17	115.60 ± 3.50	30.74 ± 5.43	5.10 ± 1.41	0.32 ± 0.08	0.56 ± 0.04	0.66 ± 0.02	0.34 ± 0.06	n. q.	n. q.
Chlorophyll *a′*	0.82 ± 0.07	31.95 ± 1.93	7.84 ± 1.52	0.70 ± 0.28	0.56 ± 0.34	1.34 ± 0.95	1.06 ± 0.08	1.78 ± 0.08	n. q.	n. q.
13^2-OH-chlorophyll *a*	0.43 ± 0.06	n. q.	n. q.	n. q.	0.70 ± 0.63	n. q.	1.96 ± 0.28	n. q.	1.36 ± 0.97	n. q.
Mg-15^2-Me-phytyl-chlorin e_6 ester		1.46 ± 0.13	0.41 ± 0.29	n. q.	7.89 ± 0.43	7.64 ± 3.04	16.18 ± 3.52	14.22 ± 0.64	6.70 ± 0.46	6.78 ± 1.28
Mg-15^2-Me-phytyl-isochlorin e_4 ester		2.91 ± 0.39	1.19 ± 0.35	3.54 ± 2.22	9.87 ± 1.07	11.68 ±2.62	14.80 ± 2.50	10.88 ± 1.58	5.64 ± 6.06	1.82 ± 1.29
Pheophytin *a*	2.74 ± 0.25	49.43 ± 5.66	58.86 ± 1.10	4.36 ± 0.73	1.83 ± 0.22	1.84 ± 0.23	0.97 ± 0.22	1.83 ± 0.46	2.14 ± 0.44	1.00 ± 0.41
Pheophytin *a′*		11.52 ± 0.81	12.73 ± 1.18	2.69 ± 0.30	n. q.	n. q.	n. q.	n. q.	n. q.	n. q.
Pyropheophytin *a*		7.30 ± 0.34	26.44 ± 2.08	23.87 ± 0.73	18.82 ± 0.06	23.00 ± 1.94	18.74 ± 0.28	15.46 ± 1.04	9.94 ± 0.78	7.04 ± 0.40
15^2-Me-phytyl-chlorin e_6 ester		19.94 ± 1.44	64.72 ± 1.78	13.45 ± 1.49	15.26 ± 1.56	17.54 ± 2.28	14.32 ± 2.00	11.96 ± 1.28	5.66 ± 1.62	5.80 ± 0.94
15^2-Me-phytyl-isochlorin e_4 ester		n. q.	n. q.	34.94 ± 5.83	19.70 ± 2.70	34.16 ± 9.56	18.84 ± 0.66	11.58 ± 0.72	6.56 ± 1.92	5.10 ± 0.60
Pheophorbide *a*	0.20 ± 0.03	n. q.	0.49 ± 0.12	n. q.	n. q.	n. q.	n. q.	n. q.	n. q.	n. q.
Pyropheophorbide *a*		n. q.	1.83 ± 0.37	n. q.	0.76 ± 0.44	0.44 ± 0.08	0.86 ± 0.48	1.60 ± 0.00	0.78 ± 0.24	3.98 ± 0.98
Total series *a*	227.90 ± 6.44	240.10 ± 2.52	205.25 ± 2.06	88.65 ± 2.36	75.72 ± 2.60	98.20 ± 4.98	88.39 ± 1.62	69.65 ± 0.89	38.78 ± 2.37	31.52 ± 0.91
Chlorophyll *b*	52.99 ± 3.88	33.46 ± 1.92	11.14 ± 0.28	0.95 ± 0.05	n. q.	1.24 ± 0.60	n. q.	n. q.	n. q.	n. q.
Chlorophyll *b′*	0.97 ± 0.07	13.11 ± 0.39	5.20 ± 1.31	0.62 ± 0.11	n. q.	n. q.	0.20 ± 0.30	n. q.	n. q.	n. q.
13^2-OH-chlorophyll *b*	n. q.	n. q.	n. q.	n. q.	1.22 ± 0.40	1.22 ± 0.42	1.60 ± 0.74	1.56 ± 0.02	0.48 ± 0.02	n. q.
Mg-15^2-Me-phytyl-rhodin g_7 ester		9.91 ± 0.64	11.85 ± 0.22	3.03 ± 0.48	6.12 ± 0.94	7.62 ± 1.68	6.86 ± 1.18	5.90 ± 1.02	2.26 ± 0.88	3.82 ± 0.42
Pheophytin *b*		n. q.	1.01 ± 1.43	n. q.	n. q.	n. q.	n. q.	n. q.	n. q.	n. q.
15^2-Me-phytyl-rhodin g_7 ester		1.44 ± 0.20	12.14 ± 1.01	1.31 ± 0.04	0.24 ± 0.16	n. q.	n. q.	n. q.	n. q.	n. q.
Total series *b*	53.96 ± 2.44	57.92 ± 1.04	41.34 ± 0.99	5.91 ± 0.25	7.58 ± 0.60	10.08 ± 1.06	8.66 ± 0.82	7.46 ± 0.72	2.74 ± 0.62	3.82 ± 0.42

[1] Data represent mean values ± SD (n = 2); [2] n.q.: not quantified.

In a previous study, Gandul-Rojas and Gallardo-Guerrero [32] noticed that the chemical treatment of olives with NaOH transformed the different chlorophyll pigments into their allomerized derivatives characterized by an open isocyclic ring, and that the degree of this transformation increased when the contact time between the fruit and the alkaline brine was prolonged. In the present study, olive fruits were treated with different concentrations of NaOH. In each sample, the fruits were maintained in the alkaline solution for the time needed to achieve the same alkali penetration in all of them—this was until the fruit stone was reached, which is characteristic for the processing of Campo Real-style table olives [15]. Under these conditions, the treatment time could be excluded as an independent variable, allowing the in-detail analysis of the influence of NaOH concentration on modifications of chlorophyll compounds. The results showed that the standard treatment (sample A), caused the formation of chlorophyll derivatives with open isocyclic ring by 12%, while the rest of the pigments (88%) maintained a closed isocyclic ring structure (Figure 4).

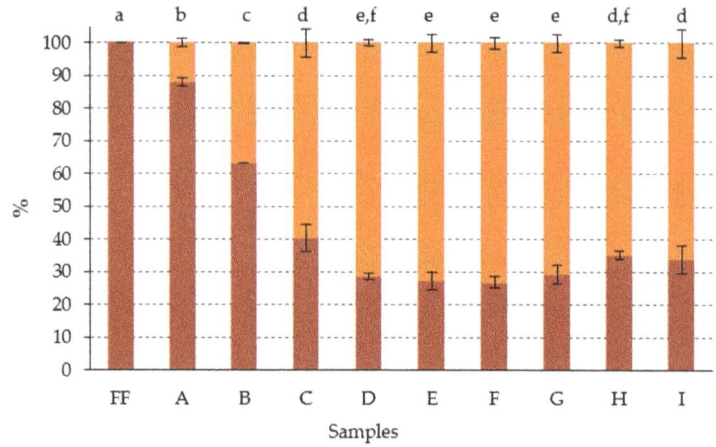

Figure 4. Percentage composition (with respect to total chlorophyll content) of chlorophyll pigments with: (•) closed isocyclic ring; (•) open isocyclic ring, in green olives. Abbreviations: FF, fresh fruit; A–I: fruits with different alkaline treatments (see Table 1 for description of samples). Data represent mean values ± SD (n = 2). Different letters above the error bars indicate significant differences according to the Duncan's multiple-range test ($p < 0.05$).

The increase of 1% in the concentration of NaOH in the alkaline brine, in spite of reducing the necessary treatment time by 1 h, caused an increase of 37% in the amount of these oxidized derivatives present in the olives (sample B). From the treatment of the fruits with alkaline brine of NaOH concentration equal or higher than 4% (samples C–I), the percentage of open isocyclic ring chlorophyll derivatives was already higher than the corresponding percentage of closed isocyclic ring derivatives, reaching around 73% in most of the samples, and without significant differences ($p < 0.05$) among them. This result suggested that there was a concentration of NaOH above which an increase in the strength of the alkaline treatment did not lead to detecting higher percentages of oxidized chlorophyll derivatives with open isocyclic ring (Figure 4), although degradation to colorless products did occur (Figure 3). Therefore, it was observed that the samples subjected to the mildest treatments (samples A–C), were those which showed the greatest differentiation, both in the total chlorophyll pigment content, and in the relationship between the chlorophyll derivatives with closed and open isocyclic ring, with no significant differences being found in this relationship from sample C.

Among the chlorophyll compounds with open isocyclic ring, two groups of derivatives could be distinguished according to the R_5 substituent of their structure (Figure 2). One of them was formed by the chlorophyll derivatives with a -COOH group in R_5, that is, those with chlorine e_6- and rhodin g_7-type structures. The other group included the derivatives in which the R_5 substitute was an H, corresponding, therefore, to compounds with isochlorin e_4-type structure. To facilitate an understanding of the results, the opening of the isocyclic ring was referred to as type O (O-ring), for the first group and type Iso (Iso-ring), for the second. Figure 5 shows the percentage composition represented by each of these groups of derivatives, with respect to the total content of chlorophyll compounds.

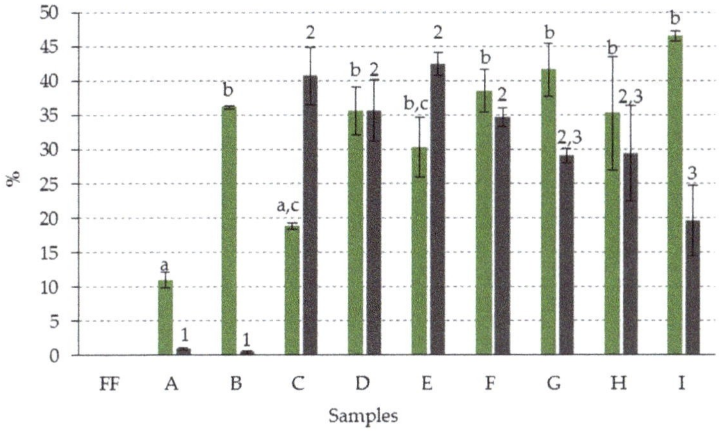

Figure 5. Percentage composition (with respect to total chlorophyll content) of chlorophyll derivatives with open isocyclic ring: (•) O-type; (•) Iso-type, in green olives. Abbreviations: FF fresh fruit; A–I: fruits with different alkaline treatments (see Table 1 for description of samples). Data represent mean values ± SD (n = 2). Different letters or numbers above the error bars indicate significant differences according to the Duncan's multiple-range test ($p < 0.05$).

The number of Iso-ring derivatives was minimal (less than 1%), in samples A and B, while that of O-ring derivatives was 11% and 36%, respectively. However, from the C sample—NaOH treatments with alkaline brine of concentration ≥4%—the presence of Iso-ring compounds increased considerably, being in some cases equal to or higher than those of O-ring (samples C–E), and reaching about 42% of the total chlorophyll compounds. However, from the E sample, a certain tendency to decrease the proportion of the Iso-ring derivatives was observed, as the NaOH concentration increased, but no parallel decrease of the O-ring derivatives was noticed. It is likely that the latter are precursors of the Iso-ring derivatives, and these, in turn, represent the step prior to the formation of colorless products, detected in the samples treated with higher NaOH concentration (Figure 3).

On the other hand, the influence of the alkaline treatment on the degradation of the chlorophyll pigments was evaluated, depending on whether they were of series *a* or *b*. To this end, the data were calculated with respect to the total content of each of the series (Figure 6). It was observed that the evolution of the formation of chlorophyll derivatives with open isocyclic ring was similar for both series, increasing as the concentration of NaOH in the alkaline brine increased, as had been previously seen globally (Figure 4). However, it was noted that, in all the samples, the percentage of oxidized derivatives corresponding to series *b* was greater than that of series *a*, reaching 100% in sample I, as opposed to the 62% quantified for series *a*. However, in none of the samples were detected Iso-ring derivatives of series *b* (Table 3) which, as mentioned above, are likely to be the step prior to the formation of colorless products. At the same time, it was found that the total content of chlorophyll derivatives of series *b* decreased in a much higher proportion than those of series *a* in most of the samples. Therefore, it could be that the Iso-ring derivatives of the series *b* were immediately transformed into uncolored compounds, making their detection impossible. Already in sample C, the series *b* derivatives decreased by about 90%, with respect to the initial fresh fruit content, fluctuating this value between 82% and 95%, in the D–I samples. In series *a*, on the other hand, the total content of chlorophyll derivatives decreased only 63% in sample C, remaining around this value until sample G. It was in samples H and I, where the degradation of derivatives of series *a* reached values close to those of series *b*, although they continued to be lower. As regards the hydroxylated derivatives (13^2-OH-chlorophyll *a* and 13^2-OH-chlorophyll *b*), a significantly higher percentage was also observed in series *b* (12%–21%) than in series *a* (1%–4%) (Table 3). All these results pointed to higher sensitivity in series *b* than series *a*, for the transformations and/or degradations of the isocyclic ring, caused by the alkaline treatment. A similar result was previously found during processing of Castelvetrano-style table olives [27].

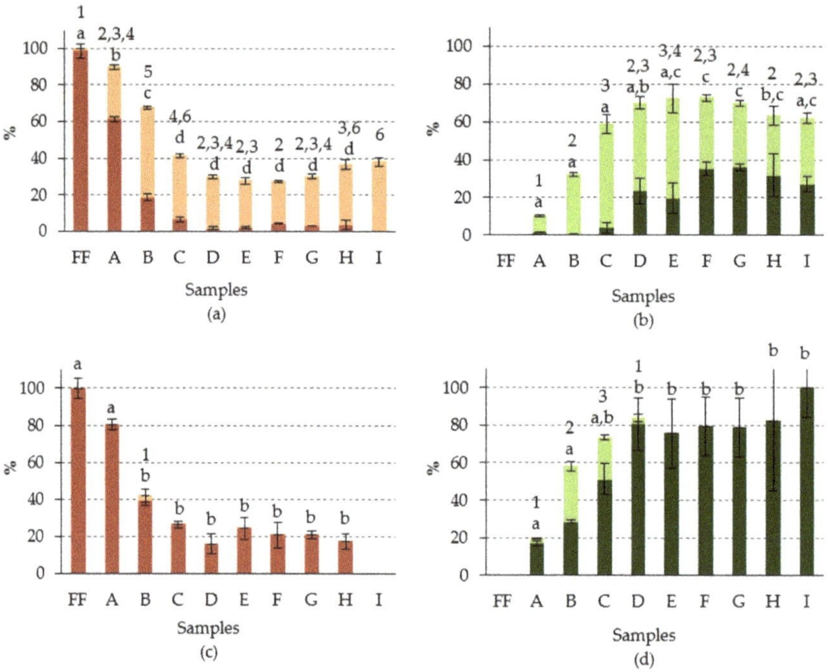

Figure 6. Percentage composition (with respect to total chlorophyll content) of chlorophyll pigments (•) with Mg and closed isocyclic ring; (•) Mg-free and with closed isocyclic ring; (•)with Mg and open isocyclic ring; (•) Mg-free and with open isocyclic ring, of (**a**,**b**): series *a*; (**c**,**d**): series *b*, in green olives. Abbreviations: FF, fresh fruit; A–I: fruits with different alkaline treatments (see Table 1 for description of samples). Data represent mean values ± SD (n = 2). Different letters or numbers above the error bars indicate significant differences (for the lower and upper data set, respectively) according to the Duncan's multiple-range test ($p < 0.05$).

However, in the case of the substitution reaction of Mg by 2 H in the porphyrin ring (pheophytinization), the result was the opposite, as expected from previous kinetic studies carried out with chlorophylls *a* and *b* [47], and in real fermented olive system [29]. This reaction was much more pronounced in all the chlorophyll compounds of series *a*, both in the chlorophyll (closed isocyclic ring), and in the derivatives with open isocyclic ring. Overall, for series *a*, Mg-free derivatives were recorded at from 37% in sample A to 90% in sample C, and varying in the rest of the samples between 61% and 80%, although without following any particular pattern. In series *b*, however, pheophytinization was not generalized, and was detected only in certain samples (A, B, C and D), and at a much lower proportion than in series *a* (Figure 6), with a maximum value of 32% recorded in sample B.

Likewise, in series *a* it was noted that part of the initial chlorophyll (closed isocyclic ring structure), which had not been transformed by the alkaline treatment into allomerized derivatives, was transformed into Mg-free derivatives. This transformation represented a greater proportion as the NaOH concentration in the alkaline brine increased, to the extent that the precursor pigment, chlorophyll *a*, was not detected in sample I (Figure 6a). On the other hand, in the derivatives with open isocyclic ring, this pattern was somewhat different. A higher percentage of Mg-free compounds was recorded in samples A to E, and the derivatives with and without Mg remained at similar percentages from sample F (Figure 6b). With respect to series *b*, in general, only the Mg-free derivative with open isocyclic ring was detected, except in the sample that had a higher percentage of this compound (sample B), in which a small amount of the Mg-free derivative with closed isocyclic ring (pheophytin *b*) was also detected (Figure 6c,d).

The alkaline treatment of the olives also caused the dephytylation reaction at C-17^3 in the chlorophyll derivatives of series *a*, with a minority presence of pyropheophorbide *a* in most samples. It highlighted that, in general, this was the only dephytylated derivative found, with the exception of sample B, in which a small amount of pheophorbide *a* was also detected (Table 3). The recorded percentages of pyropheophorbide *a* are shown in Figure 7. In samples A–F they varied between 0% and 1%, without following any particular pattern, and increased slightly to 2% in samples G and H, and more markedly in sample I, in which it reached 11%. Dephytylated chlorophyll derivatives can be formed by an enzymatic route, through the action of the endogenous enzyme chlorophyllase, or by a chemical route, through the non-specific acid or alkaline hydrolysis of esters [46]. However, according to Mínguez-Mosquera and Gandul-Rojas [46], the limited presence of oxygen in the alkaline medium may foster a certain degree of specific de-esterification of phytol by chemical action, limiting other parallel oxidation reactions at C-13. In this study, the observed increase of pyropheophorbide *a* from sample G, treated with an alkaline brine of 8% NaOH, did not seem, in principle, to be related to the activity of chlorophyllase, since, although this enzyme has optimal activity at pH 8.5 [51], strongly alkaline values promote its destabilization [52]. By contrast, since other reaction conditions were equal, the increase in NaOH concentration could foster specific chemical hydrolysis, as discussed before, especially in sample I. Nevertheless, and given the demonstrated thermal stability of chlorophyllase [51], it cannot be ruled out that the heat generated by the hydrolytic reactions inside the fruit could have fostered competition between thermal activation of chlorophyllase, and strongly alkaline pH destabilization, resulting in an initial enzymatic formation of dephytylated derivatives

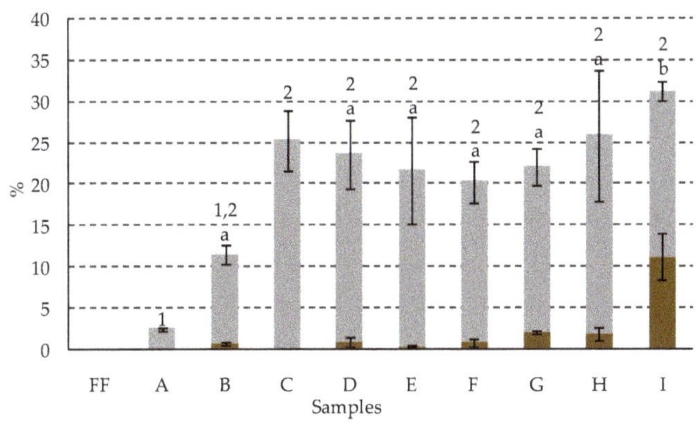

Figure 7. Percentage composition (with respect to total chlorophyll content) of: (•) pyropheophorbide *a*; (•) pyropheophytin *a*, in green olives. Abbreviations: FF, fresh fruit; A–I: fruits with different alkaline treatments (see Table 1 for description of samples). Data represent mean values ± SD (n = 2). Different letters or numbers above the error bars indicate significant differences (for the lower and upper data set, respectively) according to the Duncan's multiple-range test ($p < 0.05$).

On the other hand, the presence of pyropheophorbide *a* and pyropheophytin *a* in all the samples showed that, during the treatment of the olives with alkali, in series *a*, there was also the decarbomethoxylation reaction at C-13^2, which led to the formation of the pyroderivatives. In general, this reaction was more pronounced during the more intense alkaline treatments (Figure 7). In sample A, which was subjected to the standard alkaline treatment, only 2% pyroderivatives was produced, a percentage that increased to 11% in sample B, and up to values between 20% and 25%, without a fixed pattern or significant differences, in samples C–G. In alkaline treatments with more concentrated NaOH solutions, samples H and I, up to 26% and 31% pyroderivatives, respectively, were recorded. The formation of pyroderivatives is normally associated with the heat treatment of vegetables. Tarrado-Castellarnau et al. [48] showed an increase in the internal temperature of fruits during the alkaline treatment of green olives. The characterization of the heat transfer process led them to hypothesize that this increase could only be the result of the heat released inside the fruit, as a result of alkaline hydrolysis reactions and, to a lesser extent, the dilution of the solution with the water in the pulp. This heat generated must be the origin of the pyroderivatives, and a higher reach of the hydrolysis reactions of all the components of the olives, as the concentration of NaOH in the alkaline solution was increased, might explain the greater formation of them in olives subjected to more intense treatments.

4. Conclusions

All the treatments caused various types of reactions in the chlorophylls, oxidation with an opening of the isocyclic ring being the main one, as well as pheophytinization and degradation to colorless products. To a lesser extent, decarbomethoxylation and dephytylation reactions were detected. The increase in NaOH concentration from 2% to 5% reduced the time needed for the total penetration of alkaline brine from 7 to 4 h, but fostered greater formation of chlorophyll derivatives with open isocyclic ring, and caused a significant decrease in the chlorophyll content of the olives. However, NaOH concentrations between 6% and 10% did not lead to further reductions in the treatment time, which remained at 3 h, nor to a significant increase in oxidized compounds, although the proportion of derivatives with open isocyclic ring of isochlorin e_4 structure was modified, suggesting that these compounds might represent the stage prior to the formation of colorless products.

The chlorophyll compounds of series *b* were more sensitive than those of series *a* to the isocyclic ring oxidation reactions caused by the alkaline treatment, as well as to the degradation to colorless products. However, series *a* showed a higher degree of pheophytinization and, exclusively, decarbomethoxylation and dephytylation reactions. The first two transformations were fostered by the heat generated inside the fruit as a result of alkaline hydrolysis reactions. Dephytylation in a small proportion of chlorophylls could be the result of alkaline hydrolysis of phytol, under limited oxygen conditions, and/or thermal activation of chlorophyllase, prior to their destabilization due to a highly alkaline pH.

Therefore, a direct relationship between the degradation of chlorophyll pigments and the intensity of alkali treatment in the processing of green table olives was evidenced, with losses of more than 60% being quantified at NaOH concentration of 4% or higher. In spite of the advantage that this increase entails in reducing the necessary treatment time, the parallel negative effect on the intensity of the green color, and the functional value of the product, must be taken into account when optimizing the efficiency of the process.

Author Contributions: Conceptualization, B.G.-R.; methodology, L.G.-G. and B.G.-R.; formal analysis, M.B.-D.P.; investigation, M.B.-D.P.; resources, L.G.-G. and B.G.-R.; visualization, M.B.-D.P. and B.G.-R.; writing—original draft preparation, M.B.-D.P., L.G.-G. and B.G.-R.; writing—review and editing, M.B.-D.P., L.G.-G. and B.G.-R.; supervision, B.G.-R.; project administration, B.G.-R.; funding acquisition, B.G.-R. All authors have read and agreed to the published version of the manuscript.

Funding: This research was funded by Spanish Government [Projects AGL 2015–63890–R and AGL RTI2018–095415–B–I00], partially financed by European regional development funds (ERDF).

Acknowledgments: M.B.-D.P. thanks AGL RTI2018–095415–B–I00 for her internship contract. We thank Sergio Alcañiz García for technical assistance and Honorio Vergara Domínguez for supplying fresh olive samples.

Conflicts of Interest: The authors declare no conflict of interest.

References

1. López-López, A.; Montaño, A.; Garrido–Fernández, A. Nutrient profiles of commercial table olives: Proteins and vitamins. In *Olives and Olive Oil in Health and Disease Prevention*; Preedy, V.R., Watson, R.R., Eds.; Academic Press: Cambridge, MA, USA; Elsevier: Amsterdam, The Netherlands, 2010; Chapter 74, pp. 705–714.
2. López–López, A.; Montaño, A.; Garrido–Fernández, A. Nutrient profiles of commercial table olives: Fatty acids, sterols, and fatty alcohols. In *Olives and Olive Oil in Health and Disease Prevention*; Preedy, V.R., Watson, R.R., Eds.; Academic Press: Cambridge, MA, USA; Elsevier: Amsterdam, The Netherlands, 2010; Chapter 75; pp. 715–724.
3. Boskou, D.; Camposeo, S.; Clodoveo, M.L. Table olives as sources of bioactive compounds. In *Olive and Olive Oil Bioactive Constituents*; Boskou, D., Ed.; AOCS Press: Urbana, IL, USA, 2015; Chapter 8, pp. 217–259.
4. Charoenprasert, S.; Mitchel, A. Factors influencing phenolic compounds in table olives (Olea europaea). *J. Agric. Food Chem.* **2012**, *60*, 7081–7095. [CrossRef]
5. Gandul–Rojas, B.; Roca, M.; Gallardo–Guerrero, L. Chlorophylls and carotenoids in food products from olive tree. In *Products from Olive Tree*; Boskou, D., Ed.; InTech: Rijeka, Croatia, 2016; Chapter 5, pp. 67–97. ISBN 978-953-51-4806-7. [CrossRef]
6. Romero, C.; García, A.; Medina, E.; Ruíz–Méndez, M.V.; de Castro, A.; Brenes, M. Triterpenic acids in table olives. *Food Chem.* **2010**, *118*, 670–674. [CrossRef]
7. Fundación Dieta Mediterránea. Mediterranean Diet Pyramid: A lifestyle for Today. Available online: https://dietamediterranea.com/piramidedm/piramide_INGLES.pdf. (accessed on 1 April 2020).
8. IOC (International Olive Council). World Olive Oil and Table Olives Figures [Internet]. Available online: https://www.internationaloliveoil.org/what-we-do/economic-affairs-promotion-unit/#figures (accessed on 1 April 2020).
9. Rejano, L.; Montaño, A.; Casado, F.J.; Sánchez, A.H.; de Castro, A. Table olives: Varieties and variations. In *Olives and Olive Oil in Health and Disease Prevention*; Preedy, V.R., Watson, R.R., Eds.; Academic Press: Cambridge, MA, USA; Elsevier: Amsterdam, The Netherlands, 2010; Chapter 1, pp. 5–15.
10. Sánchez–Gómez, A.H.; García–García, P.; Rejano–Navarro, L. Elaboration of table olives. *Grasas y Aceites* **2006**, *57*, 86–94. [CrossRef]
11. IOC (International Olive Oil Council). Trade Standard Applying to Table Olives. Standard COI/OT/NC. No.1/2004. Available online: https://www.internationaloliveoil.org/what-we-do/chemistry-standardisation-unit/standards-and-methods/ (accessed on 1 April 2020).
12. Rejano Navarro, L.; Sánchez Gómez, A.H.; Vega Macías, V. New trends on the alkaline treatment–cocido– of Spanish or Sevillian Style green table olives. *Grasas y Aceites* **2008**, *59*, 197–204.
13. Lanza, B. Nutritional and sensory quality of table olives. In *Olive Germplasm–The Olive Cultivation, Table Olive and Olive Oil Industry in Italy*; Muzzalupo, I., Ed.; IntechOpen: Rijeka, Croatia, 2012; Chapter 16, p. 2840. Available online: https://www.intechopen.com/books/olive-germplasm-the-olive-cultivation-table-olive-and-olive-oil-industry-in-italy/nutritional-and-sensory-quality-of-table-olives (accessed on 1 April 2020). [CrossRef]
14. Catania, P.; Alleri, M.; Martorana, A.; Settanni, L.; Moschetti, G.; Vallone, M. Investigation of a tunnel pasteurizer for "Nocellara del Belice" table olives processed according to the "Castelvetrano method". *Grasas y Aceites* **2014**, *65*, e049.
15. De Lorenzo, C.; González, M.; Iglesias, G.; Lázaro, E.; Valiente, E.; Blázquez, N. *La aceituna de Campo Real*; Consejería de Medio Ambiente, Ed.; Dirección General de Educación y Protección Ambiental: Madrid, España, 2000; Chapter 3, pp. 25–37.
16. Luh, B.S.; Ferguson, L.; Kader, A.; Barret, D. Processing California olives. In *Olive Production Manual*; Sibbett, G.S., Ferguson, L., Eds.; University of California Agriculture and Natural Resources: Davis, CA, USA, 2005; pp. 145–153, Publication 3353.
17. Brenes, M.; García, P. Elaboración de aceitunas denominadas "Green ripe olives" con variedades españolas. *Grasas y Aceites* **2005**, *56*, 188–191.
18. Bianchi, G. Lipids and phenols in table olives. *Eur. J. Lipid Sci. Technol.* **2003**, *105*, 229–242. [CrossRef]

19. Jiménez, A.; Guillén, R.; Sánchez, C.; Fernández–Bolaños, J.; Heredia, A. Changes in the texture and cell wall polysaccharides of olive fruit during 'Spanish green olive' processing. *J. Agric. Food Chem.* **1995**, *43*, 2240–2246. [CrossRef]
20. Jaramillo Carmona, S.; de Castro, A.; Rejano Navarro, L. Traditional process of green table olives. Rationalization of alkaline treatment. *Grasas y Aceites* **2011**, *62*, 375–382. [CrossRef]
21. García–Serrano, P.; Sánchez, A.H.; Romero, C.; García–García, P.; de Castro, A.; Brenes, M. Processing of table olives with KOH and characterization of the wastewaters as potential fertilizer. *Sci. Total Environ.* **2019**, *676*, 834–839. [CrossRef]
22. Medina, E.; Romero, C.; de Castro, A.; Brenes, M.; García, A. Inhibitors of lactic acid fermentation in Spanish–style green olive brines of the Manzanilla variety. *Food Chem.* **2008**, *110*, 932–937. [CrossRef]
23. de Castro, A.; Sánchez, A.H.; Cortés–Delgado, A.; López–López, A.; Montaño, A. Effect of Spanish–style processing steps and inoculation with *Lactobacillus pentosus* starter culture on the volatile composition of cv. Manzanilla green olives. *Food Chem.* **2019**, *271*, 543–549. [CrossRef]
24. Alexandraki, V.; Georgalaki, M.; Papadimitriou, K.; Anastasiou, R.; Zoumpopoulou, G.; Chatzipavlidis, I.; Papadelli, M.; Vallis, N.; Moschochoritis, K.; Tsakalidou, E. Determination of triterpenic acids in natural and alkaline–treated Greek table olives throughout the fermentation process. *LWT Food Sci. Technol.* **2014**, *58*, 609–613. [CrossRef]
25. García, P.; Romero, C.; Brenes, B. Bioactive substances in black ripe olives produced in Spain and the USA. *J. Food Comp. Anal.* **2018**, *66*, 193–198. [CrossRef]
26. Ramírez, E.; Gandul–Rojas, B.; Romero, C.; Brenes, M.; Gallardo–Guerrero, L. Composition of pigments and colour changes in green table olives related to processing type. *Food Chem.* **2015**, *66*, 115–124. [CrossRef]
27. Gandul–Rojas, B.; Gallardo–Guerrero, L. Pigment changes during processing of green table olive specialities treated with alkali and without fermentation. *Food Res. Int.* **2014**, *65*, 224–230. [CrossRef]
28. Mínguez–Mosquera, M.I.; Garrido–Fernández, J.; Gandul–Rojas, B. Quantification of pigments in fermented manzanilla and hojiblanca olives. *J. Agric. Food Chem.* **1990**, *38*, 1662–1666. [CrossRef]
29. Mínguez–Mosquera, M.I.; Gandul–Rojas, B.; Mínguez–Mosquera, J. Mechanism and kinetic of the degradation of chlorophylls during the processing of green table olives. *J. Agric. Food Chem.* **1994**, *42*, 1089–1095. [CrossRef]
30. Mínguez–Mosquera, M.I.; Gandul–Rojas, B. Mechanism and kinetic of the degradation of carotenoids during the processing of green table olives. *J. Agric. Food Chem.* **1994**, *42*, 1551–1554. [CrossRef]
31. Mínguez–Mosquera, M.I.; Gallardo–Guerrero, L. Anomalous transformation of chloroplastic pigments in Gordal variety olives during processing for table olives. *J. Food Prot.* **1995**, *58*, 1241–1248. [CrossRef]
32. Gandul–Rojas, B.; Gallardo–Guerrero, L. Pigment changes during preservation of green table olive specialities treated with alkali and without fermentation: Effect of thermal treatments and storage conditions. *Food Res. Int.* **2018**, *108*, 57–67. [CrossRef] [PubMed]
33. Aparicio–Ruiz, R.; Riedl, K.M.; Schwartz, S.J. Identification and quantification of metallo–chlorophyll complexes in bright green table olives by high–performance liquid chromatography–mass spectrometry quadrupole/time–of–flight. *J. Agric. Food Chem.* **2011**, *59*, 11100–11108. [CrossRef] [PubMed]
34. Gandul–Rojas, B.; Roca, M.; Gallardo–Guerrero, L. Detection of the colour adulteration of green table olives with copper chlorophyllin complexes (E–141ii colorant). *LWT Food Sci. Technol.* **2012**, *46*, 311–318. [CrossRef]
35. Negro, C.; De Bellis, L.; Sabella, E.; Nutricati, E.; Luvisi, A.; Micelli, A. Detection of not allowed food-coloring additives (copper chlrophyllin-copper sulfate) in green table olives sold on the Italian market. *Adv. Hort. Sci.* **2017**, *31*, 225–233.
36. Petigara Harp, B.; Scholl, P.F.; Gray, P.J.; Delmonte, P. Quantitation of copper chlorophylls in green table olives by ultra-high-performance liquid chromatography with inductively coupled plasma isotope dilution mass spectrometry. *J. Chromatogr. A* **2020**, *1620*, 461008. [CrossRef]
37. Schiedt, H.; Liaaen–Jensen, S. Isolation and analysis. In *Carotenoids. Isolation and Analysis*; Britton, G., Liaaen–Jensen, S., Pfander, H., Eds.; Birkhäuser Verlag: Basel, Switzerland, 1995; Volume 1A, pp. 81–108.
38. Rincón–Llorente, B.; De la Lama–Calvente, D.; Fernández–Rodríguez, M.J.; Borja–Padilla, R. Table olive waste water: Problem, treatments and future strategy. A review. *Front. Microbiol.* **2018**, *9*, 1641. [CrossRef]
39. Gallardo–Guerrero, L.; Gandul–Rojas, B.; Mínguez–Mosquera, M.I. Physico–chemical conditions modulating the pigment profile in fresh fruit (*Olea europaea* var. Gordal) and favoring interaction between oxidized chlorophylls and endogenous Cu. *J. Agric. Food Chem.* **2007**, *55*, 1823–1831. [CrossRef]

40. Mínguez–Mosquera, M.I.; Gandul–Rojas, B.; Gallardo–Guerrero, L.; Roca, M.; Jarén–Galán, M. Chlorophylls. In *Methods of Analysis for Functional Foods and Nutraceuticals*, 2nd ed.; Hurst, W.J., Ed.; CRC Press, LLC: Boca Raton, FL, USA, 2008; Chapter 7, pp. 337–400.
41. Pérez–Gálvez, A.; Viera, I.; Roca, M. Chemistry in the bioactivity of chlorophylls: An overview. *Curr. Med. Chem.* **2017**, *24*, 4515–4536.
42. Gandul–Rojas, B.; Roca, M.; Mínguez–Mosquera, M.I. Chlorophyll and carotenoid degradation mediated by thylakoid–associated peroxidative activity in olives (*Olea europaea*) cv. *Hojiblanca*. *J. Plant Physiol.* **2004**, *161*, 499–507. [CrossRef]
43. Sievers, G.; Hynninen, P.H. Thin–layer chormatography of chlorophyll and their derivatives on cellulose layers. *J. Chromatogr.* **1977**, *134*, 359–364. [CrossRef]
44. Mínguez–Mosquera, M.I.; Garrido–Fernández, J. Chlorophyll and carotenoid presence in olive fruit (*Olea europaea*, L.). *J. Agric. Food Chem.* **1989**, *37*, 1–7. [CrossRef]
45. Mínguez–Mosquera, M.I.; Gandul–Rojas, B.; Montaño–Asquerino, A.; Garrido–Fernández, J. Determination of chlorophylls and carotenoids by high–performance liquid chromatography during olive lactic fermentation. *J. Chromatogr.* **1991**, *585*, 259–266. [CrossRef]
46. Mínguez–Mosquera, M.I.; Gandul–Rojas, B. High–performance liquid chromatographic study of alkaline treatment of chlorophyll. *J. Chromatogr. A* **1995**, *690*, 161–176. [CrossRef]
47. Schwartz, S.J.; Lorenzo, T.V. Chlorophylls in foods. *Crit. Rev. Food Sci. Nutr.* **1990**, *29*, 1–17. [CrossRef]
48. Tarrado–Castellarnau, M.; Domínguez–Ortega, J.M.; Tarrado–Castellarnau, A.; Pleite–Gutiérrez, R. Study of the heat transfer during the alkaline treatment in the processing of Spanish style green table olives. *Grasas y Aceites* **2013**, *64*, 415–424.
49. Louda, J.W.; Mongkhonsri, P.; Bake, E.W. Chlorophyll degradation during senescence and death-III: 3–10 yr experiments, implications for ETIO series generation. *Org. Geochem.* **2011**, *42*, 688–699. [CrossRef]
50. Viera, I.; Chen, K.; Ríos, J.J.; Benito, I.; Pérez-Galvez, A.; Roca, M. First-pass metabolism of chlorophylls in mice. *Mol. Nutr. Food Res.* **2018**, *62*, 1800562. [CrossRef]
51. Mínguez–Mosquera, M.I.; Gandul–Rojas, B.; Gallardo–Guerreo, L. Measurement of chlorophyllase activity in olive fruit (*Olea europaea*). *J. Biochem.* **1994**, *116*, 263–268. [CrossRef]
52. Lee, G.C.; Chepyshko, H.; Chen, H.H.; Chu, C.C.; Chou, Y.F.; Akoh, C.C.; Shaw, J.F. Genes and biochemical characterization of three novel chlorophyllase isozymes from brassica oleracea. *J. Agric. Food Chem.* **2010**, *5*, 8651–8657. [CrossRef]

foods

MDPI

Article

Influence of Acid Adaptation on the Probability of Germination of *Clostridium sporogenes* Spores Against pH, NaCl and Time

Antonio Valero [1,*], Elena Olague [2], Eduardo Medina-Pradas [3], Antonio Garrido-Fernández [3], Verónica Romero-Gil [4], María Jesús Cantalejo [2], Rosa María García-Gimeno [1], Fernando Pérez-Rodríguez [1], Guiomar Denisse Posada-Izquierdo [1,*] and Francisco Noé Arroyo-López [3]

[1] Department of Food Science and Technology, Agrifood Campus of International Excellence, Universidad de Córdoba, 14014 Córdoba, Spain; bt1gagir@uco.es (R.M.G.-G.); b42perof@uco.es (F.P.-R.)
[2] Department of Food Technology, Public University of Navarra, Campus de Arrosadia, E-31006 Pamplona, Spain; olagueramos@gmail.com (E.O.); iosune.cantalejo@unavarra.es (M.J.C.)
[3] Food Biotechnology Department, Instituto de la Grasa (IG-CSIC), University Campus Pablo de Olavide, Building 46, Ctra. Utrera, km 1, 41013 Seville, Spain; emedina@ig.csic.es (E.M.-P.); garfer@cica.es (A.G.-F.); fnoe@ig.csic.es (F.N.A.-L.)
[4] Technological Applications for Improvement of the Quality and Safety in Foods, R&D Division, Crta. Marbella 22. Guaro, 29108 Málaga, Spain; v.romero@oleica.es
* Correspondence: avalero@uco.es (A.V.); bt2poizg@uco.es (G.D.P.-I.)

Received: 6 November 2019; Accepted: 13 January 2020; Published: 24 January 2020

Abstract: The *Clostridium* sp. is a large group of spore-forming, facultative or strictly anaerobic, Gram-positive bacteria that can produce food poisoning. The table olive industry is demanding alternative formulations to respond to market demand for the reduction of acidity and salt contents in final products. while maintaining the appearance of freshness of fruits. In this work, logistic regression models for non-adapted and acid-adapted *Clostridium* sp. strains were developed in laboratory medium to study the influence of pH, NaCl (%) and time on the probability of germination of their spores. A *Clostridium sporogenes* cocktail was not able to germinate at pH < 5.0, although the adaptation of the strains produced an increase in the probability of germination at 5.0–5.5 pH levels and 6% NaCl concentration. At acidic pH values (5.0), the adapted strains germinated after 10 days of incubation, while those which were non-adapted required 15 days. At pH 5.75 and with 4% NaCl, germination of the adapted strains took place before 7 days, while several replicates of the non-adapted strains did not germinate after 42 days of storage. The model was validated in natural green olive brines with good results (>81.7% correct prediction cases). The information will be useful for the industry and administration to assess the safety risk in the formulation of new processing conditions in table olives and other fermented vegetables.

Keywords: *Clostridium*; logistic regression; acid-adapted strains; predictive models; table olives; fermented vegetables

1. Introduction

The *Clostridium* sp. is a foodborne pathogen that may be present in a wide variety of low acid fermented foods, being able to produce illness after its ingestion. Though there are some species for commercial use in foods such as *C. acetobutylicum*, most of them are considered spoilage or pathogenic bacteria for humans like *C. perfringens*, *C. botulinum*, *C. butyricum*, *C. tetani*, *C. difficile*, and *C. sordellii* [1].

Their spores are ubiquitously present in warm-blooded animals, and distributed in the environment, soil and water, so that they can contaminate foods during processing. In the case of *C. botulinum*, it can produce highly toxic neurotoxins, causing potentially fatal human diseases after the germination and growing of vegetative cells [2]. Therefore, to avoid the production of neurotoxin, it is essential to prevent its germination. The toxin types are classified as A, B, C, D, E, F and G. Human botulism has been mainly described with the strains of *C. botulinum* that produce toxin types A, B and E. Botulism outbreaks caused by different home-prepared or preserved foods have been widely reported in the literature, most of them due to improperly pasteurized or packed home-canned vegetables [3,4]; home-made oil condiments and sauces [5–7]; fishery products [8,9]; cheese [10,11]; or meat products [12,13]. Although it is not usual, several outbreaks in table olives have also been reported to be mainly associated with black table olives [14]. According to the last report on the trends and sources of zoonoses, zoonotic agents and food-borne outbreaks in the European Union (EU) in 2017, five strong-evidence outbreaks and 26 human cases associated with botulism were reported. Botulism cases involved hospitalization rates higher than 50%, *C. botulinum* being reported as the agent with one of the highest fatality rates [15].

The food industry requires alternative formulations with reduced acidity and salt content for canned or fermented vegetable foods, given the increasing demand by consumers for healthier and more convenient foods. However, the changes could represent a risk for the population. In the specific case of the table olive industry, salt reduction below 6% is necessary to respond to market demand. Also, excessively low pH can affect the green appearance of fruits by the degradation of chlorophylls into pheophytins. Thus, research has been oriented to study the influence of these environmental factors on the survival, growth and toxin production of *Clostridium* sp., to assess the risk associated with new packaging conditions. Temperature, pH and Knack, together with a combination of different preservatives (sodium lactate, sorbic acid, lysozyme or nisin), have been widely studied in culture media and different commodities [16–20], with most of them focused upon the growth and germination probability in different formulations, or after thermal processing.

Food safety assurance in canned or fermented, acidic foods from vegetable origins (olives, tomatoes, pickles, etc.) is typically achieved by lowering the pH below 4.6 (acidic) to prevent the proliferation of the *Clostridium* strains. However, this limit could be compromised by acid-adapted strains that could persist in contaminated products, provided the pH levels and oxygen conditions allow their germination and the subsequent production of neurotoxins. Furthermore, the survival ability of *C. botulinum* to grow, and the production of toxin in acidic environments (pH < 4.6) has been described in earlier studies, performed in both culture media and food model matrices [21–24]. Such ability can be partly explained by the implementation of a pH-inducible acid tolerance response (ATR) in sporulated bacteria at acidic pH (5.0). This tolerance produces a remarkable cell elongation [25], and increases resistance to stress at sublethal growth conditions. More recent works have studied the role of the cold shock protein-coding genes (*csp*), which are involved in growth at low temperature. Specifically, strains of *C. botulinum* having the genes *cpsB* or *cspC* develop adaptation mechanisms against NaCl, pH and ethanol stresses [26].

Therefore, a better understanding of the microbial behavior (germination and toxin production) of acid-adapted *Clostridium* sp. against environmental conditions is of particular interest for food safety assurance in the table olive industry. To this aim, microbial predictive models in foods can be effectively applied by scientists, food operators, public administration and governmental authorities, to maintain microbial quality and ensure safety [27].

The development of the probability models of *Clostridium* sp. could be useful to estimate the possibility of germination and toxin production at low infection doses, thus assisting manufacturers in the decision-making process for food quality and safety assurance [28]. Previous works have been oriented to establish food formulations for nonthermal preservation treatments by using inhibitory factors and their interactions to assess the *C. botulinum* growth probability [18–21,29,30]. Nevertheless, dedicated probability models using acid-adapted strains of *Clostridium* sp. have not been found in the literature.

In this work, logistic regression models for non-adapted and acid-adapted *Clostridium sporogenes* strains to study the influence of pH, NaCl and incubation time on the probability of germination of their spores, were developed. The factor ranges have been selected so that the model could be applied to table olive processing.

2. Materials and Methods

2.1. Strains and Culture Conditions

In the present study, *C. sporogenes* strains were used as a non-toxigenic equivalent of proteolytic *C. botulinum*, since it also causes food spoilage [16]. Thus, its use is highly recommended in challenge test studies, as microbial responses can be extrapolated to *C. botulinum* behavior [31]. Three strains of *C. sporogenes* were obtained from the Spanish Type Culture Collection (CECT), namely CECT 485 (ATCC 19404; NCTC 532); CECT 892 (ATCC 3584; NCIMB 10696); and CECT 4990 (CIP 79.39). Freeze-dried pellets were resuscitated using a small volume (2–3 mL) of Liver Broth (LB, Oxoid, Basingstoke, Hampshire, England, UK) incubating at 37 °C overnight. The use of cooked meat media for the resuscitation of sporulated bacteria has been suggested by other authors [16]. Frozen stock cultures were maintained in LB at −80 °C in Eppendorf tubes with 15% glycerol until use. Activation of vegetative cells was carried out by transferring 0.1 mL to 10 mL screw-cap tubes of LB, incubating at the same conditions above. Then, 0.1 mL of LB was transferred to 10 mL screw-cap tubes of Fluid Thioglycollate Medium (FTM, Oxoid, Basingstoke, Hampshire, England, UK), incubating at 37 °C for 48 h, until the medium became turbid. Enumeration of colonies was done by pouring 1 mL into tryptose sulphite cycloserine (TSC) agar with D-cycloserine (Oxoid, Basingstoke, Hampshire, England, UK) incubating in anaerobic jars at 37 °C for 48–72 h. Enumeration was also done using a Thoma cell counting chamber (Hirschmann Instruments™ 8100103) where the morphology of the cells and spores were visualized.

2.2. Preparation of Inoculum

To obtain the suspension of spores, 1 mL of vegetative cells in FTM was transferred to sterile 100 mL flasks of FTM to produce more vegetative cells incubating at 37 °C for 48 h. Flasks were then stored at room temperature to enhance sporulation. Sporulation was daily checked by Gram staining together with enumeration in a Thoma cell counting chamber, and spores were harvested after one week, where a high population of spores (80%) was achieved [16]. Before inoculation, the FTM flasks were heated at 87 °C for 15 min to activate sporulation and to destroy vegetative cells. Sporulated cultures were washed twice with phosphate-buffered saline (PBS) (Medicago AB, Uppsala, Sweden) by centrifugation at 4100 rpm (Jouan C4 i, Thermo Electron Corporation, France) for 10 min and finally re-suspended in saline solution (0.85%). The cocktail of *C. sporogenes* spores was prepared by mixing volumes of 4 mL from each strain suspension (1×10^9 spores/mL approx.) in a sterile test tube.

2.3. Acid Adaptation of *C. sporogenes* Strains

From the 10 mL tubes of FTM, the gradual acid adaptation of *C. sporogenes* strains was achieved by using decreasing pH levels (6.5, 6.0, 5.75 and 5.5). The pH of each prepared broth was aseptically adjusted with hydrochloric acid (1M). One ml of the grown culture at pH 7.2 (alkali) was transferred to a sterile tube of FTM at pH 6.0, incubating at 37 °C in anaerobic jars until visible turbidity was observed. Further, the same transfer was repeated to the subsequent tubes at the decreasing pH levels.

When an acid-adapted culture was obtained at pH 5.5, 1 mL of vegetative cells in FTM was transferred into sterile 100 mL flasks of FTM with their pH adjusted to 5.5. Inoculum preparation of acid-adapted spores was then carried out as explained above, using modified pH media.

2.4. Experimental Design

A full factorial design, including 32 combinations of eight pH and four NaCl levels, was achieved for non-adapted and acid-adapted strains (pre-incubated at pH 5.5). The influence of pH (4.0, 4.5, 5.0, 5.5, 5.75, 6.0 and 7.0) and NaCl (0%, 2%, 4% and 6%) was assessed. The sodium chloride (NaCl) percentage was calculated considering the salt content of the initial Differential Reinforced Clostridial Medium (DRCM, Oxoid, Basingstoke, Hampshire, England, UK) (0.5%). The pH was measured with a pH/mv-meter digit 501 (Crison, Barcelona, Spain), and its adjustment was aseptically performed using hydrochloric acid (HCl) (1M). Once modified, all media were sterilized, and subsequently, the NaCl concentrations and pH values were verified. Temperature was not initially considered as a model variable, so that it was assumed that table olives can be eventually stored at relatively high temperature conditions during summer periods. Microbial responses were recorded daily for 42-days incubation, so a total of 1344 growth/no growth data (32 pH and NaCl combinations × 42 time points) were obtained for the development of the logistic regression models.

2.5. Inoculation Procedure and Germination Assessment

For assessing the germination probability of the *C. sporogenes* cocktail, for each physiological state condition (non-adapted and acid-adapted), DRCM was used. In this culture, medium *Clostridia* can reduce sulphite to sulfide—forming iron sulfide. Iron (III) citrate is included in the formulation as an indicator of sulphite reduction. For assessing germination, microtiter plates of 10 × 10 wells each were inoculated with 300 µL medium + 100 µL of inoculum. Eight wells per condition were inoculated with two blanks (400 µL of uninoculated DRCM). Appropriate dilutions of the initial inoculum were made in such a way that a concentration of 1×10^6 spores/well was reached. Afterwards, microtiter plates were covered with a lid and sealed with paraffin. Incubation was done in 2.5 L anaerobic jars using AnaeroGen™ sachets (Thermo Scientific) for the gas generation, so that oxygen concentration was reduced below 1% and CO_2 reached 9%–13% [32]. Then, the anaerobic jars were tightly closed and incubated at 30 °C for 42 d. Germination was visually recorded daily when the medium darkened, indicating that the sulphite reduction had occurred and iron sulfide had been formed. At the end of the experiments, the microbial concentration of *C. sporogenes* was confirmed by pour plating the volume of the well (400 µL) onto Sodium polyanethol sulfonate (SPS) and tryptose sulphite cycloserine (TSC) (Oxoid, Basingstoke, Hampshire, England, UK). Positive germination was verified if there was a 1-log increase in the microbial concentration with respect to the inoculation moment. Contaminated and turbid wells which did not show blackening were discarded.

2.6. Development of Logistic Regression Models

The whole dataset was implemented in an Excel spreadsheet, and a polynomial logistic regression equation was fitted to the model data observed. Generally, this type of model contains a right-hand side term (which is a polynomial equation) and a left-hand side term, named "logit *p*", $logit\ p = ln\left(\frac{p}{1-p}\right)$ [33]. The equation used in this study was a second-order linear logistic regression model, as follows:

$$\begin{aligned} Logit\ p = \ & ln\left(\frac{p}{1-p}\right) \\ & = b_0 + b_1 * time + b_2 * pH + b_3 * NaCl + b_4 * time * NaCl + b_5 \\ & * NaCl * pH + b_6 * time * pH + b_7 * time * NaCl * pH + b_8 * pH^2 + b_9 \\ & * NaCl^2 + b_{10} * time^2 \end{aligned} \quad (1)$$

where p is the probability of germination, and b_0–b_{10} are the coefficients to be estimated. Time units were set in days.

From the observed conditions, a dataset was selected for model development (training), and internal validation was made using conditions within the model range domain. Conditions selected are represented in Table 1. The logistic regression models were fitted in R v3.4.0 (R Project for Statistical Computing) by using the *glm* function. A forward stepwise process was used by adding the significant variables ($p < 0.05$) at each step. With this procedure, a biologically consistent model was obtained, in accordance with the data observed. For assessing predictions, the cut-off value was established at 0.125, thus considering that germination was produced if there were at least 1 out of 8 positive wells.

Table 1. Experimental design for the selection of training (gray) and validation (white) conditions for the development of the logistic regression models (experimental time from 1 to 42 days).

	pH							
NaCl (%, *w/w*)	**4.00**	**4.50**	**5.00**	**5.50**	**5.75**	**6.00**	**6.50**	**7.00**
0								
2								
4								
6								

2.7. *Assessment of Model's Performance*

Once the model was obtained, its performance was evaluated using the goodness of fit statistics and predictive performance indices, which was determined by (i) the likelihood ratio test ($-2lnL$), where L is the likelihood at its optimum; (ii) Akaike Information Criterion ($AIC = -2lnL + 2k$, where k is the number of parameters in the model); (iii) the determination coefficient (R^2-Nagelkerke), which quantifies the proportion of variation explained by the logistic regression model; and iv) the Hosmer–Lemeshow (HL) statistic. The $-2lnL$ and the AIC can be used to rank models based on the same dataset, where lower values indicate better fitting models. The HL statistic indicates if the model fits the data adequately. This statistic divides the number of times in which growth occurred (observed events) into approximately ten groups (based on the predicted probabilities), and then, compares the observed and the expected number of events in the groups through a contingency table by using the Pearson coefficient. Lower values of the HL statistic indicate a better fit. The area under the Receiver Operating Characteristic (ROC) curve, c, is a measure of discrimination, obtained from a plot sensitivity (the proportion of observed events that were correctly predicted to be events), against the complement of specificity (the proportion of observed non-events that were correctly predicted to be non-events). The closer the value of c is to 1, the higher is the discrimination. For a better illustration of the adjustment of the developed model to the data observed, predicted germination probabilities at 0.125, 0.5 and 0.9 were calculated maintaining constant the pH and NaCl terms, and then were plotted in contour graphs.

2.8. *Validation of the Logistic Regression Models in Table Olive Brines*

The logistic regression models for the non-adapted and acid-adapted *C. sporogenes* strains were validated in brines from fermented table olives. Brines were obtained from directly brined green Aloreña fermentations. First, brines were centrifuged at 4000 rpm for 10 min, and the supernatant was filter-sterilized using a bacteriological filter with a pore size of 0.22 µm Ø, (Millipore filter Unit-Express plusPES, Billerica, MA, USA). Then, sterilized brines were adjusted to the different pH, and the NaCl conditions explained in Section 2.4. The studied combinations included three pH levels (5, 5.5 and 6) and two NaCl concentrations (4% and 6%). Afterwards, the adjusted brines were aseptically transferred into 7 mL sterile, screw-cap tubes and inoculated with 0.1 mL of a spore suspension of the *C. sporogenes* cocktail. Finally, the tubes were incubated anaerobically at 30 °C for 13 d. Germination of *C. sporogenes* was daily assessed through plate counting in TSC and SPS agars.

3. Results

3.1. *Performance of the Logistic Regression Models*

Logistic regression models were developed for estimating the probability of germination of non-adapted and acid-adapted *C. sporogenes* strains. Estimation of the significant coefficients together with their corresponding standard errors and *P*-values, are represented in Table 2. It should be remarked that for the pH, the ln-transformed term was used for the logistic model of non-adapted strains for improving accuracy.

Table 2. Parameter estimates of the logistic regression models for the acid-adapted and non-adapted *C. sporogenes* strains.

	Coefficient	Estimate	S.E.	Wald	df	*P*-Value	Lower C.I (95%)	Upper C.I (95%)
Acid-Adapted *C. sporogenes* Strains	*time*	−3.175	0.527	36.289	1	<0.001	−4.208	−2.142
	pH	20.898	8.639	5.851	1	0.016	3.965	37.831
	NaCl	1.291	0.410	9.903	1	0.002	0.487	2.095
	Time × pH	0.726	0.113	41.182	1	<0.001	0.504	0.948
	pH^2	−1.616	0.714	5.126	1	0.024	−3.016	−0.217
	$NaCl^2$	−0.328	0.075	19.331	1	<0.001	−0.474	−0.182
	constant	−68.996	25.954	7.067	1	0.008	–	–
Non-Adapted *C. sporogenes* Strains	*time*	−3.610	0.382	89.462	1	<0.001	−4.358	−2.862
	NaCl	−6.566	1.620	16.424	1	<0.001	−9.741	−3.390
	Time × NaCl	0.059	0.014	17.743	1	<0.001	0.032	0.087
	$NaCl^2$	−0.343	0.062	30.180	1	<0.001	−0.466	−0.221
	ln(pH) × time	2.364	0.248	91.015	1	<0.001	1.878	2.850
	ln(pH) × NaCl	4.514	0.943	22.927	1	<0.001	2.666	6.362
	constant	−3.565	0.555	41.218	1	<0.001	–	–

It should be noticed that the linear pH term was not significant ($P > 0.05$) for the model of the non-adapted strains. The performance statistics obtained indicate reasonable goodness of fit of the models obtained, mainly due to the high values of R^2-Nagelkerke (>0.921) and AIC values (Table 3). The HL statistics gave *P*-values higher than 0.05 for both models, thus indicating a good adjustment to the observed data. These values are in line with other logistic models published in the literature [18,19,34]. However, a higher degree of accuracy was obtained for the logistic model of acid-adapted strains, given the lower values of AIC and log-likelihood in comparison to that of non-adapted strains.

Table 3. Goodness of fit statistics for the logistic regression models of the acid-adapted and non-adapted *C. sporogenes* strains.

Goodness of Fit/Predictive Power	Acid-Adapted *C. sporogenes* Strains	Non-Adapted *C. sporogenes* Strains
	Coefficient	Coefficient
−2lnL [1]	119.848	195.89
AIC [2]	133.848	209.89
Hosmer-Lemeshow (df = 8)	2.065	6.044
p-value	0.979	0.642
Nagelkerke R^2	0.950	0.921

[1] log-likelihood, [2] Akaike Information Criterion.

Through the calculation of the area under the ROC curve, the corrected classified cases were calculated for model and validation data. Their percentages were estimated considering a cut-off value of 0.125 for the probability of germination (≥1/8 germinated wells). The classification percentages of observed vs. predicted conditions are shown in Table 4. The logistic models provide a certain margin of safety, since most of the misclassified cases were considered as fail-safe (i.e., germination was predicted, while no germination was observed). These findings can be translated positively into an industrial context, since by using the model, safe formulations can be designed in such a way that the germination of *C. sporogenes* is prevented.

Table 4. Classification tables of observed vs. predicted conditions of the training and validation datasets for the acid-adapted and non-adapted *C. sporogenes* strains.

	ACID-ADAPTED *C. SPOROGENES* STRAINS			
Training	Estimated Probability		Total	Correct Prediction (%)
Observed response	No germination	Germination		
No germination	292	32	324	90.12
Germination	1	683	684	99.85
Total	293	715	1008	96.83
Validation	Estimated Probability		Total	Correct Prediction (%)
Observed response	No germination	Germination		
No germination	92	12	104	88.46
Germination	0	232	232	100.00
Total	92	244	336	96.43
	NON-ADAPTED C. SPOROGENES STRAINS			
Training	Estimated Probability		Total	Correct Prediction (%)
Observed response	No germination	Germination		
No germination	311	55	366	84.97
Germination	0	642	642	100.00
Total	311	697	1008	94.54
Validation	Estimated Probability		Total	Correct Prediction (%)
Observed response	No germination	Germination		
No germination	89	26	115	77.39
Germination	0	221	221	100.00
Total	89	247	336	92.26

According to the proportion of correctly classified cases (Table 4), for the acid-adapted strains, 32 training and 12 validation conditions were misclassified as fail-safe, while only one case was considered fail-dangerous for the training dataset. For the validation dataset, all misclassified cases were fail-safe. Regarding the logistic model for the non-adapted strains, all deviations were fail-safe. For the training and validation datasets, 55 and 26 cases were misclassified, respectively. However, the average proportion of correctly classified cases was higher than 92%.

3.2. Effect of Environmental Factor on the Probability of Germination of Non-Adapted and Acid-Adapted C. sporogenes Strains

The observed responses confirmed the high sensitivity of *C. sporogenes* to low pH values, since the microorganism was not able to germinate at pH < 5.0 at any tested condition. Overall, the acid-adaptation of the strains produced a faster germination of spores at close to the limiting conditions of the pH and NaCl levels, as observed at moderately acidic pH (5.0–5.5) combined with a high (>4%) NaCl concentration. The main advantages of logistic regression models are that they can set the level of stringency required at certain environmental conditions. Contour plots representing the germination responses of both non-adapted and acid-adapted strains of *C. sporogenes* as a function of pH (5.0, 5.5 and 5.75) and incubation time (0–30 days), is shown in Figure 1. Lines of constant probabilities were then compared graphically with the experimental data at values of $p = 0.125$, $p = 0.500$ and $p = 0.900$. The homologous germination responses at NaCl concentrations of 0%, 2%, 4% and 6% as a function of pH and incubation time (0–30 days), are represented in Figure 2. A narrower transition between germination and non-germination boundaries was obtained for the acid-adapted strains. This result indicates that small changes in pH and NaCl formulations can govern the germination responses of acid-adapted *C. sporogenes* strains. The more abrupt germination/no-germination transition in the case of the acid-adapted strains produced fewer intermediate conditions, where binary responses were observed. These combinations are represented in Table 5. For non-adapted strains, intermediate conditions were observed at pH 5, 5.5 and 5.75. For the acid-adapted strains, these conditions were mainly observed at pH 5.0 and 5.5. All binary responses implied positive germination in > 1/8 wells, so that all of this model's predictions yielded probabilities higher than 0.125.

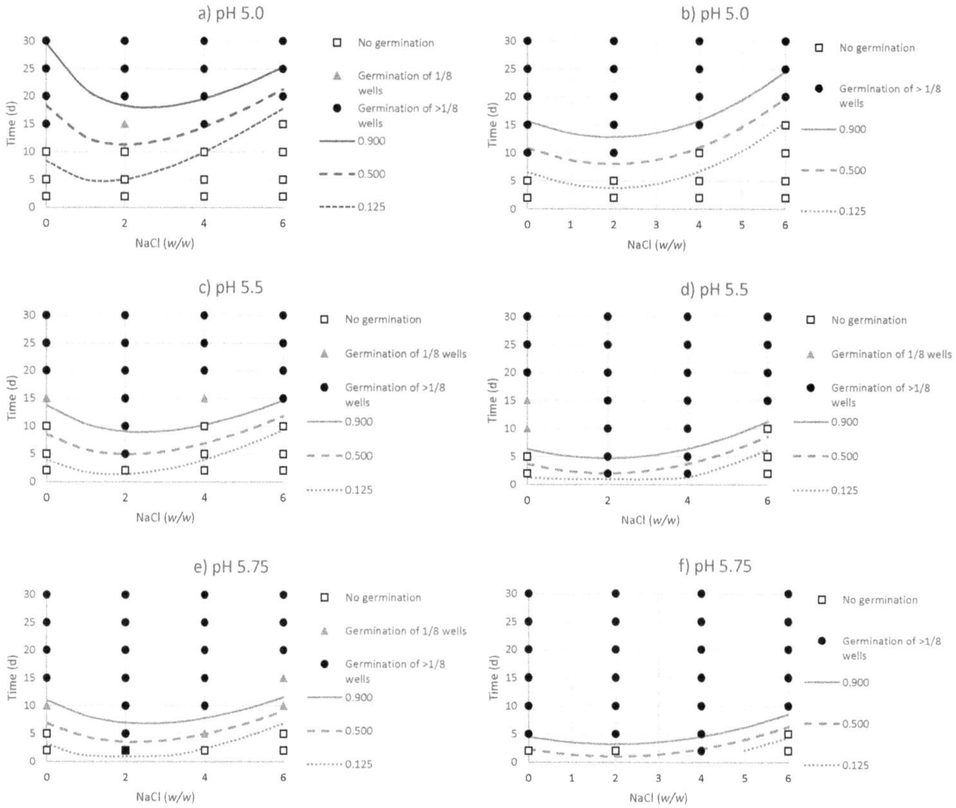

Figure 1. Contour plots for the observed germination responses and predicted probabilities (p = 0.125, p = 0.500 and p = 0.900) for the non-adapted (panels **a**,**c**,**e**) and acid-adapted strains (panels **b**,**d**,**f**) of *C. sporogenes* at pH levels 5.0, 5.5 and 5.75.

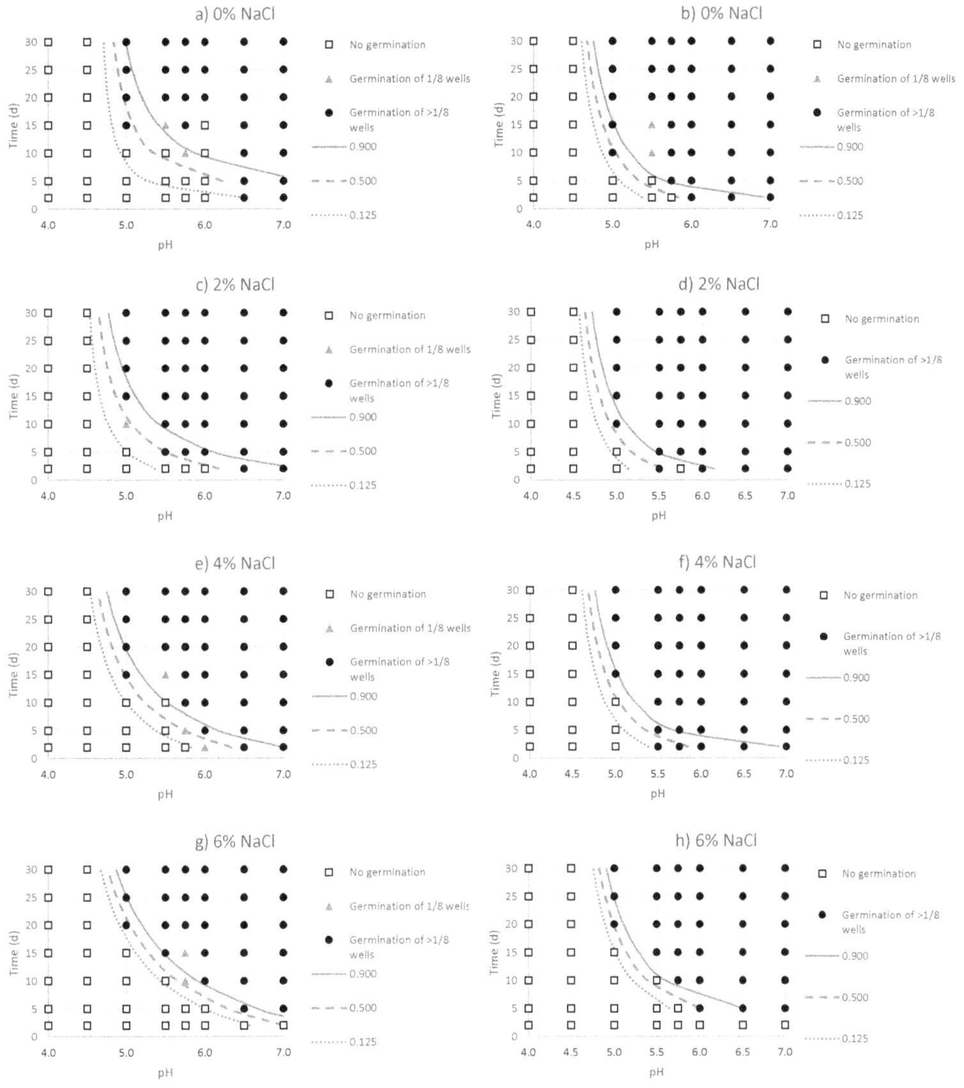

Figure 2. Contour plots for the observed germination responses and predicted probabilities (p = 0.125, p = 0.500 and p = 0.900) for the non-adapted (panels **a**,**c**,**e**,**g**) and acid-adapted strains (panels **b**,**d**,**f**,**h**) of *C. sporogenes* at NaCl concentrations of 0%, 2%, 4% and 6% *w*/*w*.

The results showed that at acidic pH values (5.0), the acid-adapted strains germinated at 10 days' incubation at NaCl concentrations ≤ 2% (Figure 1b), while the germination time of the non-adapted strains increased until 15 days at these NaCl concentrations (Figure 1a). According to the model's predictions, the growth boundary is set at pH 5.0, NaCl 3.95% and 10 d incubation for non-adapted strains, while this boundary is shifted to an increased NaCl concentration of 4.92% (pH 5 for a 10-days incubation) for the acid-adapted strains.

Table 5. Environmental conditions where a binary response was observed for *C. sporogenes* strains for the model and validation datasets after 42 d incubation at 30 °C.

Non-Adapted Strains				Acid-Adapted Strains			
pH	NaCl (%)	Germination	Dataset	pH	NaCl (%)	Germination	Dataset
5.0	0.0	5/8	Training	5.0	0.0	6/8	Training
5.0	2.0	2/8	Validation	5.0	2.0	7/8	Validation
5.0	4.0	3/8	Training	5.0	4.0	5/8	Training
5.0	6.0	4/8	Training	5.0	6.0	7/8	Training
5.5	6.0	5/8	Training	5.5	0.0	7/8	Validation
5.75	0.0	5/8	Training				
5.75	4.0	5/8	Training				
5.75	6.0	4/8	Validation				

As expected, as pH increases, higher NaCl concentrations are required to prevent germination. At pH 5.5, germination of non-adapted strains occurred after 15-days incubation, regardless of NaCl concentration (Figure 1c). For acid-adapted strains, the presence of NaCl activated the germination of spores, since at concentrations ≥ 2%, it occurred after 24 h incubation at pH 5.5 (Figure 1d). Using pH 5.5 and 6% NaCl, the models predicted germination after 9.26 and 6.20 days of incubation for the non-adapted and acid-adapted strains, respectively.

At pH 5.0, NaCl concentrations ≥ 4% delayed germination for more than 10 days (Figure 1a,b). However, by increasing the pH to 5.5, both 2% and 4% NaCl concentrations produced germination in 24 h for acid-adapted strains, while non-adapted ones delayed germination after the 15th day of incubation (Figure 1c,d).

NaCl concentrations of 2% and 4% produced a faster germination of *C. sporogenes* in comparison to conditions in the absence of NaCl and pH levels ≤ 6.0. For instance, in the case of the non-adapted strains, germination was produced, in the absence of NaCl, after 20 days of incubation at pH 6.0 (Figure 2a), but the pathogen was germinated on the 5th day in the presence of 2% NaCl (Figure 2c). For the acid-adapted strains, germination was produced at more limiting conditions, since at pH 5.5 and 2% and 4%, germination was observed after 24 h incubation (Figure 2d,f). However, when NaCl was not added, germination of all wells took place ≥ 20 days at pH 5.5 (Figure 2b).

The evolution of germination probabilities in comparison with the observed responses at representative pH and NaCl conditions are shown in Figure 3. Overall, predictions given by the logistic models indicated earlier germination for the acid-adapted strains at all assayed conditions. At pH 5.0 and 2% NaCl, the logistic model predicted germination ($p \geq 0.125$) after 5.08 and 3.73 days for the non-adapted and acid-adapted strains, respectively. The observations indicated that 7 out of the 8 wells showed germination for the acid-adapted strains, while the probability was reduced to two out of eight germinated wells for the non-adapted strains at the end of the incubation period (Figure 3a). When increasing pH and NaCl concentrations (Ph 5.5, 4% NaCl, and pH 5.75, 6% NaCl), germination responses occurred in a shorter period, since all wells of the acid-adapted strains led to positive germination after 7 days' incubation (Figure 3b,c). On the contrary, the increase in NaCl concentration gave fewer germinated wells in the case of non-adapted strains (7 and 3 out of 8 wells, respectively). The incubation times predicted by the model for the germination of acid-adapted and non-adapted strains predicted were 1.33 and 3.95 days (pH 5.5 and NaCl 4%), and 4.38 and 6.8 days (pH 5.75, NaCl 6%), respectively.

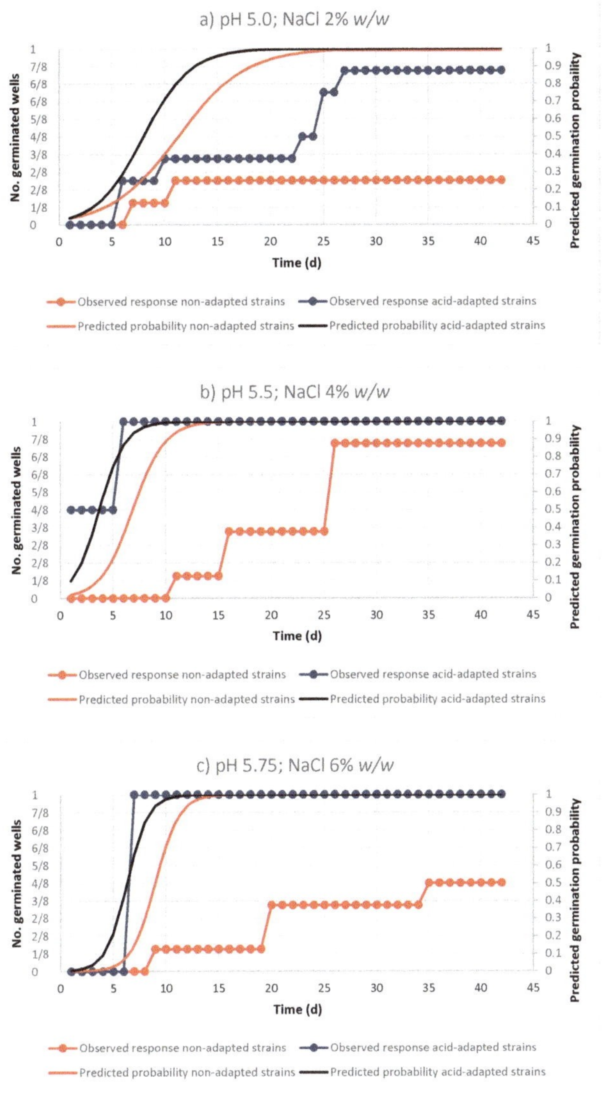

Figure 3. Evolution of predicted probabilities of germination and observed responses over experimental time for the non-adapted and acid-adapted strains of *C. sporogenes* at different combinations of pH and NaCl (panels **a**–**c**).

3.3. Germination of C. sporogenes Strains in Table Olive Brines

Validation at different pH and NaCl levels was also performed in formulated brines anaerobically stored at 30 °C for 13 d. Overall, 208 conditions for the non-adapted and the acid adapted *C. sporogenes* strains were assessed. Regarding the non-adapted strains, non-germination and germination responses of *C. sporogenes* were produced in 59 and 45 conditions, respectively. The logistic regression model was able to correctly predict microbial evolution in 87.5% of the cases, though there were 10 fail-safe (9.61%) conditions (germination was predicted by the model, but not observed) and three fail-dangerous (2.88%) (no germination predicted, but observed). The results agreed with those obtained in DRCM, since germination was observed at pH 5.0 and 2% Knack, as well as at pH 5.5 and 4% NaCl. On the contrary, when NaCl increased up to 6%, germination was only observed at pH 6.0.

For the acid-adapted strains, the logistic model was able to predict 81.73% of cases, results that were more conservative than those provided for the non-adapted strains, as confirmed by the increased number of fail-safe conditions (17, 16.34%). However, only two conditions were classified as fail-dangerous (1.92%). Such behavior can be explained by the more difficult germination of *C. sporogenes* in brines than in DRCM. Nevertheless, acid-adapted *C. sporogenes* strains were able to germinate at pH 5.0 and 4% Knack, as well as at pH 5.5 and 6% NaCl, thus confirming their higher resistance to stringent conditions when compared with the non-adapted strains.

For assessing the model's application, the predicted time required for germination ($Pred_{t\_model}$) was calculated, at a probability of 0.125, at the studied conditions, and their results compared with those observed in brines (Obs_{t_brine}) (Table 6). Predictions were conservative in most cases and provided a reasonable estimation of the germination time at different pH and NaCl conditions. Therefore, the presence of antimicrobial compounds in brines (organic acids, polyphenols, etc.) may have limited the germination of the acid-adapted *C. sporogenes* strains in table olive brines with respect to that observed in DRCM. Further studies are needed to confirm the effect of environmental factors and preservatives on the germination ability and microbial resistance of spore-forming bacteria in brines.

Table 6. Comparison between the observed time to germination in table olive brines (Obs_{t_brine}) of non-adapted and acid adapted *C. sporogenes* strains and those predicted ($Pred_{t\_model}$) by the logistic models.

Non-Adapted *C. sporogenes* Strains				**Adapted *C. sporogenes* Strains**			
pH	**NaCl (%)**	**Obs_{t_brine}**	**$Pred_{t\_model}$**	**pH**	**NaCl (%)**	**Obs_{t_brine}**	**$Pred_{t\_model}$**
5.0	2	8	5.07	5.0	2	6	3.72
5.0	4	>13	9.96	5.0	4	10	6.70
5.0	6	>13	>13	5.0	6	>13	>13
5.5	2	3	1.37	5.5	2	2	1
5.5	4	8	3.93	5.5	4	1	1.34
5.5	6	>13	9.26	5.5	6	>13	6.18
6.0	4	1	1.18	6.0	4	1	1
6.0	6	3	4.93	6.0	6	3	3.30

4. Discussion

In the present study, the acid-adaptation of *C. sporogenes* strains have influenced the subsequent germination responses as a function of different pH and NaCl conditions. Overall, acid-adapted spores produced faster germination at more limiting conditions when compared to the non-adapted ones. There are several studies in literature dealing with *C. sporogenes* behavior against various environmental conditions in different culture media using non-acid-adapted cells [18,35], as well as in food matrices such as meat products [2,16,20,31] or dairy [19,36]. However, although the growth ability of *C. botulinum* (or *C. sporogenes* as a surrogate) at low pH has been extensively reported, there are very few studies dealing with the effect of acid adaptation.

Crosthwait [37] found that acid adaptation of *C. sporogenes* in FTM and tomato serum produced germination at lower pH values (4.85) than those initially observed without any adaptation (5.4). However, it was observed that adaptation ability was maintained by continuously sub-culturing at pH 5.0. In our study, *C. sporogenes* could germinate at pH 5.0, while no germination was observed at pH 4.5 during the 42-days incubation period. Lund et al. [21] reported minimum values for pH of 4.6 to produce the growth of vegetative strains of proteolytic *C. botulinum*, though this effect was time- and strain-dependent. Other authors have confirmed these results, such as Wong et al. [24], who found spore germination and outgrowth in anaerobically-acidified media at pH < 4.6. The effect of acid pH upon the germination and subsequent growth of *C. sporogenes* or *C. botulinum* strains in culture media is variable depending on several factors, such as the inoculum size, the redox potential or the presence of antimicrobial preservatives [18,30]. It is also recognized that the physiological state and properties of spores may vary between different batches of the same strains, thus increasing the variability of the probability of germination at acidic pH.

Besides, it is reported that the addition of NaCl at high levels delays the germination and outgrowth of Clostridial strains. The relative effect of NaCl on the inhibition of *Clostridium* sp. may differ according to other factors that produce a synergistic effect or have higher significance than NaCl itself [18]. Whiting and Call [38] found that the time to the growth of proteolytic *C. botulinum* was delayed at temperatures <20 °C and pH levels <5.5, having NaCl no or little effect at concentrations ≤3%. However, when NaCl is added to food matrices, the inhibitory effect is usually enhanced. Taylor et al. [36] found that NaCl at 1.6% or 2.4% produced inhibition on *C. sporogenes* in canned butter samples. The same conclusion about the effect of NaCl was found by Knanipour et al. [19] in high moisture cheese. This result can be attributed by the effect of added food preservatives or the physical properties of foods, which can interfere with the growth of *Clostridium* sp.

Our results have confirmed previous findings in which pH and NaCl combinations could delay or inhibit the germination of spores. Montville [22] described the interaction of pH and NaCl on the growth of *C. botulinum*, reporting that germination was produced at pH 5.0 in the absence of NaCl, while concentrations up to 6% inhibited it at all of the pH levels tested. However, according to our results, the acid-adapted spores germinated faster at pH 5.0 than the non-adapted cells at NaCl concentrations of 0% and 2% (Figure 2a,b). The germination responses of acid-adapted cells were more marked at pH 5.5 when NaCl concentration ranged between 2% and 4% (Figure 2c,d), as well as at pH 5.75 and increased levels of NaCl (4% and 6%) (Figure 2e,f). The inhibitory effect of acid conditions is usually linked to the undissociated form of the acid, which dissociates into H+ and the anion in the bacterial cell. The increased concentration of protons causes a decrease in the intracellular pH, thus, disrupting cell metabolism. It is plausible that the interaction between increased NaCl concentrations and acidic pH could contribute to the increase of the turgor pressure of the cell, which in turn, may delay or prevent the germination of non-adapted spores [39]. Zhao et al. [30] reported that proteolytic *C. botulinum* did not grow at pH values < 5.5 and NaCl concentrations > 4% in a 14-day incubation period. These results match with those found in our study, since no germination was observed in 10 d at pH 5.5 and NaCl ≥ 4% for non-adapted strains and pH 5.5 and 6% NaCl for the acid-adapted ones. Sensitivity to the pH of *Clostridium* strains can produce a shift in the inhibitory pH–NaCl combinations. Montville [22] found that 6% NaCl at pH 5.5 inhibited the growth of proteolytic *C. botulinum* with intermediate pH sensitivity. Likewise, Graham et al. [40] did not obtain growth at pH less than 5.1 or 5% NaCl for non-proteolytic *Clostridium* strains. However, in our study, germination was produced at 6% NaCl and pH 5.0 for both non-adapted and acid-adapted strains (Figure 3a,b). Potential inter-strain specific differences and the use of a strain-cocktail may explain the variability in the environmental conditions allowing germination.

The effect of oxygen concentration on *Clostridium* sp. growth has been recently studied by Couvert et al. [41], finding that total inhibition for *C. sporogenes* growth is reached at the 3.26% oxygen level in the gaseous phase. Nevertheless, when other conditions are suboptimal, much lower concentrations of oxygen and lower redox potentials may be inhibitory.

The effect of the redox potential of the culture media in the presence of acid-adapted spores of *Clostridium* sp. is a matter of research for further studies, since it would allow a better understanding of the microbial behavior under suboptimal conditions.

Table 7 presents a comparison table of the observed growth responses of Clostridial strains published in earlier studies, with predicted germination probabilities found by the logistic regression models developed in this study. Though microbial responses were highly variable depending on the observation time, the strains used and the NaCl and pH combinations, predicted germination probabilities were higher for the acid-adapted strains when the outcome (p) was between 0 and 1.

Table 7. Reported growth responses of *C. botulinum* and *C. sporogenes* strains in different published studies performed in culture media and their comparison with predictions of the probability of germination of the non-adapted and acid-adapted *C. sporogenes* strains used in the present study.

Microorganisms	T (°C)	pH	NaCl (%)	Growth	Obs. Time (Days)	Reference	*p* [1] (non-Adapted)	*p* (Acid-Adapted)
C. botulinum proteolytic	30	4.7	2.5	No	>42	FSA (UK) [2]	Yes (0.97) (fs)	Yes (0.99) (fs)
C. botulinum proteolytic	30	4.8	1.5	Yes	8.96	FSA (UK)	Yes (0.13) (c)	Yes (0.13) (c)
C. botulinum proteolytic	30	5.6	5.5	Yes	>42	FSA (UK)	Yes (0.99) (c)	Yes (1.00) (c)
C. botulinum proteolytic	30	6.3	5.5	Yes	11.08	FSA (UK)	Yes (0.99) (c)	Yes (1.00) (c)
C. botulinum proteolytic	30	5.2	1.5	No	>42	FSA (UK)	Yes (0.99) (fs)	Yes (1.00) (fs)
C. botulinum proteolytic	30	5.3	4.5	Yes	12.96	FSA (UK)	Yes (0.83) (c)	Yes (0.98) (c)
C. botulinum proteolytic	30	5.1	3.5	Yes	11.00	FSA (UK)	Yes (0.48) (c)	Yes (0.86) (c)
C. botulinum proteolytic	30	7	5.5	Yes	21.17	FSA (UK)	Yes (1.00) (c)	Yes (1.00) (c)
C. botulinum proteolytic	25	4.7	0.5	Yes	11.88	FSA (UK)	No (0.07) (fs)	No (0.04) (fs)
C. botulinum proteolytic	25	5.9	3.5	Yes	8.31	FSA (UK)	Yes (0.97) (c)	Yes (0.99) (c)
C. botulinum proteolytic	25	5.7	4.5	Yes	14.05	FSA (UK)	Yes (0.99) (c)	Yes (0.99) (c)
C. botulinum proteolytic	25	5.6	5.5	Yes	29.02	FSA (UK)	Yes (1.00) (c)	Yes (1.00) (c)
C. botulinum proteolytic	25	7	4.5	Yes	8.31	FSA (UK)	Yes (1.00) (c)	Yes (1.00) (c)
C. botulinum proteolytic	35	4.7	0.5	No	>42	FSA (UK)	Yes (0.46) (fs)	Yes (0.98) (fs)
C. botulinum proteolytic [3]	30	7	0.6	Yes	1.00	Juneja et al. 1999 [4]	Yes (0.20) (c)	Yes (0.65) (c)
C. sporogenes	37	7	4	Yes	>42	*18*	Yes (1.00) (c)	Yes (1.00) (c)
C. sporogenes	37	5.5	4	Yes	>42	*18*	Yes (1.00) (c)	Yes (1.00) (c)
C. sporogenes [5]	30	5.5	2	Yes	1.00	*16*	No (0.10) (fd)	Yes (0.30) (c)
C. sporogenes [5]	29.6	5.6	6	Yes	1.00	*16*	No (0.00) (fd)	No (0.00) (fd)
C. botulinum proteolytic [6]	30	6.8	2	Yes	14.00	*21*	Yes (1.00) (c)	Yes (1.00) (c)
C. botulinum proteolytic [6]	30	5.1	2	Yes	14.00	*21*	Yes (0.78) (c)	Yes (0.96) (c)
C. botulinum proteolytic [6]	30	4.9	2	Yes	14.00	*21*	Yes (0.56) (c)	Yes (0.84) (c)
C. botulinum proteolytic [6]	30	4.9	2	No	14.00	*21*	Yes (0.41) (fs)	Yes (0.65) (fs)
C. botulinum proteolytic [6]	30	4.8	2	No	14.00	*21*	Yes (0.26) (fs)	Yes (0.35) (fs)
C. botulinum proteolytic [6]	30	4.7	2	No	14.00	*21*	No (0.07) (c)	No (0.10) (c)
C. botulinum proteolytic	30	5.5	4	No	>42	*30*	Yes (1.00) (fs)	Yes (1.00) (fs)
C. botulinum proteolytic	30	5.5	2	Yes	>42	*30*	Yes (1.00) (c)	Yes (1.00) (c)
C. botulinum proteolytic	30	6	4	Yes	>42	*30*	Yes (1.00) (c)	Yes (1.00) (c)
C. botulinum proteolytic [7]	30	5.5	3	Yes	3	*23*	Yes (0.19) (c)	Yes (0.61) (c)
C. botulinum proteolytic	30	5.0	0	Yes	30	*22*	Yes (0.91) (c)	Yes (1.00) (c)
C. botulinum proteolytic	30	5.0	3	No	30	*22*	Yes (1.00) (fs)	Yes (1.00) (fs)
C. botulinum proteolytic	30	5.5	4	Yes	30	*22*	Yes (1.00) (c)	Yes (1.00) (c)
TOTAL (c/fs/fd) [8]							68.75%/25%/6.25%	71.88%/25%/3.12%

[1] *p* (probability of germination estimated by the logistic regression models). [2] Food Standard Agency (UK). [3].Mixed strains culture of *C. botulinum* proteolytic: 62A 33 999 C11, Serotype(s): A A B B. [4] Juneja, V.K.; Whiting, R.C.; Marks, H.M.; Snyder, O.p. Predictive model for the growth of *Clostridium perfringens* at temperatures applicable to cooling of cooked meat. *Food Microbiol.* **1999**, *16*, 335-349. doi: 10.1128/AEM.70.5.2728-2733.2004. [5] Until the stationary phase was reached. Probability of germination was calculated at time 24 h. [6] Mixed strains culture of *C. botulinum* proteolytic. Type A strains ZK3, 62A, VL1, 16,037 and NCTC 3805, and proteolytic type B strains 2345, B6, NCIB 10657, 3262 and 3266. Vegetative bacteria were inoculated. [7] Mixed strains culture of *C. botulinum* proteolytic. Type A strains 17409, 62A, 25763; proteolytic type B strains 7949, 53B, and B-aphis (proteolytic type B), and type C: A028. [8] TOTAL is referred to the percentage of correct (c), fail-safe (fs) and fail-dangerous (fd) predictions classified by the non-adapted and acid-adapted logistic regression models of *C. sporogenes*.

The model's predictions were mostly in agreement with the reported responses, since more than 68% of conditions have been corrected, classified by the models. Additionally, there was a 25% of conditions (8 out of 32) classified as fail-safe by the non-adapted and acid-adapted logistic regression models of *C. sporogenes*; i.e., no germination was predicted while growth was observed for non-adapted *C. sporogenes* strains. The fail-safe predictions obtained could probably be attributed to the variability in microbial behavior against the studied environmental factors, or the physiological differences of the strains used. Inoculation level (10^6 spores/well) used in the present study may influence on the location of the germination boundary which is experimentally found at more limiting conditions when the inoculum size is large [42,43]. However, as many studies have pointed out, it is necessary to employ high inoculation levels to know the extent of a preservation system in a specific food under foreseeable conditions likely to occur in practice [18]. As the inoculation level usually used in these cases may exceed the actual contamination that could occur in food, the models' predictions tend to be fail-safe. If germination is not observed under certain combinations of factors using such inoculum size, the implementation of such formulations in foods remains safe, since the germination probability will be unlikely. However, bias to the fail-safe is more preferred than for the fail-dangerous zone, since the model can provide conservative formulations of pH and NaCl for food operators. Finally, the percentages of fail-dangerous predictions were 6.25% and 3.12% for the non-adapted and acid-adapted logistic regression models of *C. sporogenes* i.e., germination was predicted while no growth was observed.

Overall, it should be remarked that the predictions provided in Table 7 may be taken with caution, since as described above, the comparison with external literature data is subjected to different variability sources that could not be considered by the logistic regression models here developed. Further, most of these studies are referring to growth kinetics of *C. sporogenes* in different matrices, and not to germination probability, so that the comparison with our results can be limited.

The results shown in the present study could have important implications in low-acid, fermented vegetables such as table olives, in which *Clostridium* sp. may not be present during their shelf-life [44]. However, the risk of cross-contamination and its survival increase in some elaborations such as black ripe table olives (Californian style), in the case of an insufficient heat treatment due to their high pH packaging levels (>6.0), or in green table olives with a reduced NaCl content [14]. Also, some specialities, such as Aloreña de Málaga table olives, might be exposed to similar risk when, to prevent the transformation of chlorophylls to pheophytins (loss of freshness, favored in acidic medium [45]), the packaging pH levels are set close to the *Clostridium* sp growth limits. Besides, dressing such as herbs or spices are vehicles of contamination of the *Clostridium* sp. in the final product [46].

Anaerobic fermentation may produce the outgrowth and toxin production of *Clostridium* sp., but also this could happen in microaerophilic environments given the tolerance of this microorganism to low-oxygen concentrations [41]. It is widely reported that a constant monitoring of pH < 4.6 guarantees the inhibition of *Clostridium* sp. in table olives. However, as above mentioned, *C. botulinum* was reported to survive and grow at this pH level in culture media. Thus, the risk of toxin production is not negligible.

Besides, the present study demonstrated that the adaptation of strains to acidic pH produces faster germination at moderate pH (5.0–5.75) and NaCl concentrations (4%–6%) in comparison to non-adapted cells. Although no germination was observed at pH 4.5, it could be plausible that acid-adapted cells could survive at this pH and produce germination at shorter incubation periods than non-adapted ones. Further research is needed to elucidate the metabolic pathways involved in acid adaptation and the subsequent germination of *Clostridium* strains.

In summary, the logistic models developed in this study successfully describe the observed data and quantify the effect of pH, NaCl and incubation time on the probability of germination of *C. sporogenes* in a laboratory medium, with good prediction results in natural green olive brines. This study provides the first guidance to food operators and the table olive industry on the selection of alternative formulations, although further studies should be carried out to validate these results under real table olive fermentation/packaging.

Author Contributions: Conceptualization: F.N.A.-L., and A.V.; Methodology: E.M.-P., V.R.-G., F.N.A.-L., and A.V.; Software: A.V. and M.J.C.; Validation: A.V. and E.O.; Formal Analysis: A.V. and F.P.-R.; Investigation: E.M.-P., V.R.-G., and G.D.P.-I.; Resources: F.N.A.-L. and R.M.G.-G.; Data Curation: E.O. and A.V.; Original Draft Preparation: F.N.A.-L., and A.V.; Writing—Review & Editing: F.N.A.-L., A.G.-F., E.M.-P., A.V., R.M.G.-G.; Visualization: A.V. and F.N.A.-L.; Supervision: A.V. and R.M.G.-G.; Project Administration: F.N.A.-L.; Funding Acquisition: F.N.A.-L. All authors have contributed substantially to the work reported. All authors have read and agreed to the published version of the manuscript.

Funding: The research leading to these results has received funding from the Junta de Andalucía Government through the PrediAlo project (AGR-7755: www.predialo.science.com.es); the TOBE project (RTI2018-100883-B-I00) and FEDER European funds. This work has also been performed by the Research Group AGR-170 (HIBRO) of the Research Andalusian Plan.

Conflicts of Interest: The funders had no role in the design of the study; in the collection, analyses, or interpretation of data; in the writing of the manuscript, or in the decision to publish the results.

References

1. Songer, J.G. Clostridia as Agents of Zoonotic Disease. *Vet. Microbiol.* **2010**. [CrossRef]
2. Hong, Y.; Huang, L.; Byong, W. Mathematical Modeling and Growth Kinetics of *Clostridium sporogenes* in Cooked Beef. *Food Control* **2016**, *60*, 471–477. [CrossRef]
3. Barari, M.; Kalantar, E. An Outbreak of Type A and B Botulism Associated with Traditional Vegetable Pickle in Sanandaj. *Iran. J. Clin. Infect. Dis.* **2010**, *5*, 111–112.
4. Loutfy, M.R.; Austin, J.W.; Blanchfield, B.; Fong, I.W. An Outbreak of Foodborne Botulism in Ontario. *Can. J. Infect. Dis.* **2003**. [CrossRef] [PubMed]
5. Giraudon, I.; Cathcart, S.; Blomqvist, S.; Littleton, A.; Surman-Lee, S.; Mifsud, A.; Anaraki, S.; Fraser, G. Large Outbreak of *Salmonella* Phage Type 1 Infection with High Infection Rate and Severe Illness Associated with Fast Food Premises. *Public Health* **2009**. [CrossRef] [PubMed]
6. Browning, L.M.; Prempeh, H.; Little, C.; Houston, C.; Grant, K.; Cowden, J.M. An Outbreak of Food-Borne Botulism in Scotland, United Kingdom, November 2011. *Eurosurveillance* **2011**. [CrossRef] [PubMed]
7. Juliao, P.C.; Maslanka, S.; Dykes, J.; Gaul, L.; Bagdure, S.; Granzow-Kibiger, L.; Salehi, E.; Zink, D.; Neligan, R.P.; Barton-Behravesh, C.; et al. National Outbreak of Type a Foodborne Botulism Associated with a Widely Distributed Commercially Canned Hot Dog Chili Sauce. *Clin. Infect. Dis.* **2013**. [CrossRef]
8. Outbreak of Botulism Type e Associated with Eating a Beached Whale—Western Alaska, July 2002. *Morb. Mortal. Wkly Rep.* **2003**, *52*, 24.
9. Telzak, E.E.; Bell, E.R.; Kautter, D.A.; Crowell, L.; Budnick, L.D.; Morse, D.L.; Schultz, S. An International Outbreak of Type E Botulism Due to Uneviscerated Fish. *J. Infect. Dis.* **1990**. [CrossRef]
10. Franciosa, G.; Pourshaban, M.; Gianfranceschi, M.; Gattuso, A.; Fenicia, L.; Ferrini, A.M.; Mannoni, V.; De Luca, G.; Aureli, P. *Clostridium botulinum* Spores and Toxin in Mascarpone Cheese and Other Milk Products. *J. Food Prot.* **1999**. [CrossRef]
11. Aureli, P.; Di Cunto, M.; Maffei, A.; De Chiara, G.; Franciosa, G.; Accorinti, L.; Gambardella, A.M.; Greco, D. An Outbreak in Italy of Botulism Associated with a Dessert Made with Mascarpone Cream Cheese. *Eur. J. Epidemiol.* **2000**. [CrossRef] [PubMed]
12. Ghoneim, N.H.; Hamza, D.A. Epidemiological Studies on *Clostridium perfringens* Food Poisoning in Retail Foods. *Rev. Sci. Tech.* **2017**. [CrossRef] [PubMed]
13. Trotz-Williams, L.A.; Mercer, N.J.; Walters, J.M.; Maki, A.M.; Johnson, R.P. Pork Implicated in a Shiga Toxin-Producing *Escherichia coli* O157:H7 Outbreak in Ontario, Canada. *Can. J. Public Heal.* **2012**, *103*, e322–e326. [CrossRef]
14. Medina-Pradas, E.; Arroyo-López, F.N. Presence of Toxic Microbial Metabolites in Table Olives. *Front. Microbiol.* **2015**. [CrossRef] [PubMed]
15. European Food Safety Authority and European Centre for Disease Prevention and Control. The European Union Summary Report on Trends and Sources of Zoonoses, Zoonotic Agents and Food-Borne Outbreaks in 2017. *EFSA J.* **2018**. [CrossRef]

16. Dong, Q.; Tu, K.; Guo, L.; Li, H.; Zhao, Y. Response Surface Model for Prediction of Growth Parameters from Spores of *Clostridium sporogenes* under Different Experimental Conditions. *Food Microbiol.* **2007**. [CrossRef]
17. Khanipour, E.; Flint, S.H.; McCarthy, O.J.; Golding, M.; Palmer, J.; Tamplin, M. Evaluation of the Effects of Sodium Chloride, Potassium Sorbate, Nisin and Lysozyme on the Probability of Growth of *Clostridium sporogenes*. *Int. J. Food Sci. Technol.* **2014**. [CrossRef]
18. Khanipour, E.; Flint, S.H.; Mccarthy, O.J.; Golding, M.; Palmer, J.; Ratkowsky, D.A.; Ross, T.; Tamplin, M. Modelling the Combined Effects of Salt, Sorbic Acid and Nisin on the Probability of Growth of *Clostridium sporogenes* in a Controlled Environment (Nutrient Broth). *Food Control* **2016**, *62*, 32–43. [CrossRef]
19. Khanipour, E.; Flint, S.H.; Mccarthy, O.J.; Palmer, J.; Golding, M.; Ratkowsky, D.A.; Ross, T.; Tamplin, M. Modelling the Combined Effect of Salt, Sorbic Acid and Nisin on the Probability of Growth of *Clostridium sporogenes* in High Moisture Processed Cheese Analogue. *Int. Dairy J.* **2016**, *57*, 62–71. [CrossRef]
20. Huang, L.; Li, C.; Hwang, C. International Journal of Food Microbiology Growth/No Growth Boundary of *Clostridium perfringens* from Spores in Cooked Meat: A Logistic Analysis. *Int. J. Food Microbiol.* **2018**, *266*, 257–266. [CrossRef]
21. Lund, B.M.; Graham, A.F.; Franklin, J.G. The Effect of Acid pH on the Probability of Growth of Proteolytic Strains of *Clostridium Botulinum*. *Int. J. Food Microbial.* **1987**, *4*, 215–226. [CrossRef]
22. Montville, T.J. Interaction of pH and NaCl on Culture Density of *Clostridium botulinum* 62A. *Appl. Environ. Microbiol.* **1983**, *46*, 961–963. [CrossRef] [PubMed]
23. Montville, T.J. Quantitation of pH- and Salt-Tolerant Subpopulations from *Clostridium botulinum*. *Appl. Environ. Microbiol.* **1984**, *47*, 28–30. [CrossRef] [PubMed]
24. Wong, D.M.; Young-Perkins, K.E.; Merson, R.L. Factors Influencing *Clostridium botulinum* Spore Germination, Outgrowth, and Toxin Formation in Acidified Media. *Appl. Environ. Microbiol.* **1988**, *54*, 1446–1450. [CrossRef] [PubMed]
25. Jobin, M.P.; Clavel, T.; Carlin, F.; Schmitt, P. Acid Tolerance Response Is Low-PH and Late-Stationary Growth Phase Inducible in *Bacillus cereus* TZ415. *Int. J. Food Microbiol.* **2002**. [CrossRef]
26. Derman, Y.; Söderholm, H.; Lindström, M.; Korkeala, H. Role of Csp Genes in NaCl, PH, and Ethanol Stress Response and Motility in *Clostridium botulinum* ATCC 3502. *Food Microbiol.* **2015**. [CrossRef]
27. Pérez-Rodríguez, F.; Valero, A. *Predictive Microbiology in Foods*; SpringerBriefs in Food, Health, and Nutrition: New York, NY, USA, 2013. [CrossRef]
28. Bollerslev, A.M.; Nauta, M.; Hansen, T.B.; Aabo, S. A Risk Modelling Approach for Setting Microbiological Limits Using Enterococci as Indicator for Growth Potential of *Salmonella* in Pork. *Int. J. Food Microbiol.* **2017**. [CrossRef]
29. Ikawa, J.Y.; Genigeorgis, C. Probability of Growth and Toxin Production by Nonproteolytic *Clostridium botulinum* in Rockfish Fillets Stored under Modified Atmospheres. *Int. J. Food Microbiol.* **1987**, *4*, 167–181. [CrossRef]
30. Zhao, L.; Montville, T.J.; Schaffner, D.W. Time-to-Detection, Percent-Growth-Positive and Maximum Growth Rate Models for *Clostridium botulinum* 56A at Multiple Temperatures. *Int. J. Food Microbiol.* **2002**, *77*, 187–197. [CrossRef]
31. Ghabraie, M.; Vu, K.D.; Tnani, S.; Lacroix, M. Antibacterial Effects of 16 Formulations and Irradiation against *Clostridium sporogenes* in a Sausage Model. *Food Control* **2016**, *63*, 21–27. [CrossRef]
32. Beerens, H. Bifidobacteria as Indicators of Faecal Contamination in Meat and Meat Products: Detection, Determination of Origin and Comparison with *Escherichia coli*. *Int. J. Food Microbiol.* **1998**. [CrossRef]
33. Agresti, A. Building and Applying Logistic Regression Models. In *An Introduction to Categorical Data Analysis*; John Wiley & Sons, Inc.: Hoboken, NJ, USA, 2007. [CrossRef]
34. Daelman, J.; Vermeulen, A.; Willemyns, T.; Ongenaert, R.; Jacxsens, L.; Uyttendaele, M.; Devlieghere, F. Growth/No Growth Models for Heat-Treated Psychrotrophic *Bacillus cereus* Spores under Cold Storage. *Int. J. Food Microbiol.* **2013**. [CrossRef] [PubMed]
35. Yang, W.; Ponce, A. Production and Characterization of Pure *Clostridium* Spore Suspensions. *J. Appl. Microbiol.* **2009**, *106*, 27–33. [CrossRef] [PubMed]
36. Taylor, R.H.; Dunn, M.L.; Ogden, L.V.; Jefferies, L.K.; Eggett, D.L.; Steele, F.M. Conditions Associated with *Clostridium Sporogenes* Growth as a Surrogate for *Clostridium botulinum* in Nonthermally Processed Canned Butter. *J. Dairy Sci.* **2013**. [CrossRef]

37. Crosthwait, C.D. Adaptation to Growth at Low pH by *Clostridium Sporogenes*. Master's Thesis, University of Tennessee, Knoxville, TN, USA, 1979.
38. Whiting, R.C.; Call, J.E. Time of Growth Model for Proteolytic *Clostridium botulinum*. *Food Microbiol.* **1993**. [CrossRef]
39. Stewart, C.M.; Cole, M.B.; Legan, J.D.; Slade, L.; Vandeven, M.H.; Schaffner, D.W. Modeling the Growth Boundary of *Staphylococcus aureus* for Risk Assessment Purposes. *J. Food Prot.* **2001**. [CrossRef]
40. Graham, A.F.; Mason, D.R.; Peck, M.W. Predictive Model of the Effect of Temperature, pH and Sodium Chloride on Growth from Spores of Non-Proteolytic *Clostridium botulinum*. *Int. J. Food Microbiol.* **1996**. [CrossRef]
41. Couvert, O.; Divanac'h, M.L.; Lochardet, A.; Thuault, D.; Huchet, V. Modelling the Effect of Oxygen Concentration on Bacterial Growth Rates. *Food Microbiol.* **2019**. [CrossRef]
42. Skandamis, P.N.; Stopforth, J.D.; Kendall, P.; Belk, K.E.; Scanga, J.; Smith, G.C.; Sofos, J.N. Modeling the Effect of Inoculum Size and Acid Adaptation on Growth/No Growth Interface of *Escherichia coli* O157:H7. *Int. J. Food Microbiol.* **2007**, *120*, 237–249. [CrossRef]
43. Vermeulen, A.; Gysemans, K.P.M.; Bernaerts, K.; Geeraerd, H.; Debevere, J.; Devlieghere, F.; Van Impe, J.F. Modelling the Influence of the Inoculation Level on the Growth/No Growth Interface of *Listeria monocytogenes* as a Function of pH, a_w and Acetic Acid. *Int. J. Food Microbiol.* **2009**, *135*, 83–89. [CrossRef]
44. International Olive Oil Council. *Trade Standard Applying to Table Olives. RES-2/91-IV/04*; International Olive Oil Council: Madrid, Spain, 2004.
45. Gallardo-Guerrero, L.; Gandul-Rojas, B.; Moreno-Baquero, J.M.; López-López, A.; Bautista-Gallego, J.; Garrido-Fernández, A. Pigment, Physicochemical, and Microbiological Changes Related to the Freshness of Cracked Table Olives. *J. Agric. Food Chem.* **2013**. [CrossRef] [PubMed]
46. Ruiz Bellido, M.Á.; Valero, A.; Pradas, E.M.; Gil, V.R.; Rodríguez-Gómez, F.; Posada-Izquierdo, G.D.; Rincón, F.; Possas, A.; García-Gimeno, R.M.; Arroyo-López, F.N. A Probabilistic Decision-Making Scoring System for Quality and Safety Management in Aloreña de Málaga Table Olive Processing. *Front. Microbiol.* **2017**, 8. [CrossRef] [PubMed]

foods

MDPI

Article

A Preliminary Report for the Design of MoS (Micro-Olive-Spreadsheet), a User-Friendly Spreadsheet for the Evaluation of the Microbiological Quality of Spanish-Style Bella di Cerignola Olives from Apulia (Southern Italy)

Antonio Bevilacqua, Barbara Speranza, Daniela Campaniello, Milena Sinigaglia and Maria Rosaria Corbo *

Department of the Science of Agriculture, Food and Environment, University of Foggia, 71122 Foggia, Italy; antonio.bevilacqua@unifg.it (A.B.); barbara.speranza@unifg.it (B.S.); daniela.campaniello@unifg.it (D.C.); milena.sinigaglia@unifg.it (M.S.)

* Correspondence: mariarosaria.corbo@unifg.it; Tel.: +39-0881-589232

Received: 25 May 2020; Accepted: 25 June 2020; Published: 29 June 2020

Abstract: A user friendly spreadsheet (Excel interface), designated MoS (Micro-Olive-Spreadsheet), is proposed in this paper as a tool to point out spoiling phenomena in Bella di Cerignola olive brines. The spreadsheet was designed as a protected Excel worksheet, where users input values for the microbiological criteria and pH of brines, and the output is a visual code, much like a traffic light: three red cells indicate a spoiling event, while two red cells indicate the possibility of a spoiling event. The input values are: (a) Total Aerobic Count (TAC); (b) Lactic Acid Bacteria (LAB); (c) yeasts; (d) staphylococci; (e) pH. TAC, LAB, yeasts, and pH are the input values for the first section (quality), while staphylococci count is the input for the second section (technological history). The worksheet can be modified by adding other indices or by setting different breakpoints; however, it is a simple tool for an effective application of hazard analysis and predictive microbiology in table olive production.

Keywords: table olives; Bella di Cerignola; brines; microbiological quality; user-friendly spreadsheet; producers

1. Introduction

The olive tree is an iconic species in Mediterranean cultural history and diet. Its multiple uses in the food industry (olive oil and table olives) and its omnipresence in many traditional agro-systems have made this species an economic pillar and cornerstone of Mediterranean agriculture [1]. Its role as a symbol of Italy, in particular of the Region of Apulia, has increased due to the emergence of the *Xylella* outbreak. Table olives represent one of the most popular fermented foods in the Mediterranean basin and in Italy, but their production is increasing worldwide, as suggested by the International Olive Council (IOC) statistics for the period 2013-14/2017-18, with a duplication of amounts produced compared to the beginning of the millennium [2]. European Union (EU) covers 31% of world production, with Spain, Italy, and Greece being the major producing countries (97% of EU production) [2]. The olive fruit cannot be consumed directly because of the presence of oleuropein. The bitterness can be removed by alkaline treatment, or by brining/salting, fermentation, and acidification [3]. The trade standard applying to table olives describes the type of preparation of table olives that are treated, natural olives, dehydrated and/or shriveled olives, and olives darkened by oxidation; however, some traditional processes are still applied, such as the Castelvetrano system [3].

In Italy, table olive production is mainly located in Southern regions (Apulia, Sicily, etc.); however, fermentation still relies upon natural microbiota [4]. The three main techniques for table olive production used in Italy concern 82% green olives, 16% black olives and 2% processed at the cherry ripened stage [5].

The microbial ecosystem is complex and can be affected by several intrinsic (pH, water activity, diffusion of nutrients from the drupe, and concentration of phenols) or extrinsic factors (temperature, oxygen availability, and salt) [6]; if these variables are not controlled, a microbial spoilage could occur. In traditional fermentations, the modulation of salt and pH is the only way to counteract spoilage. Lactic acid bacteria (LAB) and yeasts represent the microbiota normally involved in fermentation [4], but members of *Staphylococcus* and *Pseudomonas* are detected at the beginning and throughout the process [7–9], along with some pathogens, such as *Clostridium botulinum*, *Listeria monocytogenes*, and *Staphylococcus aureus* [10–13]. Enterobacteriaceae can be also found at the beginning of fermentation and are quickly inhibited by the pH decrease by LAB [14]. LAB species encountered during table olive fermentation are *Lactiplantibacillus plantarum*, *L. pentosus*, and to a lesser extent, *L. paraplantarum* [5]. They are responsible for the rapid and safe acidification of brines [5]. Besides, different yeast species are recovered, such as *Wickerhamomyces anomalus*, *Pichia membranifaciens*, *Saccharomyces cerevisiae*, *Debaryomyces hansenii*, and *Candida boidinii* [5].

In Italy the most important varieties for table olive production are, among others, Bella di Cerignola [15], Nocellara Etnea [4], Tonda di Cagliari [16], Giarraffa [17], Termite di Bitetto, Cellina di Nardò [18], and Leccino [19], treated by either Spanish or natural styles. Bella di Cerignola (formerly Bella della Daunia, variety Bella di Cerignola, Protected Denomination of Origin-PDO) is one of the most important variety of olives in the region of Apulia. It is processed through Spanish and natural styles between October and December [8]. The fermentation of table olives still relies on natural processes led by the indigenous microbiota from the raw materials (olives, salt, water) or that are acquired during processing at factory facilities (fermenters, tanks, pipelines, pumps) and fermenter yards [20]. However, a part of the microbial diversity associated with this fermentation has been "domesticated" by the continuous replication of peculiar processing conditions and know-how. This might represent a significant contribution to "terroir" aspects [20].

The production of table olives from Bella di Cerignola through Spanish style relies upon a three-step protocol, similar to that reported by Mastralexi et al. [2] for Greek PDO table olives "Prasines Elies Chalkidikis", with some differences in the duration of each step, such as the amount of salt and lye. The three steps are as follows: (a) sorting and size grading, debittering and neutralization; (b) brining (6–10% NaCl) and fermentation (ca. 2 months); (c) storage in plastic tanks filled in brine until packaging and thermal treatment. Traditionally, fermentation is divided into four steps or phases [21]; in the first phase, Gram negative bacteria prevail, and pH decreases from 9.0 to around 6.0. This phase lasts until the growth of LAB (normally 48–72 h). The second and the third steps are characterized by the growth of LAB, along with the strong acidification of the brine. At the end of this third step, there is a potential fourth step, characterized by an increase in pH and volatile acidity, and a decrease in lactic acid [21]. The duration of this last step, usually referred to as the post-fermentation stage, is quite variable (from a month to a year), and depends on demand and market prices. If pH and NaCl are not strictly controlled, a microbial spoilage can occur due to a variety of microorganisms (*Aerobacter*, bacilli, propionibacteria, oxidative yeasts, moulds, etc.) [3,5,21]. Although new trends have been exploited in table olive production (the use of probiotics, the combinations of yeasts and LAB as starter cultures, low-salt fermentations, the study of a biogeography of olives, models to optimize the amounts of preservatives, the use of starter cultures able to degrade oleuropein) [3,22], to the best of our knowledge there are no user-friendly tools used in the post-fermentation stage in order to control the microbiological quality of batches and to help producers to perform corrective strategies.

Therefore, this paper represents a first approach to design a user-friendly worksheet that can be used by producers of Bella di Cerignola olives as a tool to focus on the microbiological stability of olives during their storage in tanks before pasteurization. The specific aims of this research were: (a) to link the idea of spoiled samples from olive producers to olive microbiology; (b) to design a simple quality management tool requiring few data; (c) to validate the tool.

2. Materials and Methods

2.1. Samples

Olives (cv. Bella di Cerignola) and brines from Spanish style processing were collected after the fermentation and/or olive storage in tanks during winter and spring from four different factories located in Cerignola (Foggia County, Southern Italy). Eighty-nine different samples were analyzed to assess microbiota, pH, NaCl amounts (brines) and sensory scores (olives); of which, 77 samples were used to design the user-friendly worksheet, and 12 for the validation.

2.2. Microbiological Analyses

Olive brines were serially diluted in saline solution (0.9% NaCl) and plated on the following media: (a) Plate Count Agar (PCA), incubated at 30 °C for 24–48 h for total aerobic count (TAC); (b) PCA, incubated at 30 °C for 24–48 h, after a heat-shock of dilutions at 80 °C for 10 min (aerobic spore-forming bacteria); (c) MRS agar, supplemented with 0.17 g/L cycloheximide (Sigma-Aldrich, Milan, Italy), incubated at 30 °C under anaerobic conditions for 48–72 h for lactic acid bacteria (LAB); (d) Pseudomonas Agar Base, added to CFC selective supplement (containing the cetrimide), incubated at 25 °C for 48–72 h for pseudomonads; (e) Baird Parker Agar Base, added with Egg Yolk Tellurite Emulsion (37 °C for 24–48 h) and Mannitol Salt Agar (37 °C for 24–48 h) for staphylococci; (f) Violet Red Bile Glucose Agar (VRBGA), incubated at 37 °C for 18–24 h for enterobacteria; (g) SPS Agar, incubated under strict anaerobic conditions at 37 °C for 24 h for clostridia; (h) Sabouraud Dextrose Agar, supplemented with 0.1 g/L chloramphenicol (C. Erba, Milan), incubated at 25 °C for 2–4 days for yeasts. All media and supplements were purchased from Oxoid (Basingstoke, UK) [8,23].

2.3. pH and NaCl Amount

The pH of brines were measured using a pH-meter Crison (Crison Instruments, Barcelona, Spain), whereas salt amounts were evaluated by means of refractometer Sper Scientific model 106 ATC (Scottsdale, AZ, USA).

2.4. Sensory Score

The producers were asked to analyze olives and brines; each producer analyzed his own samples. Before sensory evaluation, a meeting with all producers was done in order to define the sensory properties of Bella di Cerignola olives and what they meant by "good quality". The focus of this meeting was to document the "Sensory Analysis of Table Olives" [24] and the definition of negative defects (abnormal fermentation, musty, rancid, cooking effect, soapy, metallic, earthy, winery/vinegar), along with the attributes of olives (acid, bitterness, salty taste, hardness, fibrousness, crunchiness of texture).

The output of this consensus panel was to define "spoiled" samples as those with at least one of the negative attributes or if the other attributes showed negative changes. Producers were asked to score (0 vs. 1) odor, color, taste, and finally to give a score of 1 (spoiled) or 0 (non-spoiled); in case of doubts the output could be "acceptable with some problems (spoilage is beginning)". Scores were confirmed by five researchers of the laboratory of Predictive Microbiology, usually consuming green table olives.

2.5. Data Analysis

Microbiological count, pH, and salt analyses were repeated twice for each batch, and all results were analyzed through *t*-test ($p < 0.05$) using the software, Statistica for Windows (Statsoft, Tulsa, OK, USA).

2.6. Spreadsheet Design

The spreadsheet (MoS, Micro-Olive-Spreadsheet) was designed as a protected Excel spreadsheet, where users can input values for microbiological criteria and the pH of brines (Spreadsheet S1, Supplementary Materials). The input values are: (a) Total Aerobic Count (TAC); (b) Lactic Acid Bacteria (LAB); (c) yeasts; (d) staphylococci; (e) pH. TAC, LAB, yeasts, and pH are the input values for the first section (quality), while staphylococci count is the input for second section (technological history). Microbial concentrations refer to the values of the brines, as log CFU/mL (decimal logarithm).

The spreadsheet is designed based on four different cells with "If" functions. The possible outputs of "If" are "1" (true or yes) or "0" (false or no). The functions are:

Quality

- LAB < 1% TAC. The function was evaluated through the exponential values of cell count, rather than with a logarithm;
- Yeast > 5;
- pH > 4.5.

Technological process

- Staphylococci > 4.

Salt

- salt<8%.

In the cells containing the output of the "If" function, there is a conditional formatting linked to the output: if the results of function are true (or 1), the cells becomes red, while if the results are false (0) the cells become green.

3. Results

3.1. Data

Seventy-seven samples were analyzed in the first step; 12 were randomly selected and used for the validation. First, the samples were analyzed by producers and researchers to cluster them in two groups: spoiled and non-spoiled. In case of doubts (different clustering between producers and researchers), the samples were grouped as suggested by the producers. Samples were grouped as follows: 32 spoiled and 45 non-spoiled. This grouping was used for statistics to cluster microbiological data from brines, as shown in Figure 1.

Figure 1A reports the TAC count in the brines, with mean values of 7.25 ± 0.21 log CFU/mL (mean ± standard error) in the spoiled samples and 6.03 ± 0.27 log CFU/mL in the non-spoiled samples. The difference was significant (*t*-test, $p < 0.05$); LAB level was 5.79 ± 0.19 log CFU/mL in the non-spoiled samples and 4.30 ± 0.28 log CFU/mL in the spoiled samples, as shown in Figure 1B.

In order to see if it is possible to use TAC and LAB to describe spoiled and non-spoiled samples, the ratio LAB/TAC was analyzed as an index of the qualitative composition of bacterial microbiota (aerobic vs. lactic acid bacteria), as shown in Figure 1C. The ratio was evaluated by using the exponential values of cell counts to avoiding a possible "masking" exerted by logarithmic values. In spoiled samples this index was 0.26% (0.13–0.39%, 95%-confidence interval) thus suggesting that, in these batches, bacterial microbiota were mainly composed of aerobic microorganisms—in fact, their count was at least 99% higher than LAB. On the other hand, in the non-spoiled samples the ratio of LAB/TAC was 184%, with a confidence interval of ±181%.

Yeasts were 5.87 ± 0.19 log CFU/mL in the spoiled samples and 3.68 ± 0.28 log CFU/mL in the non-spoiled samples, as shown in Figure 1D. Enterobacteria were generally below the detection limit and they were found in only five samples with concentrations ranging from 1.21 to 2.09 log

CFU/mL. Clostridia, bacilli, and pseudomonads were found only in 3–6 samples (ca. 1 log CFU/mL. Staphylococci were found in most samples and were at 2.09 ± 0.45 log CFU/mL and 2.42 ± 0.38 log CFU/mL in the spoiled and non-spoiled samples, respectively; the difference was not significant (*t*-test, $p > 0.05$), as shown in Figure 1E.

NaCl content varied from 6 to 8.5%; the pH was 4.2 ± 0.3 in the non-spoiled samples and 4.6 ± 0.6 in the spoiled samples (*t*-test, $p > 0.05$).

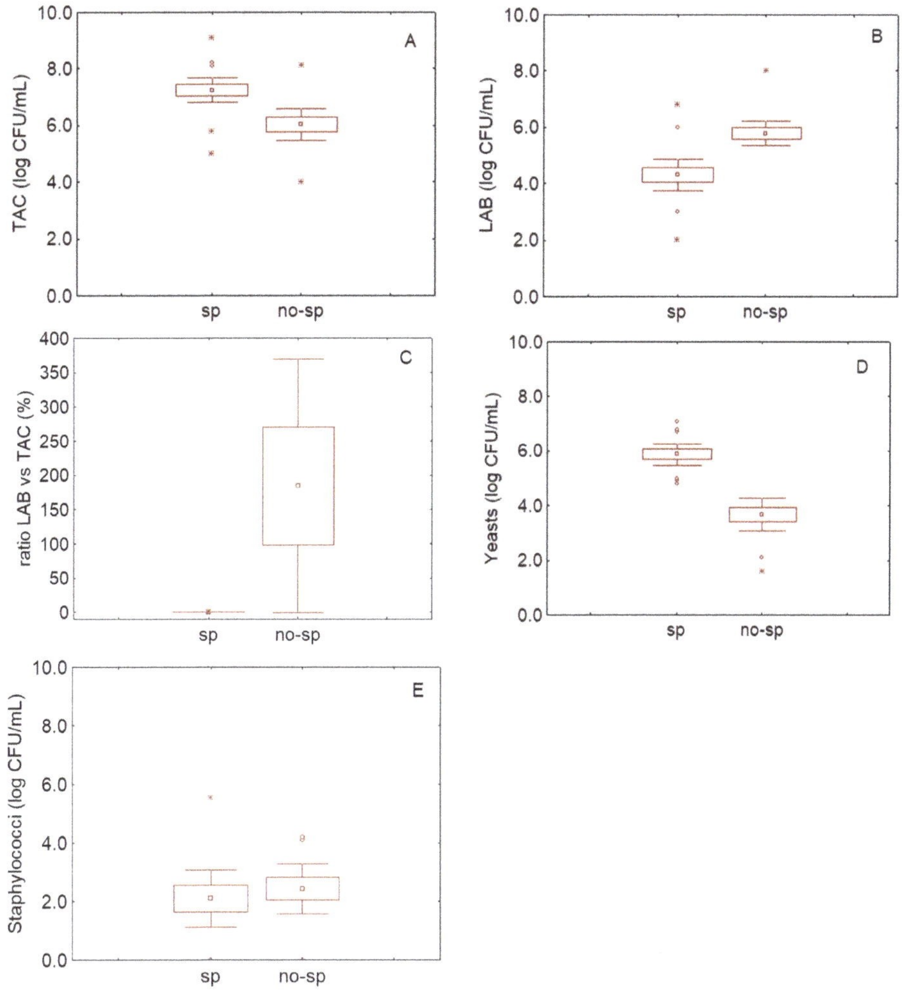

Figure 1. Total Aerobic Count (TAC) (**A**), Lactic Acid Bacteria (LAB) (**B**), ratio LAB vs. TAC (**C**), yeasts (**D**), and staphylococci in brines (**E**). sp, spoiled samples, no-sp, non-spoiled samples. □ Mean; □, Mean ± SE; ○ Outliers; * extremes; bars denote 95% confidence interval.

3.2. MoS

As a result of the microbiological analyses of the first phase, a user-friendly spreadsheet was prepared on Microsoft Excel—the spreadsheet was called MoS. The main idea was a spreadsheet where a user could input his/her own values on the quality and technological parameters in brines (TAC, LAB, yeasts, staphylococci, and pH) and preview the classification of the batch's quality.

The spreadsheet is organized in three parts: (a) microbiological quality; (b) technological history; (c) salt, as shown in Figure 2. Users can only add values in the yellow cells reading, "Please input your values here". The other sections of spreadsheet are protected.

The main outputs of the spreadsheet are a preliminary evaluation of microbiological quality and a focus on the technological history of a batch to point out an incorrect handling of olives.

Microbiological quality is based upon three criteria: the difference between TAC and LAB, yeast count, and pH. Each criterion is reported as a question, as shown in Figure 3:

1. Are LAB a small proportion of TAC in brines? This criterion was written as follows: LAB/TAC < 1%; both TAC and LAB in the equation were used as exponential values. However, user does not convert his/her values, because there is a function set in the protected cell of the "If" function, as shown in Table 1.
2. Are yeasts higher than the break-point? The threshold for yeasts in brines was set to 5 log CFU/mL.
3. The last criterion was on pH; although the results of the first phase did not show a clear difference between the spoiled and non-spoiled samples, a criterion on pH was added because of its role in the beginning of microbiological spoilage in the post-fermentation phase.

Table 1. Decision criteria for MoS (Micro-Olive-Spreadsheet) and respective equations and outcomes.

Decision Criteria (Questions)	Equations	Yes	No
	Quality		
Are LAB a small proportion of bacterial microbiota?	LAB < 1% TAC	Not acceptable	Acceptable
Is yeast concentration > breakpoint?	Yeast > 5	Not acceptable	Acceptable
Is pH > breakpoint?	pH > 4.5	Not acceptable	Acceptable
	Technological Process		
Is handling incorrect?	Staph > 4	Incorrect handling	-
Is salt too low?	Salt < 8%	Possible corrective measures	-

TAC, total aerobic count; LAB, lactic acid bacteria; staph, staphylococci. Bacterial concentrations of brines are reported as log CFU/mL.

After entering the values of TAC, LAB, yeasts, and the pH of brine, MoS answers the three questions, with one of two codes: 0 for "no", and 1 for "yes". To help users understand the impact of the answers, a visual code was added, like a traffic-light. If the answer is yes (hazard), the cell assumes a the color red, while for 0 (no risk), the cell becomes green.

Legends show the key for the correct evaluation of the results:

1. Three red cells: There is probably microbiological spoilage.
2. Two red cells: A correction strategy is required, because a spoilage could start or have started.
3. One red cell: No spoilage and no action required; however, advice was added for pH. If pH is >breakpoint, reduce it to the break-point or, better, to 4.3.

The second section of the worksheet, as shown in Figure 4, is a focus on the technological history of olives; staphylococci are generally indicator microorganisms, mainly for use in GMP (Good Manufacturing Practices).

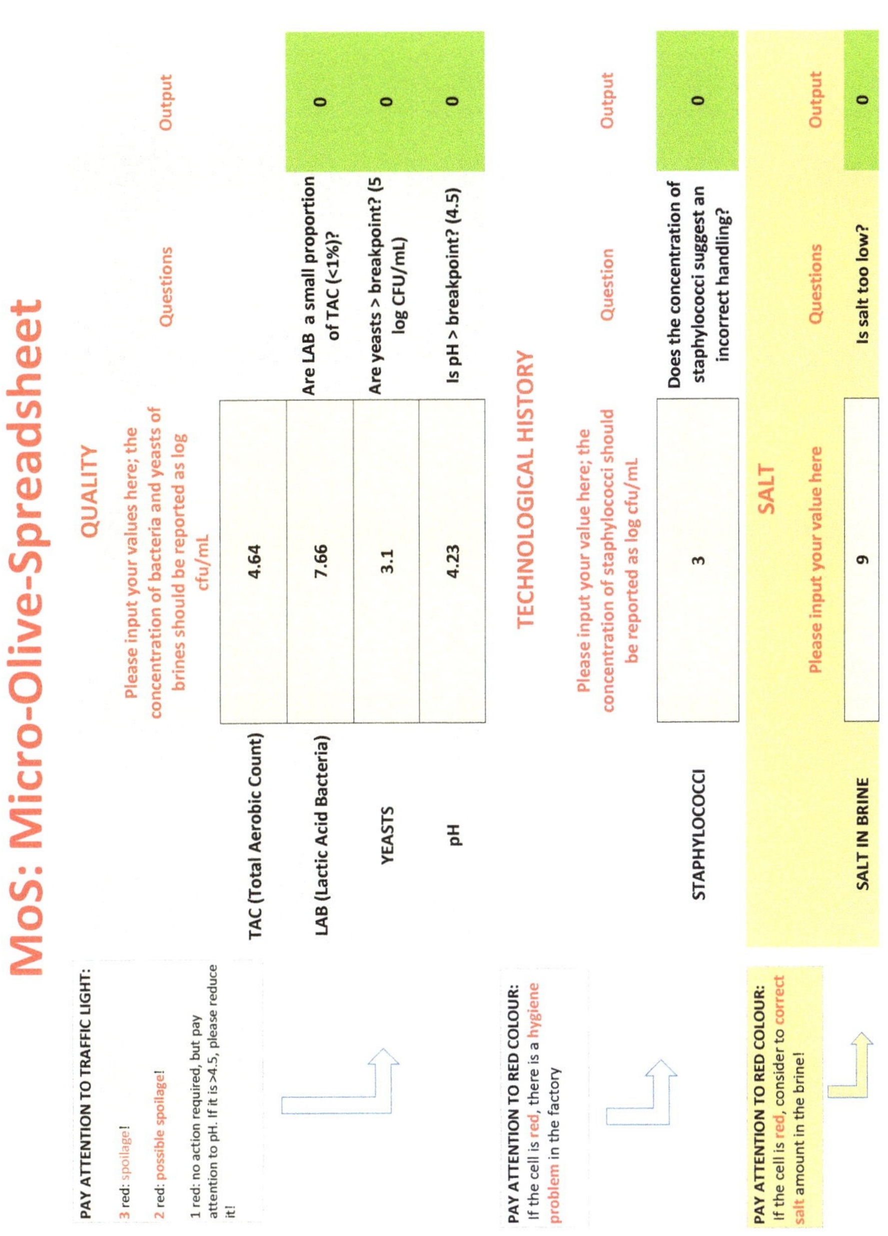

Figure 2. Input and Output of MoS.

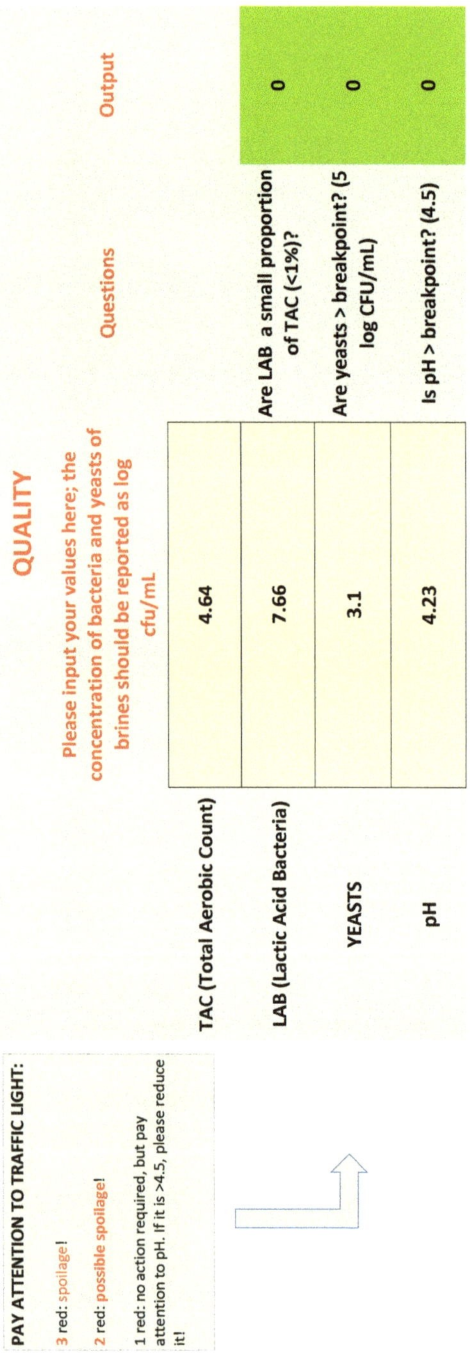

Figure 3. Section of spreadsheet for quality parameters.

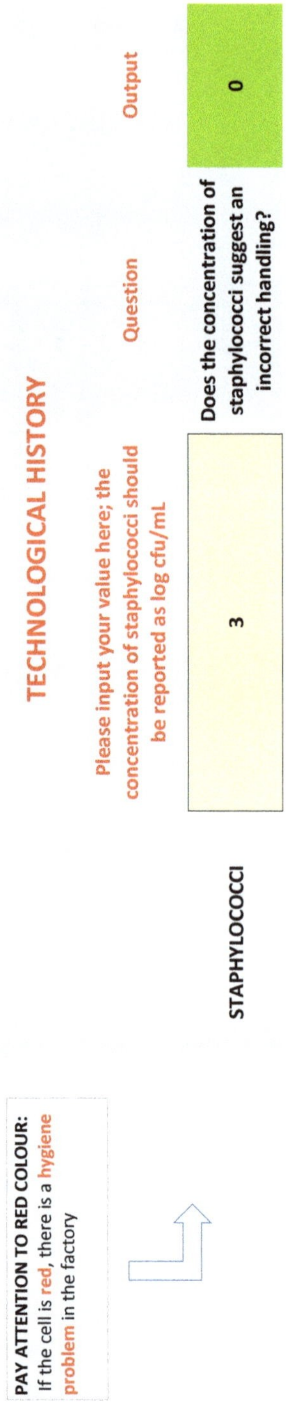

Figure 4. Section of spreadsheet on technological history.

The results of the first phase showed that staphylococci could be a significant part of the bacterial microbiota of brines. For this section, an arbitrary threshold was set, because the advice is that a high number of these microorganisms could be the result of incorrect handling; the breakpoint was set to 4 log CFU/mL. Staphylococci do not play a role in the definition of the microbiological quality, according to the criteria reported above. They are hygiene indicators and are normally transferred to olives by food handlers. The break point (4 log CFU/mL) could be modified to fit HACCP plans and regional regulations; however, the meaning for this last criterion is the following: if staphylococci count is higher than the threshold (4 log CFU/mL, as in this paper, or lower if required by other regulations) corrective measures are required because there is a serious hygiene problem in the factory.

The last section of MoS is on salt, as shown in Figure 5. The analysis of the results of the first part did not show a significant difference between spoiled and non-spoiled samples. However, a criterion on salt was added as advice for producers to perform corrective measures when NaCl <8%.

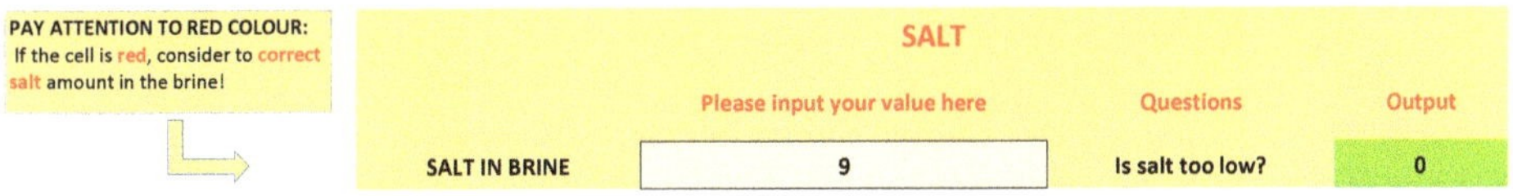

Figure 5. Section of spreadsheet for salt concentration in brine.

3.3. Validation

Twelve samples, not used for the spreadsheet design (Sections 3.1 and 3.2), were used for a preliminary validation. They were randomly selected after their collection in the factories (three per factory) and analyzed as reported above for the sensory scores and microbiology. The results are in Table 2. Figure 6 shows the output of the spreadsheet for two samples.

Five samples (1, 6, 8, 9, and 12) were recorded as acceptable or non-spoiled by the panelists (both producers and researchers). The combination of parameters on the spreadsheet (ratio of LAB/TAC, yeast count, and pH) gave the same result. Samples 2, 3, 10, and 11 were judged as not-acceptable or spoiled by the panelists. Generally, the spreadsheet returned the same output, except for sample 3, because two parameters (yeast count and pH) were in the hazard zone, while the ratio of LAB/TAC was lower than 1%. Finally, samples 4, 5, and 7 were recorded as doubtful samples by the panelists and were included in the attention class by the spreadsheet, because at least two parameters were out of range.

For all of the samples NaCl was recorded as acceptable by the producers (8–10%), enterobacteria and pseudomonads were always below the detection limit, and bacilli and clostridia were found only in two samples, but their level was very low (1–1.2 log CFU/mL).

Staphylococci were found in almost all samples, but their level was <2 log CFU/mL in brines, except for the samples 3, 5, 10, and 11, showing a count of staphylococci in the range 3.2–4.1 log CFU/mL; coagulase positive staphylococci were found in samples 3 and 4, both within range, according to the spreadsheet.

Table 2. Decision on the samples used for the validation.

Sample	Decision by Panel	Comments	TAC	LAB	Yeasts	pH	Spreadsheet
1	Acceptable	Typical odor of Spanish-style olives	2.01	6.65	4.18	4.05	Acceptable
2	Not acceptable	Off-odors	5.99	3.73	6.01	5.15	Not acceptable
3	Not acceptable	Off-odors	6.14	5.93	6.14	5.27	Attention
4	Acceptable, but the sample has some problems	Films on the surface	5.45	5.08	7.26	5.35	Attention
5	Acceptable, but the sample has some problems	Strong odor	5.53	2.23	3.46	4.67	Attention
6	Acceptable	-	4.64	7.66	3.10	4.23	Acceptable
7	Acceptable, but the sample has some problems	Fruity odor is too strong	4.82	7.63	5.87	4.85	Attention
8	Acceptable	-	4.56	6.12	4.53	4.43	Acceptable
9	Acceptable	-	4.31	5.99	3.56	4.36	Acceptable
10	Not acceptable	Films on the surface	7.80	5.79	6.01	4.67	Not acceptable
11	Not acceptable	Off-odors	5.99	2.34	5.31	4.89	Not acceptable
12	Acceptable	-	5.53	6.48	2.83	4.43	Acceptable

TAC, total aerobic count; LAB, lactic acid bacteria. The unit of counts is log CFU/mL.

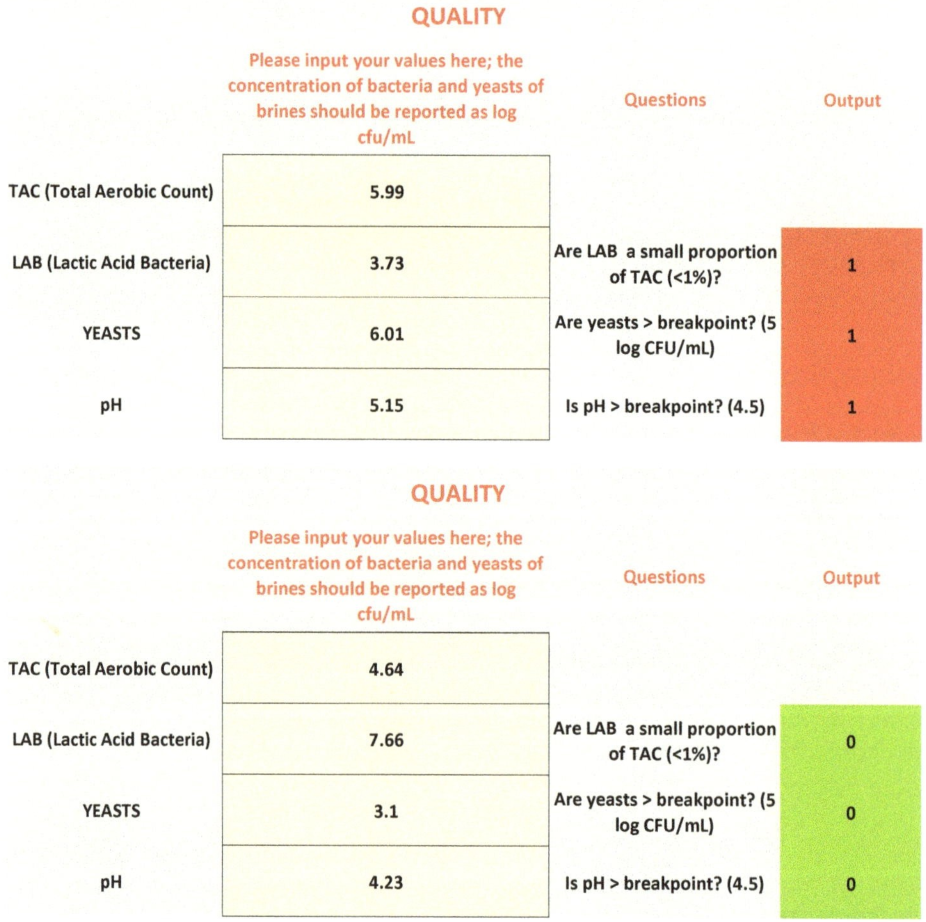

Figure 6. Traffic light for spoiled (sample 2) and acceptable samples (sample 6).

4. Discussion

The post-fermentation stage is a critical step for the production of table olives, as they are usually stored in tanks filled with brine for several months and, when the temperature increases (February and March), spoilage can occur.

Predictive microbiology is experiencing an increase in interest by olive producers, because they need tools to predict olive shelf life and/or to act with corrective measures when problems occur. An interesting example of a predictive tool is the Decision-Making System for Safety and Quality Management in Aloreña de Málaga [25]. Other applications of predictive microbiology rely upon the use of the theory of Design of Experiments or neural networks (among others [13,26]). However, during the project BiotecA we met several producers of Table Olives of Apulia and they shared with us the idea of a user-friendly spreadsheet where they could enter the values of brine to understand the microbiological scenario.

The idea is to develop this project into a tool similar to Risk Ranger [27], which is an educational and research tool developed by Australia Food Safety. Users have to answer some questions related to the technology, preparation, cooking, and storage of a food—the output is a risk rank (from 0 to 100). The benefit of this kind of tool is that it works in a spreadsheet with an Excel interface (user-friendly) and the output is understandable for non-expert users.

The first part of this research was aimed at understanding if there was a connection between the idea of spoilage by producers and the microbiological profile of olives. The preliminary results showed a link with some indices, such as yeasts, pH, and the ratio of LAB/TAC. Yeasts exert a dual role in table olives—they can cause spoilage due to the production of CO_2, bad odors and flavors, the clouding of brines or the softening of fruits [5,28,29]. In the post-fermentation stage, oxidative yeasts, such as *Pichia anomala* and *P. membranifaciens*, could prevail. In addition, film-forming yeasts (*Debaryomyces*, *Candida*, *Pichia*, and *Endomycopsis*) are often associated with pickled products and vegetable brines [30], representing the cause of olive defects and consequent product losses. Salt is the main factor able to control the yeasts and LAB during olive fermentation [31]. Yeasts dominate fermentations at salt levels > 10% NaCl; however, this process leads to a final product with a milder taste and less self-preservation characteristics [31], while salt reduction to 6–8% enables a mixed fermentation by lactic acid bacteria and yeasts that coexist until the end of fermentation, resulting in a product with better characteristics [32]. However, this practice, often used by Bella di Cerignola producers, could be responsible for the survival and/or growth of some oxidative yeasts, as reported by Fuccio et al. [5]. The analysis performed in the screening step suggested that yeasts could play a negative role in olive quality; therefore, they were set as a negative criterion in MoS. To the best of our knowledge, there is no evidence on the critical threshold of yeasts on vegetables; however, the breakpoint in MoS was set to 5 log CFU/mL because this was related to spoiled samples and is the critical threshold associated with spoilage in many foods [33].

The analyses of the first part also suggested a link between spoilage and the ratio of LAB vs. TAC. Aerobic plate counts are poor indicators of safety in some products, such as those that are fermented, which commonly show a high aerobic count. However, TAC gives information about the hygienic and sensorial quality, the adherence to good manufacturing practice, and the shelf life of the product [7,34]. The link of the ratio of LAB/TAC with the spoilage suggests that the negative effects of some aerobic bacteria could be counteracted by LAB [3,7]. TAC in table olives include *Bacillus* spp., *Aerobacter* spp., and *Pseudomonas* spp. All of these bacteria, along with some fungi, release degrading enzymes, which act on pectic substances and cellulose, hemicellulose, and polysaccharides, causing the loss of the structural integrity of the olive drupe [21,35].

pH is a key variable for olive safety and quality. An incorrect acidification of brine is a common problem for small farms where fermentation takes place without the use of starter cultures and the final pH of brine is around 5.0 [8]. Codex Alimentarius standard [36] sets the breakpoint for olive safety at 4.3; however, a pH of 4.5 is generally accepted by producers as safe, at least during storage in tanks. Another factor to control during the post-fermentation stage is salt. Although, the differences between spoiled and non-spoiled samples were not significant, salt could play a crucial role in the post-fermentation stage, mainly in spring when the increase in temperature requires additional corrective or control measures. Salt level in brines was at 6–8%; this amount could assure correct lactic fermentation and the dominance of LAB in the first stages. However, it could not be enough in post-fermentation to protect olives from abnormal fermentations [37], therefore, advice (third section) to perform corrective measure was added to MoS.

Eventually, the worksheet could be improved by adding in a revised version salt as a primary criterion; however, this criterion, as well as the advice of lowering pH to 4.3 or, better, to 4.1 to assure the microbiological stability of olives, could be the result of some changes in the habits of olive producers in Apulia.

Therefore, as a preliminary step the spreadsheet was developed for yeasts, LAB/TAC, and pH at 4.5 and the preliminary validation showed a good agreement between spoilage status, as revealed by the producers, and microbiological profile. Further investigations are required for a validation on batches from other olive producers, as well as for other olive varieties. Moreover, it was not possible to link the quality with other genera/groups of microorganisms, because in our samples they were detected occasionally.

Finally, another criterion we suggest for quality assessment is staphylococci count. In many foods the presence of staphylococci usually indicates post-processing contamination from human skin, mouths, and noses, or food handlers [38]. Due to their high salt tolerance, they can grow in table olives despite the low pH and the olive phenols may represent natural inhibitors [32].

In conclusion, this paper represents a first structured approach to design a user-friendly spreadsheet for the quick evaluation of the microbiological profile and quality of Bella di Cerignola olives during their storage in tanks. The spreadsheet is based upon four criteria (TAC, LAB, yeasts, and pH) and has two main benefits: (i) it is user-friendly; (ii) it gives an output on possible spoiling events in brines. In addition, the use of staphylocci as an indicator of microorganisms offers the possibility of analyzing and highlighting possible incorrect handling. Some issues to be addressed for an effective scaling up of the worksheet are the following: (a) The tool has been designed as an Excel spreadsheet, because this is the common software suite used by producers; however, the tool should be also designed in Apple Numbers, Open Office, or Google Sheet for a wide application; (b) It is based on lab data. The latter are not mandatory, but producers of the Apulia Region usually obtain them from experts or laboratories once or twice a month. The tool could be modified by adding screening criteria physico–chemical parameters (pH, salt, and temperature), available for producers many times per week, and microbiological counts as indices for the confirmatory classification of samples; (c) Safety is not a criterion in the tool, because producers of the Apulia region rarely have these data, since they are time-consuming and expensive, but the worksheet could be improved by pointing out some indicators (physico–chemical or microbiological) linked to safety.

The spreadsheet could be modified by adding other criteria or by setting different breakpoints, depending on regional and/or national regulations, as well as on the HACCP plan of each producer; however, it is a simple tool for an effective application of Hazard Analysis and Predictive Microbiology in table olive production and to improve a sector with many critical points.

The tool was developed for Bella di Cerignola olives, but a similar approach could be used to design a general tool for olives and to improve their performances by entering results from different seasons and places.

Supplementary Materials: The following are available online at http://www.mdpi.com/2304-8158/9/7/848/s1, Spreadsheet S1: MoS (Micro-Olive-Spreadsheet).

Author Contributions: Conceptualization, A.B., M.S., and M.R.C.; methodology, A.B.; software, A.B.; investigation, B.S. and D.C.; resources, M.S. and M.R.C.; data curation, A.B. and M.R.C.; writing—original draft preparation, A.B.; writing—review and editing, all authors; supervision, M.R.C.; funding acquisition, M.R.C. All authors have read and agreed to the published version of the manuscript.

Funding: This research was financed by "Cluster Tecnologici Regionali" Regione Puglia FSC 2007–2013, Project Bioteca QCBRAJ6.

Conflicts of Interest: The authors declare no conflict of interest.

References

1. Benítez-Cabello, A.; Romero-Gil, V.; Medinaa, E.; Sánchez, B.; Calero-Delgado, B.; Bautista-Gallego, J.; Jiménez-Díaz, R.; Arroyo-López, F.N. Metataxonomic analysis of the bacterial diversity in table olive dressing components. *Food Control* **2019**, *105*, 190–197. [CrossRef]
2. Mastralexi, A.; Mantzouridou, F.T.; Tsimidou, M.Z. Evolution of safety and other quality parameters of the Greek PDO table olives "Prasines Elies Chalkidikis" during industrial scale processing and storage. *Eur. J. Lipid Sci. Technol.* **2019**, *121*, 1800171. [CrossRef]
3. Perpetuini, G.; Prete, R.; Garcia-Gonzalez, N.; Alam, M.K.; Corsetti, A. Table olives more than a fermented food. *Foods* **2020**, *9*, 178. [CrossRef]
4. Randazzo, C.L.; Todaro, A.; Pino, A.; Pitino, I.; Corona, O.; Caggia, C. Microbiota and metabolome during controlled and spontaneous fermentation of Nocellara Etnea table olives. *Food Microbiol.* **2017**, *65*, 136–148. [CrossRef] [PubMed]
5. Fuccio, F.; Bevilacqua, A.; Sinigaglia, M.; Corbo, M.R. Using a polynomial model for fungi from table olives. *Int. J. Food Sci. Technol.* **2016**, *51*, 1276–1283. [CrossRef]
6. Nychas, G.J.E.; Panagou, E.Z.; Parker, M.L.; Waldron, K.W.; Tassou, C.C. Microbial colonization of naturally black olives during fermentation and associated biochemical activities in the cover brine. *Lett. Appl. Microbiol.* **2002**, *34*, 173–177. [CrossRef]
7. Romeo, F.V. Microbiological aspects of table olives. In *Olive Germplasm. The Olive Cultivation, Table Olive and Olive Oil Industry in Italy*; Muzzalupo, I., Ed.; InTech Publisher: Rijeka, Croatia, 2012; pp. 321–341.
8. Perricone, M.; Bevilacqua, A.; Corbo, M.R.; Sinigaglia, M. Use of *Lactobacillus plantarum* and glucose to control the fermentation of "Bella di Cerignola" table olives, a traditional variety of Apulian region (Southern Italy). *J. Food Sci.* **2010**, *75*, M430–M436. [CrossRef]
9. Bevilacqua, A.; Cannarsi, M.; Gallo, M.; Sinigaglia, M.; Corbo, M.R. Characterization and implications of *Enterobacter cloacae* strains, isolated from Italian table olives "Bella di Cerignola". *J. Food Sci.* **2010**, *75*, M53–M60. [CrossRef]
10. Panagou, E.Z.; Nychas, G.J.E.; Sofos, J.N. Types of traditional Greek foods and their safety. *Food Control* **2013**, *29*, 32–41. [CrossRef]
11. Medina-Pradas, E.; Arroyo-López, F.N. Presence of toxic microbial metabolites in table olives. *Front. Microbiol.* **2015**, *6*, 873. [CrossRef]
12. Tataridou, M.; Kotzekidou, P. Fermentation of table olives by oleuropeinolytic starter culture in reduced salt brines and inactivation of *Escherichia coli* O157:H7 and *Listeria monocytogenes*. *Int. J. Food Microbiol.* **2015**, *208*, 122–130. [CrossRef] [PubMed]
13. Bevilacqua, A.; Campaniello, D.; Speranza, B.; Sinigaglia, M.; Corbo, M.R. Survival of *Listeria monoctytogenes* and *Staphylococcus aureus* in synthetic brines. Studying the effects of salt, temperature and sugar through the approach of the Design of the Experiments. *Front. Microbiol.* **2018**, *9*, 240. [CrossRef] [PubMed]
14. Abriouel, H.; Benomar, N.; Lucas, R.; Gálvez, A. Culture-independent study of the diversity of microbial populations in brines during fermentation of naturally fermented Aloreña green table olives. *Int. J. Food Microbiol.* **2011**, *144*, 487–496. [CrossRef] [PubMed]
15. Lavermicocca, P.; Angiolillo, L.; Lonigro, S.L.; Valerio, F.; Bevilacqua, A.; Perricone, M.; Del Nobile, M.A.; Corbo, M.R.; Conte, A. *Lactobacillus plantarum* 5BG survives during the refrigerated storage bio-preserving packaged Spanish-style table olives (cv. Bella di Cerignola). *Front. Microbiol.* **2018**, *9*, 889. [CrossRef] [PubMed]
16. Comunian, R.; Ferrocino, I.; Paba, A.; Daga, E.; Campus, M.; Di Salvo, R.; Cauli, E.; Piras, F.; Zurru, R.; Cocolin, L. Evolution of microbiota during spontaneous and inoculated Tonda diCagliari table olives fermentation and impact on sensory characteristics. *LWT-Food Sci. Technol.* **2017**, *84*, 64–72. [CrossRef]
17. Randazzo, C.L.; Todaro, A.; Pino, A.; Pitino, I.; Corona, O.; Mazzaglia, A.; Caggia, C. Giarraffa and Grossa di Spagna naturally fermented table olives: Effect of starter and probiotic cultures on chemical, microbiological and sensory traits. *Food Res. Int.* **2014**, *62*, 1154–1164. [CrossRef]
18. D'Antuono, I.; Bruno, A.; Linsalata, V.; Minervini, F.; Garbetta, A.; Tufariello, M.; Mita, G.; Logrieco, A.F.; Bleve, G.; Cardinali, A. Fermented Apulian table olives: Effect of selected microbial starters on polyphenols composition, antioxidant activities and bioaccessibility. *Food Chem.* **2018**, *248*, 137–145. [CrossRef]

19. Bleve, G.; Tufariello, M.; Durante, M.; Perbellini, E.; Ramires, F.A.; Grieco, F.; Cappello, M.S.; De Domenico, S.; Mita, G.; Tasioula-Margari, M.; et al. Physico-chemical and microbiological characterization of spontaneous fermentation of Cellina di Nardò and Leccino table olives. *Front. Microbiol.* **2014**, *5*, 570. [CrossRef] [PubMed]
20. Lucena-Padrós, H.; Ruiz-Barba, J.L. Microbial biogeography of Spanish-style green olive fermentations in the province of Seville, Spain. *Food Microbiol.* **2019**, *82*, 259–268. [CrossRef]
21. Lanza, B. Abnormal fermentations in table olive processing: microbial origin and sensory evaluation. *Front. Microbiol.* **2013**, *4*, 91. [CrossRef] [PubMed]
22. Bevilacqua, A.; de Stefano, F.; Augello, S.; Pignatiello, S.; Sinigaglia, M.; Corbo, M.R. Biotechnological innovations for table olives. *Int. J. Food Sci. Nutr.* **2015**, *66*, 127–131. [CrossRef] [PubMed]
23. Campaniello, D.; Bevilacqua, A.; D'Amato, D.; Corbo, M.R.; Altieri, C.; Sinigaglia, M. Microbial characterization of table olives processed according to Spanish and Natural styles. *Food Technol. Biotechnol.* **2005**, *43*, 289–294.
24. International Olive Oil Council (IOC). Method. Sensory Analyses of Table Olives. COI/OT/MO No 1/Rev.2, November 2011. Available online: https://www.internationaloliveoil.org/what-we-do/chemistry-standardisation-unit/standards-and-methods/ (accessed on 28 June 2020).
25. Ruiz-Bellido, M.Á.; Valero, A.; Medina Pradas, A.; Romero Gil, V.; Rodríguez-Gómez, F.; Posada-Izquierdo, G.D.; Rincón, F.; Possas, A.; García-Gimeno, R.M.; Arroyo-López, F.N. A probabilistic decision-making scoring system for quality and safety management in Aloreña de Málaga table olive processing. *Front. Microbiol.* **2017**, *8*, 2326. [CrossRef] [PubMed]
26. Panagou, E.Z.; Kodogiannis, V.; Nychas, G.J.-E. Modelling fungal growth using radial basis function neural networks: The case of the ascomycetous fungus *Monascus ruber* van Tieghem. *Int. J. Food Microbiol.* **2007**, *117*, 276–286. [CrossRef]
27. Risk Ranger. Available online: https://www.cbpremium.org/RiskRanger (accessed on 15 June 2020).
28. Bevilacqua, A.; Beneduce, L.; Sinigaglia, M.; Corbo, M.R. Selection of yeasts as starter cultures for table olives. *J. Food Sci.* **2013**, *78*, M742–M751. [CrossRef]
29. Bevilacqua, A.; Corbo, M.R.; Sinigaglia, M. Selection of yeasts as starter cultures for table olives: A step-by-step procedure. *Front. Microbiol.* **2012**, *3*, 194. [CrossRef]
30. Fleming, H.P.; Mcfeeters, R.F.; Breidt, F. Staphylococcus aureus and staphyolococcal enterotoxins. In *Compendium of methods for the Microbiological examination of Foods*, 4th ed.; Downes, F.P., Ito, K., Eds.; American Public Health Association: Washington, DC, USA, 2001; pp. 387–403.
31. Leventdurur, S.; Sert-Aydın, S.; Boyaci-Gunduz, C.P.; Agirman, B.; Ben Ghorbal, A.; Francesca, N.; Martorana, A.; Erten, H. Yeast biota of naturally fermented black olives in different brines made from cv. Gemlik grown in various districts of the Cukurova region of Turkey. *Yeasts* **2016**, *33*, 289–301. [CrossRef]
32. Tassou, C.C.; Nychas, G.J.E. Inhibition of *Staphylococcus aureus* by olive phenolics in broth and in a model food system. *J. Food Prot.* **1994**, *57*, 120–124. [CrossRef]
33. Perricone, M.; Gallo, M.; Corbo, M.R.; Sinigaglia, M.; Bevilacqua, A. Yeasts. In *The Microbiological Quality of Food. Foodborne Spoilers*; Bevilacqua, A., Corbo, M.R., Sinigaglia, M., Eds.; Elsevier: Amsterdam, The Netherlands, 2017; pp. 121–132.
34. Morton, R.D. Aerobic plate count. In *Compendium of Methods for the Microbiological Examination of Foods*, 4th ed.; Downes, F.P., Ito, K., Eds.; American Public Health Association: Washington, DC, USA, 2001; pp. 63–67.
35. Golomb, B.L.; Morales, V.; Jung, A.; Yau, B.; Boundy-Mills, K.L.; Marco, M.L. Effects of pectinolytic yeast on the microbial composition and spoilage of olive fermentations. *Food Microbiol.* **2013**, *33*, 97–106. [CrossRef]
36. *CODEX/COI Codex Standard for Table Olives. CODEX STAN 66-1881. Revision 1987*; FAO: Rome, Italy, 2013.
37. Garrido-Fernández, A.; Fernández-Díez, M.J.; Adams, R.M. *Table Olives: Production and Processing*, 1st ed.; Chapman and Hall: London, UK, 1997.
38. Lancette, G.A.; Bennett, R.W. Fermented and acidified vegetables. In *Compendium of Methods for the Microbiological Examination of Foods*, 4th ed.; Downes, F.P., Ito, K., Eds.; American Public Health Association: Washington, DC, USA, 2001; pp. 521–532.

MDPI
St. Alban-Anlage 66
4052 Basel
Switzerland
Tel. +41 61 683 77 34
Fax +41 61 302 89 18
www.mdpi.com

Foods Editorial Office
E-mail: foods@mdpi.com
www.mdpi.com/journal/foods

www.ingramcontent.com/pod-product-compliance
Lightning Source LLC
LaVergne TN
LVHW070440100526
838202LV00014B/1631
9783036505145